Responses to Abiotic Stresses in Horticultural Crops

Responses to Abiotic Stresses in Horticultural Crops

Editors

Adalberto Benavides-Mendoza
Yolanda González-García
Fabián Pérez Labrada
Susana González-Morales

MDPI

Basel • Beijing • Wuhan • Barcelona • Belgrade • Novi Sad • Cluj • Manchester

Editors
Adalberto Benavides-Mendoza
Universidad Autónoma Agraria Antonio Narro
Saltillo
Mexico

Yolanda González-García
INIFAP campo experimental Todos Santos, BCS
México

Fabián Pérez Labrada
Universidad Autónoma Agraria Antonio Narro
Saltillo
Mexico

Susana González-Morales
CONAHCYT-Universidad Autónoma Agraria Antonio Narro
Saltillo 25315
Mexico

Editorial Office
MDPI AG
Grosspeteranlage 5
4052 Basel, Switzerland

This is a reprint of articles from the Special Issue published online in the open access journal *Horticulturae* (ISSN 2311-7524) (available at: https://www.mdpi.com/journal/horticulturae/special_issues/BYN1WUD87F).

For citation purposes, cite each article independently as indicated on the article page online and as indicated below:

Lastname, A.A.; Lastname, B.B. Article Title. *Journal Name* **Year**, *Volume Number*, Page Range.

ISBN 978-3-7258-1937-9 (Hbk)
ISBN 978-3-7258-1938-6 (PDF)
doi.org/10.3390/books978-3-7258-1938-6

Cover image courtesy of Adalberto Benavides-Mendoza

Contents

horticulturae

MDPI

Editorial

Response to Abiotic Stresses in Horticultural Crops

Adalberto Benavides-Mendoza [1,*], Yolanda González-García [2], Fabián Pérez-Labrada [3] and Susana González-Morales [4]

1 Department of Horticulture, Universidad Autónoma Agraria Antonio Narro, Saltillo 25315, Mexico
2 Instituto Nacional de Investigaciones Forestales, Agrícolas y Pecuarias, Centro de Investigación Regional del Noroeste, Campo Experimental Todos Santos, La Paz 23070, Mexico; gonzalez.yolanda@inifap.gob.mx
3 Department of Botany, Universidad Autónoma Agraria Antonio Narro, Saltillo 25315, Mexico; fabian.perezl@uaaan.edu.mx
4 CONAHCYT-UAAAN, Universidad Autónoma Agraria Antonio Narro, Saltillo 25315, Mexico; sgonzalezmo@conahcyt.mx
* Correspondence: adalberto.benavides@uaaan.edu.mx

1. Introduction

Horticultural production systems provide multiple benefits: economic, social, and health. However, the climate crisis poses a significant potential impact on horticultural production, due to changes in weather patterns and various environmental stresses [1]. Despite the expansion of protected horticulture in recent decades, crops grown in greenhouses, on mulches, and under shade nets are not immune to the adverse effects of these stresses; furthermore, a substantial portion of horticultural crops, in terms of area and production volume, are cultivated in open fields. Therefore, research on the impact of abiotic factors on the productivity, quality, and yield of horticultural crops is crucial [2].

Recently, the study of abiotic stress in horticultural crops has gained significant attention, reflecting the growing need to understand and mitigate the negative effects of environmental stressors on crop productivity and quality. Abiotic stresses such as drought, salinity, extreme temperatures, and heavy metal contamination are major limiting factors in horticultural production. These stresses can adversely affect plant growth, development, and yield, leading to substantial economic losses and threats to food security. Recent developments in the field of horticultural plant stress have underscored the intricate interplay between environmental stressors and plant physiological responses, leading to innovative strategies for improving crop resilience and productivity [3].

The study of stress in crops can be approached from different points of view. One angle to begin from is the consideration of elements that make up the agroecosystem, such as soils and water. This can be used in the field of plant populations or in the physiology and biochemistry of individual plants. Similarly, analyzing agronomic management or using biostimulants constitutes a valuable approach to understanding and improving crop responses to stress. All these elements were covered by the articles published in this Special Issue, which demonstrates the scope of the efforts made by the scientific community to address the problem of horticultural crop stress. Below is a review of the published studies in the Special Issue.

Citation: Benavides-Mendoza, A.; González-García, Y.; Pérez-Labrada, F.; González-Morales, S. Response to Abiotic Stresses in Horticultural Crops. *Horticulturae* **2024**, *10*, 815. https://doi.org/10.3390/horticulturae10080815

Received: 3 July 2024
Accepted: 16 July 2024
Published: 1 August 2024

2. Overview of Published Articles

In horticultural production systems, soil serves as the fundamental medium for plant growth and development, and its physicochemical properties are crucial external factors that influence these two aspects. In the first paper of the Special Issue, Gu et al. (Contribution 1) studied the growth of *Zanthoxylum armatum* in five different types of soil, namely, red soil (RS), yellow soil (YS), acidic purple soil (ACPS), alkaline purple soil (ALPS), and alluvial soil, which exhibited distinct physical and chemical characteristics due to variations in their formation processes and nutritional conditions. The results indicated that the morphological indices of the *Z. armatum* seedlings grown in the alluvial soil were greater than those in the other four soils. Because of the relatively high nutrient levels in the plant organs, alluvial soil and red soil may be advantageous for cultivating *Z. armatum*. The above findings aid our understanding of how this plant adopts various nutrient acquisition strategies under different soil conditions.

The application of microbial biostimulants is a valuable and increasingly used activity in soils and horticultural plants. Plant growth-promoting rhizobacteria are capable of inducing a tolerance to abiotic and biotic stresses. A report by Andy et al. (Contribution 2) assessed the potential of two isolated rhizobacterial strains of *Bacillus* possessing PGPR capabilities. The different tests on the production capacity of metabolites and enzymes with a biostimulant capacity under conditions of stress, due to water deficit and in the presence of heavy metals, indicated the potential of these strains to improve the response of plants to stress.

Horticultural species that produce metabolites with medicinal value constitute a valuable sector of the industry. These species respond to stress factors by modifying their composition and varying their quality and medicinal potential. In a study reported by Honório et al. (Contribution 3), *Annona emarginata* was subjected to three water levels (flooding, field capacity, and drought) for two periods of time (stress and recovery), which significantly modified the primary metabolism and impacted the accumulation of metabolites with medicinal value; drought promoted a higher concentration of total alkaloids and flooding lead to a decrease in total alkaloids and an increase in the liriodenine concentration. This study revealed that different kinds of stress can constitute tools for controlling the composition of medicinal plants. Conversely, following the same topic of medicinal plants, Vinogradova et al. (Contribution 4) presented a review with a complete overview of the impact on health, therapeutic potential, and production of medicinal plants when they are subjected to heavy metals, whether from natural sources or anthropogenic pollution. This topic is relevant considering the great horticultural importance of medicinal plants and the growing popularity of herbal remedies.

Cold damage can reduce the value of climacteric fruits and cause losses in the commercial chain when they are handled postharvest. Sensitivity to low-temperature damage varies between species and between varieties within species. However, the mechanism that causes the difference in sensitivity is poorly understood. In the study published by Zhan et al. (Contribution 5), two types of peach fruits (cold insensitive and cold sensitive) were selected for an analysis of the mechanisms of chilling injury in fruits with varying levels of chilling sensitivity. This analysis utilized lipidomic and transcriptome data and dynamic changes in plant hormones. In cold-insensitive peach fruits, the endogenous ABA and dilactosyl diacylglycerol contents significantly increased during low-temperature storage. In contrast, cold-sensitive peach fruits accumulated higher levels of ethylene, phospholipids, and ABA glucose esters than CM fruits, explaining the observed severe symptoms of the cold stress.

Heavy metals represent a significant stress factor for crops, a problem exacerbated by pollution from industrial waste and hydrocarbon combustion. The relationship between tomato growth and the ability to accumulate cadmium (Cd) in different organs was reported by Zhang et al. (Contribution 6). The authors found that while roots absorb and transport metal, they do not serve as storage organs. Additionally, plant growth did not influence the Cd concentration in tomato tissues. The highest Cd concentration was found in the leaves, followed by the stems and then the fruits. The results are valuable for projects focused on selecting cultivars with a low rate of metal transport to fruits. Similarly, Pérez-Labrada et al. (Contribution 7) described how the biostimulants citric acid and humic substances modify the absorption of Fe in tomato plants using different physiological, biochemical, and gene expression variables. Both biostimulants favorably modified the variables under study and improved the response of the plants that grew in calcareous soil.

Water deficit and salinity are increasingly common stresses affecting horticultural activity. Therefore, the use of and research on tolerant fruit-producing species are considered essential. Kenanoğlu et al. (Contribution 8) investigated the impact of two different pitaya species and different biostimulants applied for seed priming on the tolerance of plants to salinity. The pitaya species was a determining factor in tolerance, while oxalic acid and mepiquat chloride were the best inducers of tolerance. This study is relevant for the genetic improvement of this species. With respect to salinity, Širić et al. (Contribution 9) described the impact of TiO_2 nanoparticles and biochar on Chinese spinach under salinity stress. The authors found a synergistic effect in the two biostimulants, indicating that the combination of nanomaterials and conventional biostimulants is an area that should receive greater attention from researchers.

Studying the adaptive mechanisms of halotolerant species is important for developing strategies to improve crop responses to soil salinization. Joshi et al. (Contribution 10) described the mechanisms used by two halophytic species of the genus *Suaeda* to tolerate high levels of Na^+, Cl^-, and heavy metals. Changes in the ionic composition of the cells, the accumulation of osmolytes, and the improvement in antioxidant metabolism were part of the metabolic adjustments made by the plants.

One of the challenges of climate change and the destruction of ecosystems is biodiversity conservation. Micropropagation is a technique that allows the propagation of plant material from rare or endangered species. Chokheli et al. (Contribution 11) published an article describing the development of an optimization method to obtain the appropriate medium for propagating rare species.

3. Conclusions

The scope of the Special Issue is broad, and includes different approaches to studying and improving the response of different horticultural species to stresses from factors such as low temperature, salinity, water deficit, heavy metals, and micronutrient deficiency. However, some important stresses, such as high temperature and high irradiance, which are common in many places where horticultural species are grown, were not included.

Despite the above, the Special Issue presents a complete overview of information for those interested in research, the development of biostimulants, or horticultural production in the field.

Conflicts of Interest: The authors declare no conflicts of interest.

List of Contributions:

1. Gu, T.; Ren, H.; Wang, M.; Qian, W.; Hu, Y.; Yang, Y.; Yu, T.; Zhao, K.; Gao, S. Changes in Growth Parameters, C:N:P Stoichiometry and Non-Structural Carbohydrate Contents of *Zanthoxylum armatum* Seedling in Response to Five Soil Types. *Horticulturae* **2024**, *10*, 261. https://doi.org/10.3390/horticulturae10030261.
2. Andy, A.K.; Rajput, V.D.; Burachevskaya, M.; Gour, V.S. Exploring the Identity and Properties of Two Bacilli Strains and Their Potential to Alleviate Drought and Heavy Metal Stress. *Horticulturae* **2023**, *9*, 46. https://doi.org/10.3390/horticulturae9010046.
3. Honório, A.B.M.; De-la-Cruz-Chacón, I.; da Silva, G.C.; Mimi, C.O.; Campos, F.G.; da Silva, M.R.; Boaro, C.S.F.; Ferreira, G. Differential Tolerance of Primary Metabolism of Annona Emarginata (Schltdl.) H. Rainer to Water Stress Modulates Alkaloid Production. *Horticulturae* **2024**, *10*, 220. https://doi.org/10.3390/horticulturae10030220.
4. Vinogradova, N.; Glukhov, A.; Chaplygin, V.; Kumar, P.; Mandzhieva, S.; Minkina, T.; Rajput, V.D. The Content of Heavy Metals in Medicinal Plants in Various Environmental Conditions: A Review. *Horticulturae* **2023**, *9*, 239. https://doi.org/10.3390/horticulturae9020239.
5. Zhan, W.; Wang, Y.; Duan, W.; Li, A.; Miao, Y.; Wang, H.; Meng, J.; Liu, H.; Niu, L.; Pan, L.; et al. Preliminary Analysis, Combined with Omics of Chilling Injury Mechanism of Peach Fruits with Different Cold Sensitivities during Postharvest Cold Storage. *Horticulturae* **2024**, *10*, 46. https://doi.org/10.3390/horticulturae10010046.
6. Zhang, X.; Zhang, C.; Zhang, Y. Tomato Accumulates Cadmium to a Concentration Independent of Plant Growth. *Horticulturae* **2023**, *9*, 1343. https://doi.org/10.3390/horticulturae9121343.

7. Pérez-Labrada, F.; Benavides-Mendoza, A.; Juárez-Maldonado, A.; Solís-Gaona, S.; González-Morales, S. Effects of Citric Acid and Humic-like Substances on Yield, Enzyme Activities, and Expression of Genes Involved in Iron Uptake in Tomato Plants. *Horticulturae* **2023**, *9*, 630. https://doi.org/10.3390/horticulturae9060630.
8. Kenanoğlu, B.B.; Mertoğlu, K.; Sülüşoğlu Durul, M.; Korkmaz, N.; Çolak, A.M. Maternal Environment and Priming Agents Effect Germination and Seedling Quality in Pitaya under Salt Stress. *Horticulturae* **2023**, *9*, 1170. https://doi.org/10.3390/horticulturae9111170.
9. Širić, I.; Alhag, S.K.; Al-Shuraym, L.A.; Mioč, B.; Držaić, V.; Abou Fayssal, S.; Kumar, V.; Singh, J.; Kumar, P.; Singh, R.; et al. Combined Use of TiO_2 Nanoparticles and Biochar Produced from Moss (*Leucobryum glaucum* (Hedw.) Ångstr.) Biomass for Chinese Spinach (*Amaranthus dubius* L.) Cultivation under Saline Stress. *Horticulturae* **2023**, *9*, 1056. https://doi.org/10.3390/horticulturae9091056.
10. Joshi, A.; Rajput, V.D.; Verma, K.K.; Minkina, T.; Ghazaryan, K.; Arora, J. Potential of Suaeda Nudiflora and Suaeda Fruticosa to Adapt to High Salinity Conditions. *Horticulturae* **2023**, *9*, 74. https://doi.org/10.3390/horticulturae9010074.
11. Chokheli, V.A.; Bakulin, S.D.; Ermolaeva, O.Y.; Kozlovsky, B.L.; Dmitriev, P.A.; Stepanenko, V.V.; Kornienko, I.V.; Bushkova, A.A.; Rajput, V.D.; Varduny, T.V. Investigation of Growth Factors and Mathematical Modeling of Nutrient Media for the Shoots Multiplication In Vitro of Rare Plants of the Rostov Region. *Horticulturae* **2023**, *9*, 60. https://doi.org/10.3390/horticulturae9010060.

References

1. Alotaibi, M. Climate Change, Its Impact on Crop Production, Challenges, and Possible Solutions. *Not. Bot. Horti Agrobot. Cluj-Napoca* **2023**, *51*, 13020. [CrossRef]
2. Singh, P.; Singh, S.; Dubey, R.S. Climate Change Impacts on Agriculture: Crop Productivity and Food Security. In *Climate Change and Sustainable Development*; Fulekar, M.H., Dubey, R.S., Eds.; CRC Press: London, UK, 2023; p. 26, ISBN 978-1-00-320554-8.
3. Salse, J.; Barnard, R.L.; Veneault-Fourrey, C.; Rouached, H. Strategies for Breeding Crops for Future Environments. *Trends Plant Sci.* **2024**, *29*, 303–318. [CrossRef] [PubMed]

horticulturae

MDPI

Article

Exploring the Identity and Properties of Two *Bacilli* Strains and their Potential to Alleviate Drought and Heavy Metal Stress

Aruna Kumari Andy [1,2], Vishnu D. Rajput [3], Marina Burachevskaya [4] and Vinod Singh Gour [2,*]

1 Directorate of Research, Sam Higginbottom University of Agriculture, Technology and Sciences, Allahabad 211007, India
2 Amity Institute of Biotechnology, Amity University Rajasthan, Jaipur 303002, India
3 Academy of Biology and Biotechnology, Southern Federal University, 344006 Rostov-on-Don, Russia
4 Soil Chemistry and Ecology Laboratory, Faculty of Natural Sciences, Tula State Lev Tolstoy Pedagogical University, Lenin Avenue, 125, 300026 Tula, Russia
* Correspondence: vinodsingh2010@gmail.com

Abstract: Naturally available plant growth-promoting rhizobacteria (PGPR) have 1-aminocyclopropane-1-carboxylic acid (ACC) deaminase enzymes, and are capable of processing the plant-borne ACC by converting it into α-ketobutyrate and ammonia. Thus, the PGPRs help in the depletion of ethylene levels, and enhance abiotic stress tolerance in plants. In the present study, two rhizobacterial strains, i.e., *Bacillus cereus* and *B. haynesii*, isolated from *Vigna mungo* and *Phaseolus vulgaris*, were used. These strains were taxonomically identified by 16S rDNA sequencing as *B. cereus* and *B. haynesii*, with NCBI accession numbers LC514122 and LC 514123, respectively. The phylogeny of these strains has also been worked out based on homology, with data available on NCBI GenBank. The strains were screened for their plant growth-promoting traits, and quantified in the same way. The enzymatic activity and molecular weight of the ACC deaminase obtained from both bacterial strains have also been determined. An in vitro drought tolerance study was done by using PEG 6000. These bacterial strains exhibited higher ACC deaminase activity (~5 to 6 μmol/mL), exopolysaccharide yield (15 to 18 mg/10 mL protein), and indole acetic acid (27–32 μg/mL). These characteristics indicate that the bacterial strains under present study may be helpful in enhancing the drought tolerance of the crops with enhanced yield. *Bacillus cereus* has been found to be a tolerant strain to As, Ba, and Ni, based on the plate assay method, and so it has the potential to be used as biofertilizer in fields affected by these metals.

Keywords: PGPR characterization; ACC deaminase; exopolysaccharides; molecular diversity

Citation: Andy, A.K.; Rajput, V.D.; Burachevskaya, M.; Gour, V.S. Exploring the Identity and Properties of Two *Bacilli* Strains and their Potential to Alleviate Drought and Heavy Metal Stress. *Horticulturae* **2023**, 9, 46. https://doi.org/10.3390/horticulturae9010046

Academic Editors: Adalberto Benavides-Mendoza, Yolanda González-García, Fabián Pérez Labrada and Susana González-Morales

Received: 30 November 2022
Revised: 22 December 2022
Accepted: 23 December 2022
Published: 2 January 2023

1. Introduction

The human population keeps on increasing year by year, and the total population is expected to reach nine billion by 2050. This increase is accompanied by rapid urbanization and industrialization, which inevitably leads to environmental stress and damage. The consequences are higher evaporation and transpiration, which eventually cause global warming [1]. This leads to reduced agriculture land and soil fertility. To attain food security, agricultural production has been increased using synthetic fertilizers and chemicals at a very high rate, which has damaged soil health and, consequently, resulted into abiotic stress and ultimately decreased the quantity and quality of crops, considerably [2].

Meanwhile, the genetic potential of crop yield is hindered by natural abiotic stresses [3], among which drought is one of the worst constraints for agricultural production. The mechanism of drought tolerance has been explored in several crops, and a cluster of gene network has been identified to regulate the plant–water relationship [4]. Drought affects the physiology, morphology, and biochemistry of plants, and it also reduces soil fertility. These factors ultimately lead to a reduction in crop productivity, effecting global economics as well.

Salt tolerance in crops, which has been attained by genetic modifications to improve crop production, helps plants grow in saline conditions [5]. Studies indicate that microorganisms such as plant growth-promoting rhizobacteria (PGPR) have the potential to restore the degraded soil, and can enhance crop productivity even under diverse stress environmental conditions [6,7].

Micro-organisms are microscopic living creatures also found in diverse extreme conditions, and mainly include bacteria, cyanobacteria, algae, fungi, yeast, actinomycetes, and myxomycetes. Almost all existing natural materials are decomposed by these microbes. Such micro-organisms can transform naturally available organic matter into plant nutrients and make them freely available to the plants. Bacteria and mycorrhizal fungi present in agricultural fields help plants uptake water and minerals [2]. Many of these microbes protect the crops from various diseases and nutrient deficiency, and enrich the soil by acquiring enhanced available P and N. They can also increase the water retention capacity of soil, and thereby enhance soil fertility. PGPRs are known to be helpful for plants to grow even in abiotic stress conditions, as they have the potential to fix nitrogen, solubilize phosphorus, and sequester iron by siderophore production, and produce different phytohormones (auxins, gibberellins, cytokinins) [8]. Further, PGPRs are also able to produce ACC deaminase, which degrades 1-aminocyclopropane- 1-carboxylic acid (ACC) into α-ketobutyrate and ammonia, which inhibits ethylene production and ultimately help root growth, and thereby support plants to grow well even under stress conditions [9,10]. The association of PGPRs with soil helps in the formation of aggregates of soil particles, thereby increasing soil aeration.

Microbiological approaches have also attracted researchers, due to their potential to remove/sequester heavy metals and relieve plants from the pollution of these metals. Various biological approaches (bioleaching, bio-stimulation, bioventing, composting, bio-augmenting, land forming, and bioremediation) have been explored by soil scientists using microbes (single or in combination) to enhance soil health and increase yield. Such eco-friendly strategies received public acceptance at a large scale [11,12].

In several developing nations of the world, pulses and vegetables belonging to the leguminous family stand as the major sources of dietary proteins for millions of people, and play an important role in mitigating protein malnutrition [13], especially for the poor residing in parts of South Asia and sub-Saharan Africa; here, chickpea remains a most important source [14].

In the past decades, increased drought frequency and severity have been noticed in different parts of the world [15]. It is pertinent that we develop a strategy to make the crops drought-tolerant, or facilitate crop growth using biofertilizers to enhance yield, and thereby ensure food security. Chickpea is one of the major crops globally grown with other crops such as finger millet, and groundnut is branded as a forgotten crop, though it provides a good income to poor farmers. In chickpea production, South Asia contributes about 72%, while only 28% is contributed by the rest of the world [16].

The literature reveals that chickpea produces ethylene under drought and salt stress, which leads to a negative impact on plant growth. The impact of drought on chickpea production had reduced the yield by 33% globally. PGPRs could help chickpeas to tolerate the stress by secreting ACC deaminase [17], as noted earlier by Khan et al. (2019), where they identified that some PGPRs, such as *Bacillus subtilis*, *Bacillus thuringiensis*, and *Bacillus megaterium*, helped chickpeas to cope with drought stress [18].

The potential of our isolated rhizobacterial strains possessing PGPR capabilities were assessed. Out of 25 isolates, seven isolates were characterized for Gram's staining, biochemical characteristics [Methyl red (MR) test, Voges–Proskauer (VP) test, Catalase test], and PGPR potential (Phosphate solubilization, Siderophore production, Exopolysaccharide yield, ACC deaminase production, Nitrogen fixation) [19]. In the present manuscript, the investigation and results are described for two bacterial isolates, AV-7 from our previous study [19], and AV-12, a new strain which is reported for the first time.

Based on our present study, a new isolate AV-12 and a previously reported isolate AV-7 are further characterized based on biochemical assays such as IMViC tests (Indole test, Methyl red (MR) test, Voges–Proskauer (VP) test, catalase test). Our study also includes other important biochemical characteristics such as the citrate utilization test, phosphate solubilisation, amylase hydrolysis, and carbohydrate fermentation test, where D-glucose, sucrose, and maltose are utilized. Then, these strains were studied for their PGPR (qualitative and quantitative) potentials, which mainly included IAA production, ammonia production, HCN production, siderophore production, and ACC deaminase. These bacterial strains have also been studied for exopolysaccharide production under normal and induced water-stress conditions. The bacterial strains have been taxonomically identified using their 16S rRNA sequence.

2. Materials and Methods

Soil samples were collected from rhizospheric zones of *Vigna mungo* and *Phaseolus vulgari*, grown in agriculture farm (cultivated with many varieties of crops including different legumes) of Sam Higginbottom University of Agriculture, Technology and Sciences (SHUATS), Allahabad with GPS location of 25°24′36″ N 81°51′11″ E. The soils were analyzed for their physical and chemical properties including the presence of heavy metals. Heavy metals were analyzed by Atomic Absorption Spectrophotometer [19].

The bacterial strains were studied for Gram staining (Figure 1), and preliminary qualitative biochemical tests were performed as described previously [19]. The experimental methodologies did not add in details here, as full details can be obtained from Andy et al. [19].

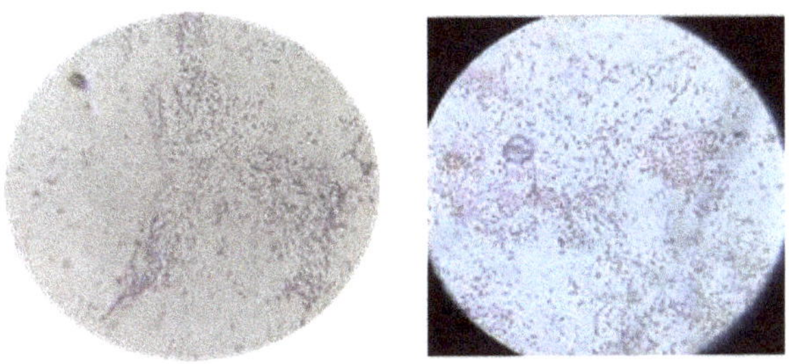

Bacillus cereus *Bacillus haynesii*

Figure 1. Slides observed under light microscope (100× oil immersion resolution) of two strains after Gram staining.

Biochemical characteristics

The bacterial isolates were characterized biochemically by various assays, namely indole test, methyl red test, Voges–Proskauer test, catalase test, Simmons citrate test, carbohydrate hydrolysis test, carbohydrate fermentation test [20], and amylase hydrolysis test [21].

Plant Growth Promoting Rhizobacterial traits evaluation

Phosphate solubilization efficiency of the bacterial isolates was studied by using Pikovskaya medium, which contained inorganic phosphate as main ingredient. The isolates were inoculated into the media and incubated at 30 °C for 24 h [22] (Figure S1). Then, the final OD was read at 600 nm, and the obtained values of cultured isolates were compared with the non-inoculated control values. The amount of phosphate released by the bacterial isolates was studied using the standard curve of potassium di-hydrogen phosphate as a source of P in the range of 10–100 mg/mL.

HCN production by the bacterial isolates was determined by adapting the method proposed by Lorck [23]. In this method, nutrient agar media was amended with glycine (4.4 g/L), and then the isolates were streaked in it. Then, the Petri plates were placed with Whatman filter paper, previously soaked in 2% sodium carbonate in picric acid (0.5%) solution. After sealing the plates with parafilm, these were incubated for 4 days at 28 °C. The filter paper changed from greenish-yellow to reddish-brown in the testing procedure and was concluded to be HCN-positive (Figure S1). The concentration of HCN was determined by spectrophotometric measurements at 424 nm, with standard curve of KCN.

Indole-3-acetic Acid (IAA) production by bacterial isolate has been estimated in nutrient broth containing 0.5 mg/mL L-tryphtophan. The cultures were incubated for 24 h at 30 °C. After incubation, when 1 mL of bacterial culture was mixed with 2 mL of Salkowski's reagent (7.5 mL 0.5 M$FeCl_3 \cdot 6H_2O$, 150 mL concentrated H_2SO_4, 250 mL distilled water), it resulted in appearance of reddish-orange colour. It indicates production of IAA by the cultured strain (Figure S2) [24]. For quantitative estimation of IAA, optical density was recorded at 530 nm by spectrophotometer and the concentration of IAA was calculated using standard curve of IAA (Hi-media) constructed in the range of 10–100 μg/mL.

For quantitative estimation of siderophore, CAS-shuttle assay (CAS: Chrome Azurol S agar) was performed [25]. Siderophore removes the iron from the dye complex, causing reduction in the intensity of blue colour, which was recorded at 630 nm. In this test, minimal medium was used as blank and percentage siderophore units were calculated.

Bacterial isolates were tested for their potential to produce ammonia in peptone water using Nessler's reagent [26] (Figure S2). Bacterial isolates were grown in peptone water broth for 48 h at 28 °C temperature; development of brown to yellow colour indicated production of ammonia, and its optical density was measured at 450 nm using spectrophotometer for its quantification. The concentration of ammonia was estimated from the standard curve of ammonium sulphate in the range of 1–10 μmol/mL.

2.1. Identification of Bacterial Strains Based on 16s rRNA Sequence

The total genomic DNA of the bacterial isolates was extracted and quantified at 260/280 nm by using spectrophotometer and analyzed on agarose gel, and the same was used in PCR amplification as a template for the 16S rDNA. The universal primers, 27F (5′- AGA GTT TGA TCC TGG CTC AG-3′) and 1492R (5′-ACG GCT ACC TTG TTA CGA CTT-3′), were used for amplification of 16S rDNA region as described by Weisburg et al. [27]. The PCR reactions were performed in 20 μL volumes containing 2 μL of the genomic DNA. The PCR master mix was having 0.16 mM dNTP Mix, 20 pmol of forward and reverse primers, and 0.75U Taq DNA polymerase (MBI, Fermentas, Lithuania). Two drops of mineral oil were overlaid on mixes contained in PCR tubes. Amplification was carried out in a thermal cycler (Eppendorf Master cycler nexus 230 V/50–60 Hz) [28]. The PCR conditions used in this study were as follows: an initial denaturation at 95 °C for 6 min, 40 cycles of denaturation at 95 °C for 30 s, annealing at 50 °C for 1 min, and extension at 72 °C for 1 min. Final extension was performed at 72 °C for 10 min. Gel electrophoresis with agarose (2%) and ethidium bromide was used to analyze the obtained PCR products. The DNA bands were observed by Gel Documentation System (Zenith, Gel Documentation System; Model No. Gel.LUMINAX-312) (Figure 2).

The PCR product was sequenced based on Sanger's Method [29]. BioEdit (an online) tool was used for alignment of obtained forward and reverse sequence, then the query sequences were identified, considering *E* value a $<1 \times 10^{-5}$ and maximum hits (99 or 100%) with a species in the reference NCBI database. Nucleotide BLAST and MEGA X were used for construction of phylogenetic tree by employing Maximum Parsimony (MP) program using the fast minimum evolution method [30].

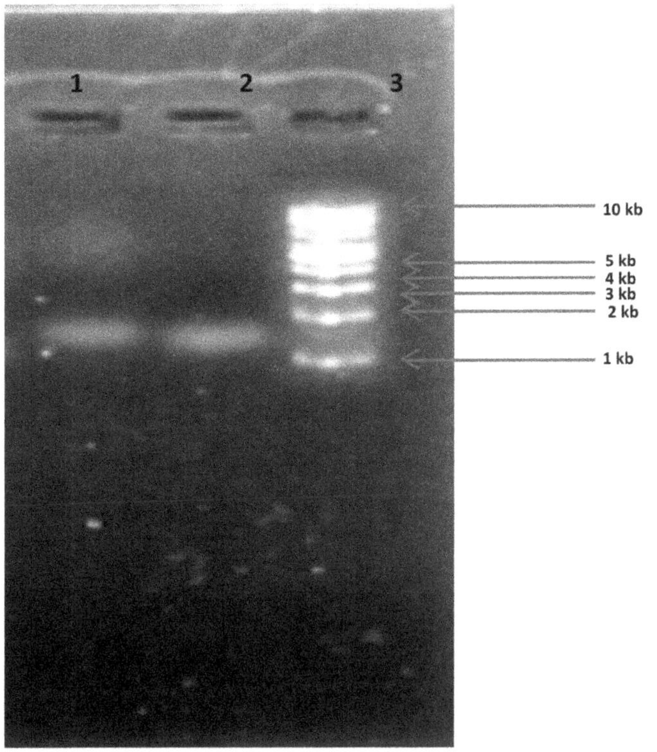

Figure 2. 16s rRNA gene amplification for bacterial isolates, where Lane 1 well was loaded with the *Bacillus cereus* sample, Lane 2 well was loaded with *Bacillus haynesii* sample, and as a check, a 10 Kb marker was loaded in Lane 3 well.

2.2. Screening ACC Deaminase Activity from Isolated Rhizobacteria

2.2.1. ACC Deaminase Activity (Qualitative)

For qualitative estimation of ACC deaminase, the bacterial isolates were serial diluted in LB (Luria–Bertani) medium, which contains 10 g of peptone, 5 g of yeast extract, and 10 g of NaCl per liter. This medium was supplemented with agar (1.5%). The appropriate dilutions (0.1 mL) of the sample bacterial isolates were cultured on LB agar medium and incubated at 28 °C for 24 h.

After incubation, distinct colonies (morphologically) were identified, and the same were observed for ACC deaminase production and its activity simultaneously on sterile Dworkin and Foster (DF) minimal salts media [31] (Figure S1). DF salts minimal media was prepared as per standard protocol and $(NH_4)_2SO_4$ was replaced by 3 mM ACC. Under this condition, bacteria will secrete ACC deaminase and use ACC as a sole source of N. Therefore, the bacteria which survive on this medium will definitely have ACC deaminase activity.

2.2.2. ACC Deaminase Quantified from Selected Bacterial Isolates

ACC deaminase activity was quantified using the method described by Penrose and Glick [32], where α-ketobutyrate produced by activity of ACC deaminase has been calculated by spectroscopic observations [32].

2.2.3. Measurement of ACC Deaminase (Enzyme) Activity

ACC deaminase converts ACC into α-ketobutyrate, and so, indirectly, ACC deaminase was estimated by estimating α-ketobutyrate production at 540 nm by spectroscopic method [32].

2.2.4. Standard Curve of α-Ketobutyrate

The standard solutions of α-ketobutyrate were prepared with 1–10 μmol concentrations. 300 μL of 2, 4-dinitrophenylhydrazine was added to each tube and gently vortexed. Then, the tubes were incubated at 30 °C for 30 min, then 2 mL of 2N sodium hydroxide was added to each tube, and then the absorbance was read at 540 nm, and standard curves were constructed accordingly.

2.3. Portrayal of Partially Purified ACC Deaminase Enzyme

Purification of ACC deaminase enzyme was performed by ammonium sulfate precipitation method followed by dialysis [33], where silica-based column chromatography was used. 10 g of silica gel was weighed and mixed into sodium phosphate buffer until it is completely dissolved. Then, the gel is poured into Pasteur pipette by using beaker (10 mL). The column was left until the silica gel settles down completely. Finally, when the buffer and gel become separated, the buffer was allowed to flow down the column drop by drop. The solvent level was monitored for its flows through the silica gel and left until its level reaches 3 mm layer, which indicated the sample was completely poured out. At the end of the process, pure sample in different concentrations were collected.

2.4. Quantification of Partially Purified Protein by Bradford Method

Bradford reagent was prepared by dissolving 100 mg Coomassie Brilliant Blue G-250 in 50 mL 95% ethanol, and then 100 mL 85% (*w*/*v*) phosphoric acid was added. Solution was diluted to 1 L. When the dye was completely dissolved, it was filtered through Whatman No. 1 paper. Standard solutions were prepared containing a range of 5 to 100 micrograms protein (albumin) in 100 μL volume. Then, 5 mL dye reagent was added in each solution and incubated for 5 min. Absorbance at 540 nm was measured using spectrophotometer and standard curve was plotted. Amount of protein present in the sample was calculated by comparing absorbance by sample with the standard curve.

2.5. Determination of Molecular Weight of ACC Deaminase

The molecular weight of ACC deaminase was determined by using SDS PAGE, where the bands of ACC deaminase were compared with ladder of 1 Kb [34].

2.6. Estimation of Exopolysaccharide (EPS)

The bacterial isolates were grown in nutrient broth containing 5% of sucrose as carbohydrate source. 10 mL of culture suspension was collected after 6 days and centrifuged at 30,000 rpm for 45 min. Then, thrice the volume of chilled acetone was added to the supernatant. EPS was separated from the mixture in the form of a slimy precipitate. Precipitate was collected on a filter paper. The precipitates were allowed to dry overnight at 50 °C temperature. Then, by gravimetric method, weight of EPS was recorded [35].

2.7. Protein Content in EPS: PGPRs Potentials at Various Levels of Drought

For estimating protein in EPS media with different water potentials (−0.05, −0.15, −0.30, and −0.49 MPa) in trypticase, soya broth (TSB) was prepared by adding appropriate quantity of polyethylene glycol (PEG 6000) [9], then 1% overnight-raised bacterial cultures were inoculated in the screw-capped tubes. These tubes were incubated at 120 rpm for 24 h at 28 °C. The cultures grown at minimum (−0.05 MPa) and maximum (−0.49 MPa) stress level were analyzed for their ability to produce EPS, for which the culture grown was centrifuged for 25 min at 20,000 rpm and the supernatant was collected. Intracellular polysaccharides extraction was possible, so it was ruled out by adding 2, 4 Diphenylalanine (DPA) reagent for the presence of DNA in the collected supernatant. Nitrocellulose membrane (0.45 μm) was used to filter supernatant, and then it was extensively dialyzed with chilled water (4 °C). To remove insoluble material (if present) from obtained dialysate, it was centrifuged at 20,000 rpm for 25 min by mixing with ice-cold (3 volumes) absolute ethanol and incubated at 4 °C overnight. The mixture was centrifuged at 10,000 rpm for

15 min to obtained EPS. Further, the precipitate was purified by repeated dialysis and precipitation process again, after suspended in water. The growth in cultures was estimated by measuring the OD at 590 nm. The total protein content in the precipitated EPS was determined by comparing with the standard curve of Bovine serum albumin (BSA) by Bradford's assay.

2.8. Heavy Metal Tolerance Test by Plate Method

To study the tolerance of As, Ba, and Ni in the bacterial strains, the bacterial strains were cultured on the nutrient agar media amended with 14.74 mg/L and 29.48 mg/L of As, 6.28 mg/L and 12.56 mg/L of Ba, and 2.963 mg/L and 5.926 mg/L of Ni separately. Arsenic (III) oxide (As_2O_3), barium chloride ($BaCl_2 \cdot 2H_2O$), and nickel (II) chloride ($NiCl_2 \cdot 6H_2O$) were used for this study [19]. The bacterial strain which could survive on amended media with heavy metal has been considered as tolerant strain.

Each experiment has been repeated thrice and the average value is presented in this manuscript.

3. Results

3.1. Soil Properties

The soil samples were analyzed for both physical and chemical properties. The physical properties of the soil included pH (7.6), and electric conductivity (0.20). The soil texture was sandy loam and had 53% sand, 27% silt, and 20% clay. Soil bulk density was found to be 1.26 Mg/m^3 and soil particle density was determined to be 2.86 Mg/m^3. The soil has been found to be non-saline. The chemical analysis of soil indicate presence of 0.68% organic carbon, 54 kg/ha nitrogen, 28.8 kg/ha phosphorous, and 230 kg/ha potassium. Thus, the soil had high organic carbon, low N, high P, and moderate K content. The presence of these nutrients allowed proliferating bacteria belonging to families *Enterococcaceae*, *Bacillaceae*, *Coccidae*, *Pseudomonadaceae*, and *Pseudomonadaceae aerobe*. The soil was also studied for heavy metals like As, Ba, and Ni, out of which As and Ba were found to be below detectable levels and Nickel was found to be 0.66 ppm.

3.2. Bacterial Properties

The morphological features, biochemical characteristics, PGPR characteristics, and molecular characteristics of the two bacterial isolates are summarized in Table 1. The stain AV-7 has been found to be Gram-negative cocci with round and smooth colonies. The bacterium has been found to be positive for various biochemical tests except indole test and citrate utilization (Table 1). It was not capable of fermenting D-glucose and sucrose in carbohydrate fermentation test (Table 1). The bacterial strain AV-12 has been found similar to AV-7 with white convex elevated circular colonies. However, this bacterial strain has potential to utilize D-glucose and sucrose in carbohydrate fermentation test (Table 1).

Based on morphological and Gram's staining results, the bacterial strains were identified to be *Bacillus species*.

3.3. Identification of PGPRs by 16s rRNA Sequence

The quantitative analysis of isolated genomic DNA of both the strains was recorded as 7.57 µg/mL for AV-12 and 7.145 µg/mL for AV-7. The 16S rRNA sequences were found to be 1377 bp for AV-12 and 1340 bp for AV-7. AV-12 and AV-7 were the isolates from rhizospheric soils of *Vigna mungo* and *Phaseolus vulgaris*, respectively [19]. The 16S rRNA sequence for both bacterial strains are submitted to DNA Database Bank of Japan (DDBJ), which is a NCBI collaborator. The isolates were assigned with accession no. LC514122 and LC514123 for *Bacillus cereus* (AV-12) and *Bacillus haynesii* (AV-7), respectively. The obtained partial 16S rRNA sequences from both bacterial isolates were compared with existing sequences database from NCBI GenBank and found to be related with other reported strains (Figure 3) [30].

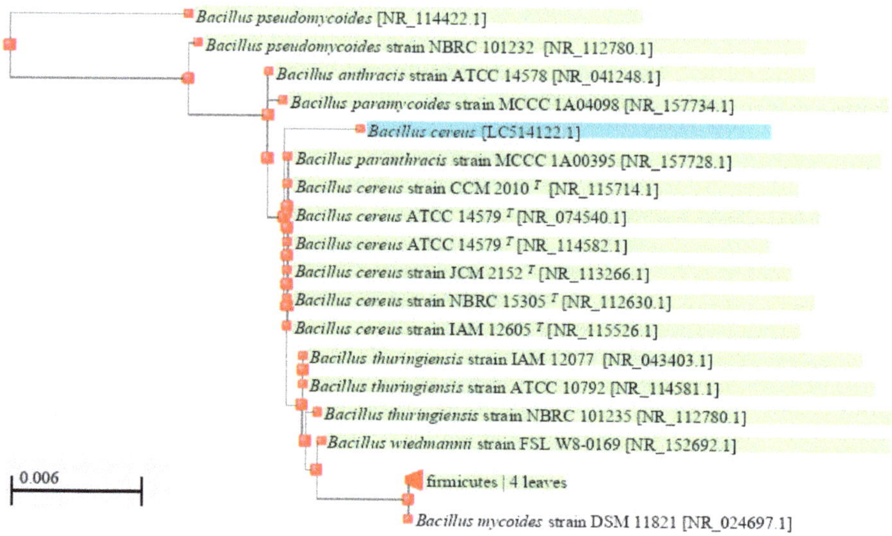

(A)

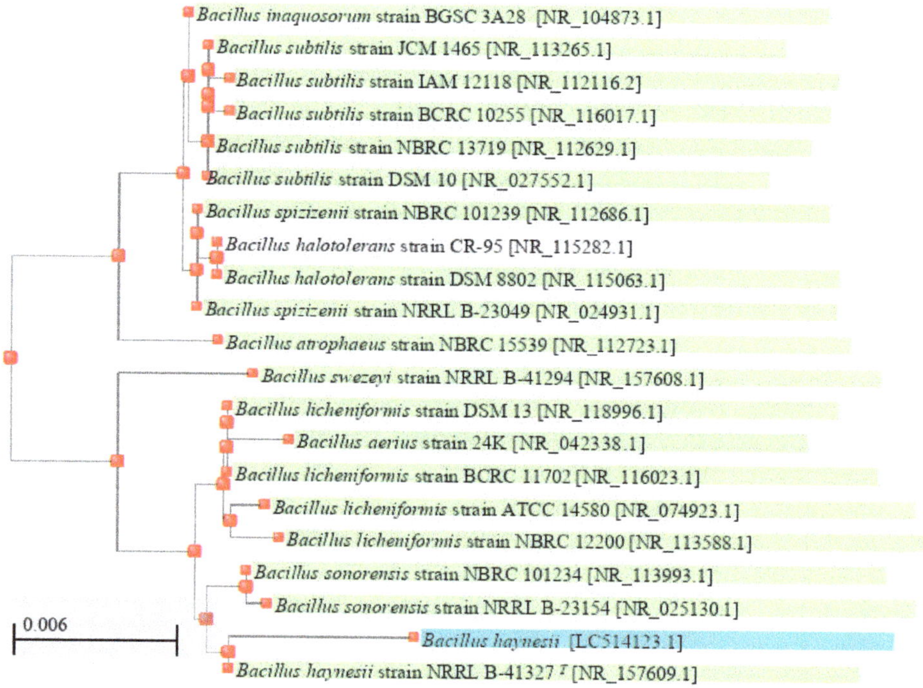

(B)

Figure 3. (**A**) Phylogenetic tree (NCBI BLAST): Tree Method used in constructing this picture was Fast minimum evolution with maximum sequence distance as 0.75. Sea-green-colour-indicated query (*Bacillus cereus*) was searched for 16S rRNA partial sequences from Bacteria and Archaea, bottle-green-colored strains are from type material, and red colour is an indication for Firmicutes. The Type strains are superscripted with T. (**B**) Phylogenetic tree (NCBI BLAST): Tree Method used in constructing this picture was Fast minimum evolution with maximum sequence distance as 0.75. Sea-green-colour-indicated query (*Bacillus haynesii*) was searched for 16S rRNA partial sequences from Bacteria and Archaea, bottle-green-coloured strains are from type material, and red colour is an indication for Firmicutes. The Type strains are superscripted with T.

Table 1. Source of rhizosphere soil and morphological, biochemical, and molecular characteristics of bacterial isolates.

Characteristics	**AV-12**	**AV-7**
Soil sample source	*Vigna mungo*	*Phaseolus vulgaris*
	Morphology	
Cell morphology	Round and Smooth colonies	Round and Smooth colonies#
Gram reaction	Gram-Negative	Gram-Negative #
Shape of organism	Bacilli (rod)	Cocci #
Spore formation	Observed	Observed
Arrangement of cells	Cells form clusters	Cells form clusters
	Culture	
Colony colour	White	White
Elevation	Convex	Convex
	Biochemical tests	
Indole test	Negative	Negative
Methyl red test	Positive	Positive #
VP test	Positive	Positive #
Catalase test	Positive	Positive #
Citrate utilization test	Positive	Positive
Phosphate solubilization	Positive	Positive #
Amylase hydrolysis	Positive	Positive
Carbohydrate production test	Positive	Positive
Carbohydrate fermentation		
D-Glucose	Negative	Positive
Sucrose	Negative	–
Maltose	Negative	Positive
	PGP Traits	
Phosphate solubilization (mg/mL)	0.063	1.881
Ammonia production (µmol/mL)	0.518	0.413
HCN production (µmol/mL)	21.30	30.58
Siderophore production (%)	48.71	42.30
Exopolysaccharide yield (mg/10mL)	18	15
ACC deaminase production (µmol/mL)	5.484	6.008
	Molecular	
BLAST Comparison (16S rDNA)	*Bacillus cereus*	*Bacillus haynesii*
Accession Number	*LC514122*	*LC514123*

These results have already reported in our previous study [19].

3.4. *Screening ACC Deaminase Activity from Rhizobacteria Isolates*

3.4.1. ACC Deaminase Qualitative Test

Both the AV-7 and AV-12 rhizobacterial strains were found to produce ACC deaminase in Petri plate method.

3.4.2. Quantification of ACC Deaminase

ACC deaminase activity for *B. cereus* and *B. haynesii* were estimated and presented in Table 2. *B. cereus* and *B. haynesii* produced 5.484 μM/mL and 6.008 μM/mL concentration of α- ketobutyrate, respectively.

Table 2. Performance of bacterial isolates under in vitro drought and heavy metal stress.

Properties	***B. cereus***	***B. haynesii***
Cell density (Number/mL)	108×10^6	108×10^6
EPS (mg/mg protein) No Stress	2.26	1.29
EPS (mg/mg protein) Stress	5.88	4.84
ACC deaminase activity (μM αKB/mg/min)	12.6	11.0
Total protein in crude extract (mg)	2.87	1.98
After purification		
ACC deaminase activity (μM/mg/min)	3.33	2.85
Mol weight of ACC deaminase (35-42 kDa)	35	40
Heavy metal stress at two concentrations of each heavy metal		
Arsenic (As) [14.74 mg/L]	Resistant	Sensitive
Arsenic (As) [29.48 mg/L]	Resistant	Sensitive
Barium (Ba) [6.28 mg/L]	Resistant	Sensitive
Barium (Ba) [12.56 mg/L]	Resistant	Sensitive
Nickel (Ni) [2.963 mg/L]	Resistant	Sensitive
Nickel (Ni) [5.926 mg/L]	Resistant	Sensitive

3.4.3. Measurement of ACC Deaminase Activity

The catalytic activity of ACC deaminase was found to be 12.6 μM αKB/mg/min for *B. cereus* and 11.0 μM αKB/mg/min for *B. haynesii.*

3.4.4. Characterization of Partially Purified ACC Deaminase Enzyme

The total protein was precipitated using ammonium sulphate to obtain partially purified segment of the ACC deaminase enzyme using chromatography. Initially, the total protein in crude extract was estimated to be 2.87 mg/mL for *B. cereus* and 1.98 mg/mL for *B. haynesii.* The ACC deaminase activity of crude protein extract was found to be 12.6 and 11.8 μM α keto butyrate/mg/min for *B. cereus* and *B. haynesii*, respectively (Table 2). The crude protein was further purified using silica gel column chromatography and obtained partially purified enzyme, whose specific activity was noted as 3.33 μM/mg/min and 2.85 μM/mg/min for *B. Cereus* and *B. haynesii*, respectively (Table 2).

3.4.5. Quantification of Partially Purified Protein Extraction by Bradford Method

Bacillus cereus and *B. haynesii* were tested for protein content in the partially purified enzyme. The average protein was quantified by using α-ketobutyrate standard curve and the concentrations were found to be 0.372 μg/mL for *B. cereus* and 0.272 μg/mL for *B. haynesii.*

3.4.6. Determination of Molecular Weight of ACC Deaminase

ACC deaminase is a multimeric sulfhydryl enzyme, and each subunit has molecular mass ranging 35–42 kDa approximately and is always tightly bound with pyridoxal phosphate (PLP) as co-factor [19]. Molecular weight of purified ACC deaminase enzyme of one subunit from *B. Cereus* and *B. haynesii* were found to be 35 kDa and 40 kDa, respectively (Figure 4).

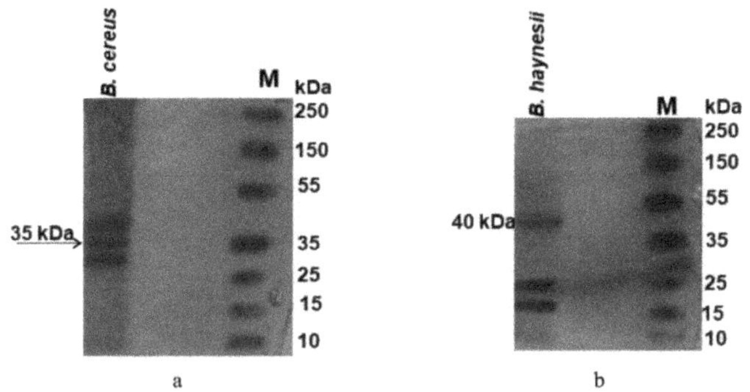

Figure 4. Purified 1-aminocyclopropane-1-carboxylate (ACC) deaminase derived from (**a**) *Bacillus cereus strain* (35 KDa) and (**b**) *Bacillus haynesii* (40 KDa).

3.5. Quantification of EPS

EPS yield was recorded after 6 days of inoculation of the bacterial strains and the results indicated that *B. cereus* and *B. haynesii* produced 18 mg/10 mL and 15 mg/10 mL EPS, respectively (Table 2).

Protein content in EPS

EPS contains protein, carbohydrates, uronic acid, and little amount of DNA. The protein and other content in the EPS help the PGPRs to stick on the surface of the roots of the crop [6]. To study the efficacy of the PGPRs under varied stress conditions, 5% PEG 6000 (−0.05 MPa); 10% PEG 6000 (−0.15 MPa); 15% PEG 6000 (−0.30 MPa), and 20% PEG 6000 (−0.49 MPa), to produce protein content in EPS; the protein content in the EPS from both the bacterial strains has been compared. The results revealed that *B. cereus* produced 2.27 μg/mg, 3.73 μg/mg, 4.61 μg/mg, and 5.87 μg/mg protein in its EPS at −0.05 MPa, −0.15 MPa, −0.30 MPa, and −0.49 MPa water potential, respectively. Similarly, *B. haynesii* also produced increased amount of protein content in EPS with a decrease in osmotic potential. It produced 1.31 μg/mg, 2.07 μg/mg, 3.47 μg/mg, and 4.82 μg/mg of protein in EPS at −0.05 MPa, −0.15 MPa, −0.30 MPa, and −0.49 MPa water potential, respectively.

3.6. Heavy Metal Tolerance

Out of these two rhizobacterial isolates, only *B. cereus* (AV-12) was capable of growing on nutrient agar supplemented with As, Ba, and Ni (Table 2). Therefore, AV-12 can tolerate these heavy metals *in vitro*, but the rest of the isolates including *B. haynesii* (AV-7) could not grow on the provided media and remained sensitive towards these heavy metals.

4. Discussion

In the present study, the comparative analysis of 16S rRNA gene sequence indicated that the bacterial isolate AV-12 is *B. cereus*. This bacterium belongs to: Species: *B. cereus*, Genus: *Bacillus*, Family: *Bacillaceae*, Order: *Caryophanales*, Class: *Bacilli*, and Phylum: *Firmicutes*. The scientific name of this bacterium was assigned by Frankland and Frankland (1887). The type strains are ATCC 14579, CCM 2010, JCM 2152, BCRC 10603, BCRC 11026, IAM 12605, NRRL B-3711, DSM 31, NBRC 15305, CGMCC 1.3760, LMG 6923, CECT

148, NCIB 9373, NCTC 2599, CECT 5050, CCUG 7414, CIP 66.24, IFO 15305, HAMBI 1887, IAM 14174, KCTC 3624, NCCB 75008, NCAIM B.02078, NCIMB 9373, VKM B-504, VTT E-93143, and NCFB 1771. Hence, during 16S rRNA nucleotide sequence BLAST performance, ATCC 14579 was selected as a reference sequence, whose similarity with AV-12 was found to be 99.64%, and other strains showed similarity ranging from 99.64% to 95.36% in the phylogenetic tree. *B. cereus* group comprises of other closely related species, namely: *B. cereus sensu stricto* (referred to herein as *B. cereus*), *B. anthracis*, *B. thuringiensis*, *B. mycoides*, *B. pseudomycoides*, and *B. cytotoxicus*.

In a similar way, the other bacterial isolate AV-7 was identified as *B. haynesii*. This bacterium belongs to: Species: *B. haynesii*, Genus: *Bacillus*, Order: Caryophanales, Family: Bacillaceae class: Bacilli, and phylum: Firmicutes. The scientific name of this bacterial strain was assigned by Dunlap et al. (2017) [36]. The type strains are NRRL B-41327, and CCUG 70178. Hence, during 16S rRNA nucleotide sequence BLAST performance, NRRL B-41327 (*B. haynesii* Dunlap et al., 2017, Accession No. 3EC4C1) was selected as a reference sequence, whose similarity was found to be 99.18%, and 20 other *B. cereus* group type strains showed similarity ranging from 98.964 to 97.61% in the phylogenetic tree.

Diverse organisms with a wide range of ACC deaminase activity act as PGPR. A low level of ACC deaminase activity, namely approximately $\geq$20 nmol α-ketobutyrate/mg/h, is sufficient to permit a bacterium to grow on ACC and to act as a PGPR, and can support root elongation in crops [37]. Another similar study was carried out with 841 rhizobacterial isolates, isolated from 74 chickpea plants, out of which 743 isolates were determined to be *Bacillus* and *Pseudomonas* by using taxonomically selective and enrichment isolation protocols. Out of these 743 isolates, 19 isolates were closely related to *Bacillus* spp. and could produce ACC deaminase activity ranging from 0.35 to 7.32 μmol α KB/h, and it was sufficient to alleviate drought stress [37]. In another study, Duan et al. (2021) isolated seven bacterial isolates from grapevine rhizosphere. Out of these seven isolates, two (DR3 and DR6) isolates were found to produce high amount of protein in EPS, which ranged from 41.18 to 60.11 μmol α-KB/(mg/Protein/h) and showed drought tolerance potentials [38]. A wide range of bacteria belonging to the genera *Bacillus*, *Burkholderia*, *Pseudomonas* and *Variovorax* showed ACC deaminase production ability which reduced ethylene production under various abiotic stress conditions.

A study was conducted in chickpeas by Kumar et al. (2016), who studied the potential of two PGPR (*Bacillus amyloliquefaciens* and *Pseudomonas putida*) to alleviate drought stress in chickpeas, as these strains had multiple PGP traits including ACC deaminase activity when performed in vitro. In this study, PGPRs colonization in chickpea rhizosphere was visualized with gfp labelling. The in vivo experiments disclosed that a combined application (in consortia) of these microbes ameliorated negative effects of drought stress in chickpeas, as it was evident by increased biomass. The increased colonization, enhanced ACC deaminase activity, increased chlorophyll content, and proline accumulation were observed in chickpeas, which indicates that the bacterial consortium was able to alleviate drought stress [39].

In a similar study, Sharma et al. (2013) isolated 47 bacterial isolates from rhizospheric soils of chickpeas in Punjab (India), and they found that 25 and 22 strains were of *Bacillus* and *Pseudomonas*, respectively. Further, they found that out of these only ten strains could use ACC as sole source of N. They reported that *Bacillus* isolate 23-B and *Pseudomonas* 6-P in combination with *Mesorhizobium ciceris* can help the plant to mitigate the water stress for *Cicer arietinum* (Kabuli L-552 and Desi GPF-2) and enhanced all growth parameters even under water stress conditions [17].

It was noted that the molecular weight of ACC deaminase from *Pseudomonas* strains are in a range of 35 kDa to 41 kDa [40]. The molecular mass of the whole ACC deaminase enzyme isolated from *Pseudomonas* sp. UW4 was found to be 168 kDa [41]. The molecular mass of ACC deaminase has been reported in various studies such as *Methylobacterium nodulans* ORS2060 (Homotetramer 144 kDa) [42], *Methylobacterium radiotolerans* JCM2831 (Homotetramer 144 kDa) [43], and *Amycolatopsis methanolica* 239 (Homotetramer 144 kDa) [43].

Water retention and cementing properties are the unique qualities of exopolysaccharide which have vital roles in stabilizing soil aggregates, forming biofilm, and enhancing nutrient flow up to plant roots. *Bacillus cereus* and *B. haynesii,* isolated in this study, also produced fairly good amounts of exopolysaccharides, due to which they can result as promising bioinoculants in various stress conditions, especially drought. EPS production in significant amounts have been reported in several studies. According to Sandhya et al. [44], out of 81 bacterial strains isolated from alfisols, vertisols, inceptisols, oxisols, and aridisols, 26 isolates could tolerate the maximum level of drought stress (−0.73 MPa) and produced EPS. Further, they observed that *P. putida* GAP-P45 had higher drought tolerance capacity [44].

Heavy metal tolerance was studied in *Bacillus* sp., which was collected and isolated from heavy-metal-polluted soil. This bacterial isolate was helpful in the detoxification of trivalent and tetra valent chromium [45]. Similarly, *Bacillus subtilis* was found to be instrumental in alleviating stress caused by high concentration of Cd in soil for carrots [46]. The inoculation of this bacterial strain could immobilize Cd through bioaugmentation, and enhance the plant shoot and root growth 16% and 55%, respectively [46]. *Acidithiobacillus caldus* and *Sulfobacillus thermotolerans* were reported for their heavy metal bioleaching properties for Cu, Cd, Pb, Zn, Mn, Hg, and As [47]. Some other endophytic bacterial isolated from *Tridax procumbens* were reported for heavy metal detoxification including Zn, Pb, and As [48].

In the present study, efforts have been made to isolate the rhizobacterial strains from the rhizospheric soil collected from two legume crops. Two bacterial strains were characterized based on morphology and biochemical properties. Both, strains were Gram-negative. These bacterial isolates were evaluated for their potential as PGPR. *Bacillus cereus* and *B. hyanesii* had potential to produce acid during fermentation and are capable of producing catalase; these tests reveal their capacity to scavenge reactive oxygen species. These strains have potential to solubilize P and produce siderophore. *Bacillus cereus* can withstand heavy metals (As, Ba, and Ni), while *B. haynesii* was sensitive to said heavy metals. These strains can be used to promote growth of the crops cultivated in the soil lacking available P. These two strains also were able to tolerate drought under in vitro conditions. These strains also produce exopolysaccharides and these were also capable of producing the stress-releasing enzyme ACC deaminase. Looking at the various characteristics of these rhizobacterial isolates, they can be considered a potential PGPR bacteria to support the growth and yield of the legumes. In comparison to the use of synthetic hazardous chemicals in agriculture, the use of PGPRs is considered as the best alternative, as an eco-friendly and sustainable approach for agriculture. Applications of PGPRs as biofertilizers have the chance of increase agriculture productivity. Further, field study of these strains with crops will reveal their potential as a future bio-fertilizer [49].

5. Conclusions

The present study reveals that both bacterial strains (*B. cereus* and *B. haynesii*) produce EPS and ACC deaminase in considerable amounts, sufficient for mending abiotic stress, and showed their potential towards plant-growth-promoting properties. The molecular weight for ACC deaminase monomers was found to be 35 kDa and 40 kDa for *B. cereus* and *B. haynesii*, respectively. The heat stress tolerance potential of the bacterial strains has been further evaluated by looking into EPS-based protein production at various drought stress levels. The results indicate that the *B. cereus* bacterial strain has better protein yield than *B. haynesii* strain in both normal and induced water-stress conditions. Hence, from the present study, it can be concluded that both bacterial strains have PGPR potential. However, the *B. cereus* strain has shown better performance in vitro for PGPR properties with reference to ACC deaminase activity and also for protein quantity in EPS, both in normal and drought conditions, proving its potentiality as a biocontrol agent, biofertilizer, or as part of a bio-inoculum, especially as it can be one of the best solutions for abiotic stresses. The *Bacillus cereus* strain also showed heavy metal resistance to arsenic, barium, and nickel heavy metals, whereas *Bacillus haynesii* was sensitive towards these heavy metals. Further studies are required to look into the performance of these bacterial strains in association with chickpeas, initially in a pot experiment and later in field trials.

Supplementary Materials: The following supporting information can be downloaded at: https://www.mdpi.com/article/10.3390/horticulturae9010046/s1, Figure S1: PGPR traits (qualitative): A: Control; B: Bacillus cereus; C: Bacillus haynesii. First row: Phosphate solubilization; Second row: HCN production; Third row: ACC deaminase production. Figure S2: PGPR traits (qualitative): A: Production of ammonia- both the isolates showed positive response B: Production of IAA- both the bacterial isolates were positive for the test.

Author Contributions: V.S.G.: Concept, experimental design, and help in writing manuscript. A.K.A.: Conducted the experiments and collected the data, compiled the data, and wrote the manuscript. V.D.R. and M.B.: Provided critical suggestions, and edited and improved the manuscript. All authors have read and agreed to the published version of the manuscript.

Funding: This research received no external funding.

Institutional Review Board Statement: Not applicable.

Informed Consent Statement: Not applicable.

Data Availability Statement: Not applicable.

Acknowledgments: The authors would like to thank the dean of JIBB, SHUATS for conducting the experiment in their laboratory. V.D.R. acknowledges the support by the laboratory «Soil Health» of the Southern Federal University with the financial support of the Ministry of Science and Higher Education of the Russian Federation, agreement No. 075-15-2022-1122. M.B. would like to recognize the Grant of the Ministry of Education and Science of the Russian Federation for the development of youth laboratories, within the framework of the implementation of the Tula State Lev Tolstoy Pedagogical University program "Priority 2030" under Agreement No. 073-03-2022-117/7.

Conflicts of Interest: The authors declare no conflict of interest.

References

1. Barnawal, D.; Singh, R.; Singh, R.P. Role of Plant Growth Promoting Rhizobacteria in Drought Tolerance: Regulating Growth Hormones and Osmolytes. In *PGPR Amelioration in Sustainable Agriculture*; Singh, A.K., Kumar, A., Singh, P.K., Eds.; Woodhead Publishing: Sawston, UK, 2019; pp. 107–128. [CrossRef]
2. Sahu, P.K.; Jayalakshmi, K.; Tilgam, J.; Gupta, A.; Nagaraju, Y.; Kumar, A.; Hamid, S.; Singh, H.V.; Minkina, T.; Rajput, V.D.; et al. ROS generated from biotic stress: Effects on plants and alleviation by endophytic microbes. *Front. Plant Sci.* **2022**, *13*, 1042936. [CrossRef] [PubMed]
3. Tuteja, N.; Sopory, S.K. Chemical signaling under abiotic stress environment in plants. *Plant Signal. Behav.* **2008**, *3*, 525–536. [CrossRef] [PubMed]
4. Kour, D.; Khan, S.S.; Kaur, T.; Kour, H.; Singh, G.; Yadav, A.; Yadav, A.N. Drought adaptive microbes as bioinoculants for the horticultural crops. *Heliyon* **2022**, *8*, e09493. [CrossRef]
5. Ashraf, M.; Akram, N.A. Improving salinity tolerance of plants through conventional breeding and genetic engineering: An analytical comparison. *Biotechnol. Adv.* **2009**, *27*, 744–752. [CrossRef] [PubMed]
6. Ahmed, B.; Shahid, M.; Syed, A.; Rajput, V.D.; Elgorban, A.M.; Minkina, T.; Bahkali, A.H.; Lee, J. Drought tolerant enterobacter sp./Leclercia adecarboxylata secretes indole-3-acetic acid and other biomolecules and enhances the biological attributes of *Vigna radiata* (L.) R. Wilczek in water deficit conditions. *Biology* **2021**, *10*, 1149. [CrossRef]
7. Upadhyay, S.K.; Rajput, V.D.; Kumari, A.; Espinosa-Saiz, D.; Menendez, E.; Minkina, T.; Dwivedi, P.; Mandzhieva, S. Plant growth-promoting rhizobacteria: A potential bio-asset for restoration of degraded soil and crop productivity with sustainable emerging techniques. *Environ. Geochem. Health* **2022**. [CrossRef]
8. Kour, D.; Rana, K.L.; Sheikh, I.; Kumar, V.; Yadav, A.N.; Dhaliwal, H.S.; Saxena, A.K. Alleviation of drought stress and plant growth promotion by *Pseudomonas libanensis* EU-LWNA-33, adrought-adaptive phosphorus-solubilizing bacterium. *Proc. Natl. Acad. Sci. India Sect. B Biol. Sci.* **2020**, *90*, 785–795. [CrossRef]
9. Ali, S.Z.; Sandhya, V.; Rao, L.V. Isolation and characterization of drought-tolerant ACC deaminase and exopolysaccharide-producing fluorescent *Pseudomonas* sp. *Ann. Microbiol.* **2014**, *64*, 493–502. [CrossRef]
10. Vacheron, J.; Desbrosses, G.; Bouffaud, M.L.; Touraine, B.; Moenne-Loccoz, Y.; Muller, D.; Legendre, L.; Wisniewski-Dye, F.; Prigent-Combaret, C. Plant growth-promoting rhizobacteria and root system functioning. *Front. Plant Sci.* **2013**, *4*, 356. [CrossRef]
11. Fulekar, M.H.; Sharma, J.; Tendulkar, A. Bioremediation of heavy metals using biostimulation in laboratory bioreactor. *Environ. Monit. Assess.* **2012**, *184*, 7299–7307. [CrossRef]

12. Pande, V.; Pandey, S.C.; Sati, D.; Bhatt, P.; Samant, M. Microbial interventions in bioremediation of heavy metal contaminants in agroecosystem. *Front. Microbiol.* **2022**, *13*, 824084. [CrossRef] [PubMed]
13. Choudhary, A.K.; Kumar, S.; Patil, B.S.; Bhat, J.S.; Sharma, M.; Kemal, S.; Ontagodi, T.P.; Datta, S.; Patil, P.; Chaturvedi, S.K.; et al. Narrowing yield gaps through genetic improvement for Fusarium wilt resistance in three pulse crops of the semi-arid tropics. *SABRAO J. Breed. Genet.* **2013**, *45*, 341–370.
14. Araujo, S.S.; Beebe, S.; Crespi, M.; Delbreil, B.; Gonzalez, E.M.; Gruber, V.; Lejeune Henaut, I.; Link, W.; Monteros, M.J.; Prats, E.; et al. Abiotic stress responses in strategies used to cope with environmental challenges. *Crit. Rev. Plant Sci.* **2015**, *34*, 237–280. [CrossRef]
15. Fahad, S.; Bajwa, A.A.; Nazir, U.; Anjum, S.A.; Farooq, A.; Zohaib, A.; Sadia, S.; Nasim, W.; Adkins, S.; Saud, S.; et al. Crop Production under Drought and Heat Stress: Plant Responses and Management Options. *Front. Plant Sci.* **2017**, *8*, 1147. [CrossRef] [PubMed]
16. Available online: http://www.fao.org/faostat/en/#data/QC (accessed on 7 September 2022).
17. Sharma, P.; Khanna, V.; Kumar, P.I. Efficacy of aminocyclopropane-1-carboxylic acid (ACC)-deaminase-producing rhizobacteria in ameliorating water stress in chickpea under axenic conditions. *Afr. J. Microbiol. Res.* **2013**, *7*, 5749–5757. [CrossRef]
18. Khan, N.; Bano, A.; Rahman, M.A.; Guo, J.; Kang, Z.; Babar, M. Comparative physiological and metabolic analysis reveals a complex mechanism involved in drought tolerance in chickpea (*Cicer arietinum* L.) induced by PGPR and PGRs. *Sci. Rep.* **2019**, *9*, 2097. [CrossRef]
19. Andy, A.K.; Masih, S.A.; Gour, V.S. Isolation, screening and characterization of plant growth promoting rhizobacteria from rhizospheric soils of selected pulses. *Biocatal. Agric. Biotechnol.* **2020**, *27*, 101685. [CrossRef]
20. Sharma, K. *Manual of Microbiology: Tools & Techniques*; Ane Books: Delhi, India; pp. 141–171.
21. Bhaskara, R.K.V.; Ashwini, K.; Gaurav, K.; Karthik, L. Optimization, production and partial purification of extracellular á-amylase from *Bacillus* sp. marini. *Arch. Appl. Sci. Res.* **2011**, *3*, 33–42.
22. Pikovskaya, R.I. Mobilization of phosphorus in soil in connection with vital activity of some microbial species. *Mikrobiologiya* **1948**, *17*, 363–370.
23. Schwyn, B.; Neilands, J. Universal chemical assay for the detection and determination of siderophores. *Anal. Biochem.* **1987**, *160*, 47. [CrossRef]
24. Gordon, S.A.; Weber, R.P. Colorimetric estimation of indole acetic acid. *Plant Physiol.* **1951**, *26*, 192–195. [CrossRef] [PubMed]
25. Cappuccino, J.C.; Sherman, N. *Microbiology: A Laboratory Manual*, 3rd ed.; Benjamin/Cummings Pub. Co.: New York, NY, USA, 1992; pp. 125–179.
26. Lorck, H. Production of hydrocyanic acid by bacteria. *Physiol. Plant* **1948**, *1*, 142–146. [CrossRef]
27. Weisburg, W.G.; Barns, S.M.; Pelletier, D.A.; Lane, D.J. 16S ribosomal DNA amplification for phylogenetic study. *J. Bacteriol.* **1991**, *173*, 697–703. [CrossRef] [PubMed]
28. Jain, D.; Kachhwaha, S.; Jain, R.; Kothari, S.L. PCR based detection of cry genes in indigenous strains of Bacillus thuringiensis isolated from the soils of Rajasthan. *Indian J. Biotechnol.* **2012**, *11*, 491–494.
29. Deharvengt, S.J.; Petersen, L.M.; Jung, H.S.; Tsongalis, G.J. Chapter 13—Nucleic acid analysis in the clinical laboratory. In *Contemporary Practice in Clinical Chemistry*, 4th ed.; Academic Press: Cambridge, MA, USA, 2020; pp. 215–234.
30. Tamura, K.; Stecher, G.; Peterson, D.; Filipski, A.; Kumar, S. MEGA6: Molecular evolutionary genetics analysis version 6.0. *Mol. Biol. Evol.* **2013**, *30*, 2725–2729. [CrossRef]
31. Dworkin, M.; Foster, J.W. Experiments with some microorganisms which utilize ethane and hydrogen. *J. Bacteriol.* **1958**, *75*, 592. [CrossRef]
32. Penrose, D.M.; Glick, B.R. Methods for isolating and characterizing ACC deaminase-containing plant growth-promoting rhizobacteria. *Physiol. Plant* **2003**, *118*, 10–15. [CrossRef]
33. Maxton, A.; Singh, P.; Masih, S.A. ACC deaminase-producing bacteria mediated drought and salt tolerance in *Capsicum annuum*. *J. Plant Nutr.* **2018**, *41*, 574–583. [CrossRef]
34. Jia, Y.J.; Kakuta, Y.; Sugawara, M.; Igarashi, T.; Oki, N.; KisAKi, M.; Shoji, T.; Kanetuna, Y.; Horita, T.; Matsui, H.; et al. Synthesis and degradation of 1-aminocyclopropane-1-carboxylic acid by *Penicillium citrinum*. *Biosci. Biotechnol. Biochem.* **1999**, *63*, 542–549. [CrossRef]
35. Chug, R.; Mathur, S.; Kothari, S.; Gour, V.S. Maximizing EPS production from *Pseudomonas aeruginosa* and its application in Cr and Ni sequestration. *Biochem. Biophys. Rep.* **2021**, *26*, 100972. [CrossRef]
36. Dunlap, C.A.; Schisler, D.A.; Perry, E.B.; Connor, N.; Cohan, F.M.; Rooney, A.P. *Bacillus swezeyi* sp. nov. and *Bacillus haynesii* sp. nov., isolated from desert soil. *Int. J. Syst. Evol. Microbiol.* **2017**, *67*, 2720–2725. [CrossRef] [PubMed]
37. Alemneh, A.A.; Zhou, Y.; Ryder, M.H.; Denton, M.D. Is phosphate solubilizing ability in plant growth-promoting rhizobacteria isolated from chickpea linked to their ability to produce ACC deaminase? *J. Appl. Microbiol.* **2021**, *131*, 2416–2432. [CrossRef] [PubMed]
38. Duan, B.; Li, L.; Chen, G.; Su-Zhou, C.; Li, Y.; Merkeryan, H.; Liu, W.; Liu, X. 1-Aminocyclopropane1-carboxylate deaminase-producing plant growth-promoting rhizobacteria improve drought stress tolerance in grapevine (*Vitis vinifera* L.). *Front. Plant Sci.* **2021**, *12*, 706990. [CrossRef]

39. Kumar, M.; Mishra, S.; Dixit, V.; Kumar, M.; Agarwal, L.; Chauhan, P.S.; Nautiyal, C.S. Synergistic effect of Pseudomonas putida and *Bacillus amyloliquefaciens* ameliorates drought stress in chickpea (*Cicer arietinum* L.). *Plant Signal. Behav.* **2016**, *11*, e1071004. [CrossRef] [PubMed]
40. Glick, B.R.; Nascimento, F.X. Pseudomonas 1-Aminocyclopropane-1-carboxylate (ACC) Deaminase and Its Role in Beneficial Plant-Microbe Interactions. *Microorganisms* **2021**, *9*, 2467. [CrossRef]
41. Hontzeas, N.; Zoidakis, J.; Glick, B.R.; Abu-Omar, M.M. Expression and characterization of 1-aminocyclopropane-1-carboxylate deaminase from the rhizobacterium *Pseudomonas putida* UW4: A key enzyme in bacterial plant growth promotion. *Biochim. Biophys. Acta* **2004**, *1703*, 11–19. [CrossRef]
42. Fedorov, D.N.; Ekimova, G.A.; Doronina, N.V.; Trotsenko, Y.A. 1-Aminocyclopropane-1-carboxylate (ACC) deaminases from *Methylobacterium radiotolerans* and *Methylobacterium nodulans* with higher specificity for ACC. *FEMS Microbiol. Lett.* **2013**, *343*, 70–76. [CrossRef]
43. Ekimova, G.A.; Fedorov, D.N.; Doronina, N.V.; Trotsenko, Y.A. 1-aminocyclopropane-1-carboxylate deaminase of the aerobic facultative methylotrophic actinomycete *Amycolatopsis methanolica* 239. *Microbiology* **2015**, *84*, 584–586. [CrossRef]
44. Sandhya, V.; Ali, S.Z.; Grover, M.; Reddy, G.; Venkatesvarlu, B. Alleviation of drought stress effects in sunflower seedlings by exopolysaccharides producing *Pseudomonas putida* strain GAP-P_{45}. *Biol. Fertil. Soils* **2009**, *46*, 17–26. [CrossRef]
45. Syed, S.; Chinthala, P. Heavy metal detoxification by different *Bacillus* species isolated from solar salterns. *Scientifica* **2015**, *2015*, 319760. [CrossRef]
46. Wang, T.; Sun, H.; Jiang, C.; Mao, H.; Zhang, Y. Immobilization of Cd in soil and changes of soil microbial community by bioaugmentation of UV-mutated *Bacillus subtilis* 38 assisted by biostimulation. *Eur. J. Soil Biol.* **2014**, *65*, 62–69. [CrossRef]
47. Panyushkina, A.E.; Babenko, V.V.; Nikitina, A.S.; Selezneva, O.V.; Tsaplina, I.A.; Letarova, M.A.; Kostryukova, E.S.; Letarov, A.V. *Sulfobacillus thermotolerans*: New insights into resistance and metabolic capacities of acidophilic chemolithotrophs. *Sci. Rep.* **2019**, *9*, 15069. [CrossRef] [PubMed]
48. Govarthanan, M.; Mythili, R.; Selvankumar, T. Bioremediation of heavy metals using an endophytic bacterium *Paenibacillus* sp. RM isolated from the roots of *Tridax procumbens*. *3 Biotech* **2016**, *6*, 242. [CrossRef]
49. Basu, A.; Prasad, P.; Das, S.N.; Kalam, S.; Sayyed, R.Z.; Reddy, M.S.; El Enshasy, H. Plant Growth Promoting Rhizobacteria (PGPR) as Green Bioinoculants: Recent Developments, Constraints, and Prospects. *Sustainability* **2021**, *13*, 1140. [CrossRef]

horticulturae

MDPI

Article

Investigation of Growth Factors and Mathematical Modeling of Nutrient Media for the Shoots Multiplication In Vitro of Rare Plants of the Rostov Region

Vasiliy A. Chokheli [1,*], Semyon D. Bakulin [2], Olga Yu. Ermolaeva [1], Boris L. Kozlovsky [1], Pavel A. Dmitriev [1], Victoriya V. Stepanenko [1], Igor V. Kornienko [1,3], Anastasia A. Bushkova [1], Vishnu D. Rajput [1] and Tatiana V. Varduny [1]

1 Academy of Biology and Biotechnology, Southern Federal University, Rostov on Don 344090, Russia
2 Department of Botany, Moscow Timiryazev Agricultural Academy, Russian State Agrarian University, Plant Breeding and Seed Technology, Moscow 127550, Russia
3 Southern Scientific Center of the Russian Academy of Sciences, Paleogeography Laboratory, Rostov-on-Don 344006, Russia
* Correspondence: vachokheli@sfedu.ru

Abstract: Micropropagation is an effective way to preserve the gene pool of threatened plants. This study is devoted to the mathematical modeling of nutrient media and the study of the effect of *m*T (meta-topoline) on the multiplication of shoots of *Hedysarum grandiflorum*, *Hyssopus cretaceus*, and *Matthiola fragrans* in vitro in comparison with benzylaminopurine (BAP) and kinetin (KT). Initiation was performed on an MS medium with 0.5 mg/L BAP. For shoots multiplication, MS, B5, and WPM media were used with the addition of *m*T, BAP, KT. For *H. grandiflorum*, the multiplication coefficient of shoots was highest on medium B5 with the addition of mT at a concentration of 1 mg/L—2.90 shoots per plant, for *H. cretaceus*—B5 + 0.5 mg/L *m*T, and for *M. fragrans*—B5 + 1 mg/L KT. A positive effect of *m*T on *H. grandiflorum* and *M. fragrans* in vitro was found. The efficiency of using KT for *H. cretaceus* shoot multiplication is shown. The effectiveness of the B5 nutrient medium for *H. grandiflorum* and *M. fragrans* was determined. The positive effect of WPM for *H. cretaceus* micropropagation has been demonstrated. It is not recommended to use the MS media for micropropagation of these plant species.

Keywords: red list; 6-benzylaminopurine; meta-topoline; kinetin; micropropagation; phytohormones

Citation: Chokheli, V.A.; Bakulin, S.D.; Ermolaeva, O.Y.; Kozlovsky, B.L.; Dmitriev, P.A.; Stepanenko, V.V.; Kornienko, I.V.; Bushkova, A.A.; Rajput, V.D.; Varduny, T.V. Investigation of Growth Factors and Mathematical Modeling of Nutrient Media for the Shoots Multiplication In Vitro of Rare Plants of the Rostov Region. *Horticulturae* **2023**, 9, 60. https://doi.org/10.3390/horticulturae9010060

Academic Editors: Adalberto Benavides-Mendoza, Yolanda González-García, Fabián Pérez Labrada and Susana González-Morales

Received: 24 November 2022
Revised: 23 December 2022
Accepted: 26 December 2022
Published: 4 January 2023

1. Introduction

Nowadays, one of the most popular and highly effective methods of preserving the gene pool of rare and endangered plant species is micropropagation. This method not only make it possible to preserve endangered plant species in vitro but also to study their genetic, physiological, anatomical, and morphological aspects of biology, to identify and isolate secondary metabolites that are used in medicine, as well as to produce an amount of material sufficient for breeding or reproduction of regenerating plants for further sale [1–3]. The improvement of old and the search for new techniques is necessary due to introducing new plant species into in vitro conditions. The selection of the optimal phytohormonal composition of nutrient media can contribute not only to an increase in the degree of multiplication but also to the effective production of callus, and the creation of bioreactor cultures, tissue cultures, embryos, and pollen [3].

The flora of the Rostov region is rich in rare and endangered plant species. Nowadays, more than 273 species of plants and fungi with a threatened status have been included in the Red List of the region, and 45 plant species from this list are included in the Red List of the Russian Federation [4]. For threatened flora species of the Rostov region, such as *Hedysarum grandiflorum* Pall. and *Hyssopus cretaceus* Dubj., data on the methods of micropropagation vary, and no microclonal propagation schemes have been developed for *Matthiola fragrans* Bunge.

The studies on the micropropagation of these and related species recommend use of benzylaminopurine (BAP) as the only cytokinin [5,6]. Growth regulators such as kinetin (KT) [7], 2-isopentyladenine [8], and tidiazuron [9], are rarely used. Many plants, especially threatened ones, are characterized by difficulties in in vitro cultivation. Such phenomena may be caused by seed dormancy, difficulties in sterilizing explants, low rates of shoot multiplication and rhizogenesis, and problems with histogenesis and morphogenesis of plants [10]. The success of in vitro cultivation of plants and their further acclimatization strongly depends on the choice of a suitable phytohormonal growth regulator. It is known that BAP in plant tissues is metabolized to stable toxic compounds—6-benzylaminopurine-9-glycosides (9-B-glucopyransosyl-benzyladenine), which accumulate in tissues at the base of plants, inhibiting their further development [11].

An alternative to BAP is N6-(3-hydroxybenzyl)adenine (meta-topoline, *m*T), a less toxic phytohormone that improves further processes of root formation in vitro and acclimatization. The *m*T strengthens the processes of multiplication and rhizogenesis, non-irrigation of tissues, and increases the success of acclimatization of plants [12,13]. Successful cultivation of plants in tissue culture depends not only on the choice of growth stimulants and their concentration, but also depends on the composition of the components of the medium. In this regard, the necessary stage is the mathematical planning of the nutrient medium [14].

The purpose of this research is to mathematically plan and develop effective methods of multiplication of plant species in vitro to determine the influence of BAP and KT on the *m*T growth and development of plants.

2. Materials and Methods

Hedysarum grandiflorum is an herbaceous perennial with ornamental, forage, and medicinal properties [4,15]. *Hedysarum cretaceus* is a perennial herb, and an obligate inhabitant of chalk outcrops. It has decorative, medicinal, ethiromalenic, and honey-bearing properties [4,16]. *M. fragrans* is an herbaceous perennial inhabitant of chalk outcrops, and an ornamental plant [4]. In the Red List of the Rostov region, *H. grandiflorum* and *H. cretaceus* have the status of 3 b, d—rare species with narrow ecological confinement associated with specific growing conditions, having a limited range, part of which is located in the Rostov region. *Matthiola fragrans* has the status of 3 b—a rare species with a narrow ecological amplitude associated with a specific substrate for growth. These species are listed in the Red Book of the Russian Federation under the status 3—rare species [4]. They experience a strong negative impact from the destruction of natural habitats, low competitiveness, and anthropogenic disturbances of the habitat. All three species are plants that prefer chalky soil substrates [4,17].

Seeds collected at the Nursery of Rare and Endangered Plants of the Botanical Garden of the Southern federal university (BG SFedU) were used as explants for the introduction of research objects into the culture.

For *H. cretaceous* and *M. fragrans*, the seed material was sterilized by washing the seeds in a mixture of 70% ethyl alcohol solution and 3% hydrogen peroxide solution in a ratio of 1:1 for 10 min, followed by washing the diaspores in distilled water for 10 min four times. *Hedysarum grandiflorum* seeds, after washing in a mixture of alcohol and hydrogen peroxide, were immersed in 96% ethanol for 1 s, and then burned with a gas burner flame. This is necessary to get rid of the hard seed. The seeds were germinated on a Murashige-Skoog nutrient medium with the addition of BAP at a concentration of 0.5 mg/L [18].

When developing a nutrient medium for the cultivation of rare plants, it is necessary to carry out mathematical planning of the main components (factor gradations); this is the type of phytohormone, its concentration, and mineral-organic base (nutrient medium). The course of planning is based on the following algorithm: (1) selection of 3 or more different bases for nutrient media (MS, B5, etc.), including fractional bases (1/2 MS, 1/3 MS, etc.); (2) selection of phytohormone (BAP, *m*T, etc.); (3) selection of phytohormone concentration in large steps (0.5 mg/L): 0 (hormone-free control medium), 0.5 mg/L, 1 mg/L, 1.5 mg/L, 2 mg/L. If necessary, you can take large concentrations. This is what concerns the animation of escapes. For rhizogenesis, it is recommended to use the following concentrations of phytohormones (in step of 0.2 mg/L): 0 (hormone-free control medium), 0.2 mg/L, 0.4 mg/L, 0.6 mg/L, 0.8 mg/L, 1 mg/L. If necessary, you can also take large concentrations. After setting up all the experiments, the data is calculated by multivariate analysis of variance (ANOVA) and the influence of the factor is determined.

The second part of the experimental development on nutrient media planning is to reduce the step (0.1 mg/L) of the phytohormone concentration on a certain nutrient medium. For example, if the effective basis turned out to be a WPM medium with a phytohormone mT concentration equal to 1.5 mg/L, then it is worth reducing the step in one direction (1.4 mg/L, 1.3 mg/L) and increasing the step in the other direction (1.6 mg/L, 1.7 mg/L). As a result, the optimal nutrient medium for this genotype of a rare plant is obtained.

Our study presents the first part of the experiment and shows the influence of various factors on 3 species from 3 genera belonging to 3 different families: *Fabaceae* family (*Hedysarum grandiflorum*), Lamiaceae family (*Hyssopus cretaceus*), and the Brassicaceae family (*Matthiola fragrans*).

MS, Gamborg and Eveleg media, Woody Plant Medium [19], with the adding of several phytohormones in different concentrations, were used to stimulate the multiplication of shoots (Table 1).

Table 1. Variants of experimental media for stimulating the animation of shoots of research objects *in vitro*.

	MS	B5	WPM
Control	-	-	-
BAP	0.5 1.0 1.5 2.0	0.5 1.0 1.5 2.0	0.5 1.0 1.5 2.0
*m*T	0.5 1.0 1.5 2.0	0.5 1.0 1.5 2.0	0.5 1.0 1.5 2.0
KT	0.5 1.0 1.5 2.0	0.5 1.0 1.5 2.0	0.5 1.0 1.5 2.0

Plant passages were carried out in a laminar box using sterilized instruments (scissors, tweezers, dissecting needles). The pH of the nutrient media was adjusted to a value of 6.0 using a 1 M KOH solution. The culture media was sterilized in an autoclave MLS-3751L (Sanyo) at a temperature of 121 °C and pressure of 1.5 atmospheres for 30 min. The plants were cultivated at a constant temperature of 25 °C and a 16-h photoperiod.

The sample for each variant of the experiment was 20 plants for each media combination. The main parameter to assess the success of the media combinations revealed in the course of the study was the multiplication coefficient of the shoots. The quadratic error was determined for the obtained parameter values, and a comparative analysis was performed using the Student's t-test at $p = 0.05$ [20]. Statistical analysis of the values of the proliferation coefficient was carried out using the method of multivariate analysis of variance ANOVA using programs Statistica 13.3 and Microsoft Excel.

3. Results

3.1. Sterilization

The results of the use of the described scheme of sterilization of seeds of *H. grandiflorum*, *H. cretaceus*, and *M. fragrans* are shown in the diagram (Figure 1).

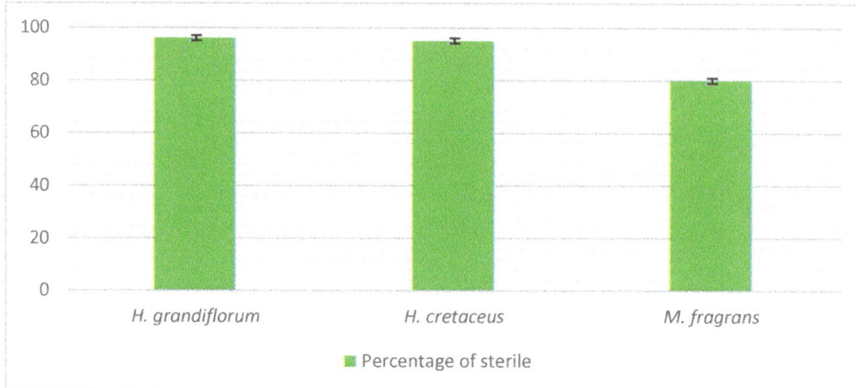

Figure 1. Results of application of methods of sterilization of seeds of objects of research.

It was managed to achieve, respectively, (mean ± standard error), 96.00 ± 2.77%, 95.00 ± 3.08%, and 80.00 ± 5.65% of the seed sterility level.

3.2. Multiplication

The average values of the multiplication coefficient of shoots of the studied plant species on different nutrient media were revealed. Standard errors for the obtained values are determined (Tables 2–4).

Table 2. Average values of *H. grandiflorum* multiplication coefficient on different nutrient media.

		MS	B5	WPM
Control		1.00 ± 0.00 *	1.00 ± 0,00 *	1.7 ± 0.22 *
*m*T	0.5	1.00 ± 0.00 *	2.10 ± 0.16 *	1.7 ± 0.23 *
	1.0	1.00 ± 0.00 *	2.90 ± 0.34	1.3 ± 0.11*
	1.5	1.00 ± 0.00 *	2.40 ± 0.15	2.00 ± 0.21*
	2.0	1.00 ± 0.00 *	2.45 ± 0.23	1.30 ± 0.13*
BAP	0.5	1.75 ± 0.19 *	1.25 ± 0.01 *	1.60 ± 0.18 *
	1.0	2.35 ± 0.26	1.60 ± 0.11 *	1.20 ± 0.09 *
	1.5	1.20 ± 0.14 *	1.30 ± 0.11 *	1.65 ± 0.20 *
	2.0	1.55 ± 0.14 *	1.70 ± 0.19 *	1.35 ± 0.15 *
KT	0.5	1.35 ± 0.14 *	1.55 ± 0.14 *	1.25 ± 0.10 *
	1.0	1.10 ± 0.07 *	1.90 ± 0.1 *	1.20 ± 0.09 *
	1.5	1.15 ± 0.08 *	1.55 ± 0.14 *	1.35 ± 0.13 *
	2.0	1.40 ± 0.13 *	2.65 ± 0.23	1.35 ± 0.11 *

* The values with a significant statistical difference from the highest value of the parameter according to the Student's t-criterion at $t_{teor} = 2.09$, $\alpha = 5\%$, $p = 0.05$ are emphasized.

Table 3. Average values of *H. cretaceus* multiplication coefficient on different nutrient media.

		MS	B5	WPM
Control		1.00 ± 0.00 *	1.45 ± 0.18 *	1.80 ± 0.25 *
*m*T	0.5	1.00 ± 0.00 *	4.20 ± 0.42 *	3.00 ± 0.32 *
	1.0	1.15 ± 0.08 *	2.05 ± 0.20 *	3.00 ± 0.36 *
	1.5	1.60 ± 0.23 *	1.90 ± 0.20 *	2.25 ± 0.19 *
	2.0	1.20 ± 0.16 *	1.85 ± 0.20 *	3.50 ± 0.52
BAP	0.5	1.40 ± 0.15 *	1.60 ± 0.20 *	3.20 ± 0.46
	1.0	2.60 ± 0.22 *	1.75 ± 0.20 *	1.95 ± 0.22 *
	1.5	2.20 ± 0.21 *	1.30 ± 0.15 *	1.45 ± 0.15 *
	2.0	2.40 ± 0.23 *	1.65 ± 0.21 *	1.00 ± 0.00 *
KT	0.5	2.25 ± 0.19 *	2.60 ± 0.20 *	2.00 ± 0.24 *
	1.0	2.10 ± 0.20 *	1.85 ± 0.23 *	2.30 ± 0.24 *
	1.5	2.85 ± 0.33 *	2.25 ± 0.25 *	2.45 ± 0.26 *
	2.0	2.35 ± 0.31 *	1.80 ± 0.20 *	2.60 ± 0.22 *

* The values with a significant statistical difference from the highest value of the parameter according to the Student's t-criterion at t_{teor} = 2.09, α = 5%, p = 0.05 are emphasized.

Table 4. Average values of *M. fragrans* multiplication coefficient on different nutrient media.

		MS	B5	WPM
Control		1.95 ± 0.14 *	2.90 ± 0.28	3.05 ± 0.22
*m*T	0.5	2.65 ± 0.27 *	3.25 ± 0.36	3.15 ± 0.30 *
	1.0	2.75 ± 0.37	1.95 ± 0.22 *	2.55 ± 0.23 *
	1.5	2.85 ± 0.34	1.95 ± 0.22 *	2.10 ± 0.22
	2.0	3.05 ± 0.38	2.40 ± 0.26 *	3.35 ± 0.39 *
BAP	0.5	1.00 *	2.90 ± 0.45 *	1.40 ± 0.18 *
	1.0	1.00 *	2.40 ± 0.37 *	1.75 ± 0.22 *
	1.5	1.00 *	2.35 ± 0.27 *	1.75 ± 0.22 *
	2.0	1.00 *	2.30 ± 0.33 *	1.85 ± 0.24 *
KT	0.5	1.85 ± 0.20 *	1.65 ± 0.20 *	2.65 ± 0.25 *
	1.0	1.90 ± 0.14 *	3.75 ± 0.40 *	1.90 ± 0.27 *
	1.5	1.80 ± 0.29 *	2.30 ± 0.24 *	2.65 ± 0.30 *
	2.0	2.60 ± 0.36 *	2.75 ± 0.33 *	2.40 ± 0.24 *

* The values with a significant statistical difference from the highest value of the parameter according to the Student's t-criterion at t_{teor} = 2.09, α = 5%, p = 0.05 are emphasized.

The highest average value of the multiplication coefficient of *H. grandiflorum* shoots *in vitro* was found on nutrient medium B5 + 1 mg/L mT—2.90 ± 0.34 shoots per plant (Table 2, Figure 2). A comparative analysis using the *t*-test indicates the reliability of this result compared to most parameter values when using other nutrient media options. The differences with the use of 1 mg/L of BAP on MS medium and 2 mg/L of KT on mineral base B5 are unreliable. The range of the multiplication coefficient when using *m*T varied from 1.00 to 2.90 ± 0.34 shoots per plant. The use of BAP resulted in obtaining average values of the multiplication coefficient in the range from 1.20 ± 0.14 to 2.35 ± 0.26 shoots per plant. The variation of the values of the analyzed parameter on media with KT ranged from 1.10 ± 0.07 to 2.65 ± 0.23. It was when using the B5 nutrient medium that both *m*T and other growth regulators used proved to be more effective than on other mineral bases. When using the MS mineral base, the multiplication coefficient reached its maximum in the variant with the addition of 1 mg/L of BAP—2.35 ± 0.26 per plant. The maximum value of the multiplication coefficient on the B5 medium was recorded using *m*T at a concentration of 1 mg/L—2.90 ± 0.34 shoots per plant. At the same time, the coefficient values are distributed more evenly on the WPM medium than in variants with other mineral bases.

The highest average value of the multiplication coefficient of *H. cretaceus* shoots in vitro was found on nutrient medium B5 + 0.5 mg/L *m*T—4.20 ± 0.42 shoots per plant (Table 3, Figure 3). A comparative analysis using the *t*-test indicates the reliability of this result compared to most parameter values when using other nutrient media options. The differences with the use of 2 mg/L *m*T and 0.5 mg/L BAP on a mineral basis WPM are unreliable. Variants of experiments with *m*T showed relatively high values of the multiplication coefficient on B5 and WPM media (Table 3). BAP demonstrates a less stimulating effect on all media (from 1.00 to 3.20 ± 0.46 shoots per plant). The maximum value of the multiplication coefficient when using BAP is observed at its concentration of 0.5 mg/L in the WPM medium. When using KT as a growth regulator, the multiplication coefficient varied less than when using mT and BAP—from 1.80 ± 0.20 to 2.85 ± 0.33 shoots per plant. There is a linear increase in the multiplication coefficient when using KT in a WPM environment.

Figure 2. The effect of different types of growth regulators on plants of *H. grandiflorum* shoots *in vitro*.

Figure 3. The effect of different concentrations of *m*T on plants of *H. cretaceus* shoots *in vitro*.

The highest average value of the multiplication coefficient of *M. fragrans* shoots in vitro was found on nutrient medium B5 + 1 mg/L KT (Figure 4). On all variants of nutrient media with the addition of mT, relatively high values of the multiplication coefficient are observed—from 1.95 ± 0.22 to 3.35 ± 0.39 shoots per plant (Table 4). The maximum value is fixed at WPM + 2 mg/L mT. Significant differences were revealed in most cases of comparison with media using BAP, where the spread of results varied from 1.00 to 2.90 ± 0.45 shoots per plant. At the same time, the maximum value of the parameter is fixed at the lowest concentration of BAP on medium B5—0.5 mg/L. When using this mineral base, relatively high values with BAP were observed, although they did not give significant differences when compared with most other experimental options. The multiplication coefficient on nutrient media with the addition of KT varied from 1.65 ± 0.20 (B5 + 0.5 mg/L KT) to 3.75 ± 0.40 (B5 + 1 mg/L KT). The last value of the coefficient turned out to be the maximum for the entire experiment. Based on comparative analysis, the use of KT at a concentration of 1 mg/L on B5 medium does not differ significantly from the use of mT at most concentrations, as well as BAP at a concentration of 0.5 mg/L on all types of mineral bases studied.

When using KT as a growth regulator, regardless of the type of mineral base, hormone, and concentration, the plants looked green, devoid of signs of chlorosis, vitrification, and necrosis. When using BAP, vitrification of plants was observed. Such growth features were recorded more intensively when using the MS mineral base, both together with BAP and mT, but to a lesser extent, at a concentration of *m*T 2 mg/L.

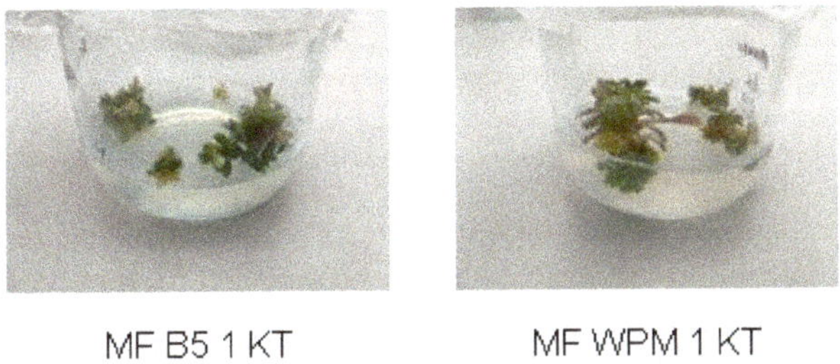

Figure 4. The effect of different media with *m*T on plants of *H. cretaceus* shoots *in vitro*.

3.3. Statistical Analysis

Statistical comparisons were made between all variants of nutrient media for each studied species using the Student's *t*-test at $p = 0.05$ (Tables 2–4).

Multivariate analysis of the values of the multiplication coefficient was performed. The factors were: mineral base type (MS, B5, WPM), growth regulator (*m*T, BAP, KT), and growth regulator concentration, mg/L (0.0; 0.5; 1.0; 1.5; 2.0). Numerical results of the analysis are presented in Appendix A Table A1, Table A2, Table A3, and graphical data are shown in Figure 5.

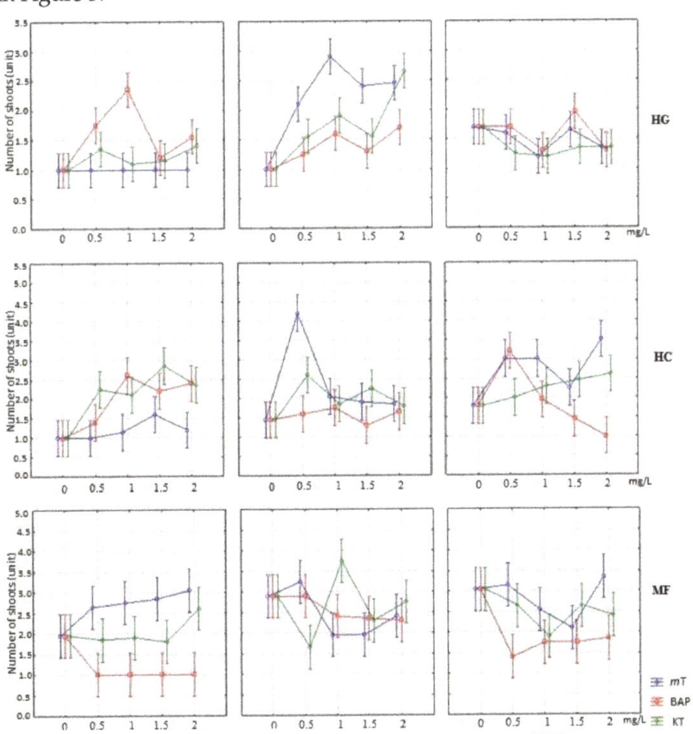

Figure 5. The combined effect of mineral base, growth regulator, and growth regulator concentration on the average number of *H. grandiflorum* (HG), *H. cretaceous* (HC), and *M. fragrans* (MF) shoots *in vitro*. Confidence interval = 0.95.

4. Discussion

4.1. Seed Sterilization

Roasting in the flame had a scarifying effect on the seeds of *H. grandiflorum*, which have a hard-seed surface. The firing efficiency was also shown earlier by us using the example of *H. cretaceum* seeds [21]. The results of sterilization of *H. grandiflorum* seeds were higher compared to the results given in some other studies on the micropropagation of *H. grandiflorum* and other species of the genus, where diacid, 70% alcohol solution, sodium hypochlorite, and Domestos were used as sterilizing agents [6,22,23].

Sterilization of *H. cretaceus* seeds according to the described scheme allowed the obtainment of 95% sterile seeds. This methodic sterilization of primary explants was successfully tested by us earlier on seeds of another species—*H. angustifolius* [3]. The chosen technique is not inferior in its effectiveness to other methods of sterilization of seeds of this species using sodium hypochlorite [9,24], silver nitrate [25], "Lysoformin 3000" [26], 70% ethanol [9,26], mercury chloride [27], chloramine B [25], and Tween 20 [24].

The result of using the chosen sterilization technique makes it more effective for *M. fragrans* seeds than using sodium hypochlorite, as indicated in the literature [28]. To achieve a higher percentage of sterile seeds, it may be necessary to slightly reduce the exposure time of seeds in sterilizers.

4.2. Multiplication of Shoots

4.2.1. Hedysarum Grandiflorum

Interestingly, for the micropropagation of other members of the genus *Hedysarum*, it is recommended to use BAP, sometimes KT, as suitable cytokinins in a fairly wide range—from 0.1 to 10.0 mg/L, often together with various auxins [5,29–33]. For several representatives of the genus *Hedysarum* (*H. cumuschtanicum, H. cephalotes, H. songoricum, H. ferganense, H. neglectum, H. semenovii, H. sultanovae, H. montanum, H. plumosum*), it is proposed to use the nutrient medium of Gamborg and Eveleg (B5) with the addition of BCI and BAP in concentrations of 0.25 mg/L and 0.1 mg/L, respectively, as a substrate for multiplication of shoots. Variants of the B5 nutrient medium with the addition of 0.1 mg/L of KT and BAP separately also proved to be effective. Medium B5 with the addition of 0.25 mg/L of BCI had the greatest stimulating effect on the process of rhizogenesis of regenerating plants of all studied plants [34].

Visually, the healthiest, developed *H. grandiflorum* plants looked on medium B5 using *m*T and KT as growth regulators. The *m*T effectively stimulates multiplication, and KT—elongation of shoots. At the same time, the use of nutrient media with the addition of BAP causes vitrification of plants when using all mineral bases. With an increase in the concentration of BAP, the degree of vitrification increased. When using the mineral base of WPM, the plants looked weak, and chlorosis was often detected. This may be caused by an insufficient amount of nitrogen in the composition of the medium compared with other variants of mineral bases [18,19,35].

MS mineral base is known for its high content of both nitrate and ammonium nitrogen. We assume that ammonium nitrogen in high concentrations can block the proliferation and multiplication of *H. grandiflorum* shoots *in vitro*. There are known data on the effect of the ratio of nitrate and ammonium on the organogenesis of various plants *in vitro*. Thus, Bennett et al. [36] found that with the complete removal of ammonium nitrate from the nutrient medium, the rooting efficiency and survival rate during the adaptation of *Eucalyptus globulus* plants increase. A decrease in the concentration of ammonium nitrogen compared to nitrate increased the fresh weight of potato plants *in vitro*, made the pH level of the medium more stable, but did not increase the efficiency of plant multiplication [37]. An increase in the concentration of nitrate–nitrogen to ammonium led to a better degree of micro-cloning of garlic plants [38].

Numerical results of the analysis of variance (Appendix A Table A1) show that the degree of multiplication of *H. grandiflorum* shoots in vitro slightly depends on the selected regulator. The type of mineral base and the concentration of growth regulators have a greater effect on changing this parameter. It can be seen from the graphs in Figure 5 that the greatest variation in values is observed when using phytohormones at a concentration of 1 mg/L. More definite patterns of the dependence of the effectiveness of animation on the selected types of hormone (Figure 2) and the mineral basis of the nutrient medium can be traced when using growth regulators at concentrations of 0.0 and 2.0 mg/L. Transitional situations are observed due to the use of hormones in concentrations of 0.5 and 1.5 mg/L. Interestingly, in most variants of the experiment, the effect of phytohormones was more noticeable on the B5 medium and to a lesser extent when using MS and WPM mineral bases. Application of *m*T had no effect on the multiplication of shoots in combination with the mineral base of MS.

The data obtained are consistent with the literature information on the effectiveness of using the B5 medium or its components for microcloning of representatives of the Fabaceae family [5,39,40]. For the embryogenesis of *Albizzia lebbeck*, a modified nutrient medium B5 was used, with the addition of phytohormones 2,4-D and Kin at concentrations of 0.5 mg/L and 2 mg/L, respectively [39]. For in vitro germination of seeds of *Vigna subterranea* as a mineral base, medium B5 was used, along with media such as MS, SH, and CHU [40].

4.2.2. Hysoppus Cretaceus

Low concentrations of cytokinins, up to 1 mg/L, are indicated as effective in studies on micropropagation of *Hyssopus* representatives [3,8,9,41,42]. Despite the effectiveness of *m*T in stimulating multiplication, plants on all media with its addition acquired chlorosis, and from the middle of the third week of cultivation, tissue necrosis. A similar situation was observed on all media with BAP, which is why vitrification and stunting also developed in plants. The absence of such adverse effects was observed when using KT as growth regulators, which additionally stimulated the elongation of shoots of the studied plants. BAP is often recommended together with auxins for in vitro micropropagation of shoots of other members of the genus—*H. officinalis*, *H. angustifolius* [9,25].

The actual data of the dispersion analysis show that in the case of *H. cretaceus*, the type of mineral medium, the choice of the growth regulator, and its concentration have both individually and together an impact on the effectiveness of the multiplication of shoots (Appendix A Table A2). According to the graphs in Figure 5, a similar pattern of action on the animation of shoots is observed for all used growth regulators at concentrations of 1.0 and 1.5 mg/L (Figure 3). Here, *m*T is more effective in the B5 media, BAP, and KT—in the MS media. At the same time, both BAP and KT reduce their effectiveness of action at given concentrations on the B5 medium. When using mineral medium B5, there is an increase in the efficiency of the use of all phytohormones at a concentration of 0.5 mg/L, especially *m*T. On the WPM medium, at this concentration, the use of BAP shows itself best of all. At a concentration of growth regulators of 2 mg/L, a similar efficiency dynamic is observed between *m*T and KT, whereas BAP in the range from MS to WPM becomes less effective for multiplying shoots. A low concentration of cytokinins also has a more effective effect on *H. angustifolius* multiplication in vitro [3].

Unfortunately, the above scientific literature does not explain the detrimental effect of *meta*-topoline on the growth and development of *H. cretaceous* regenerating plants *in vitro*. Being a "mildly" acting cytokinin, KT in the case of *H. cretaceus* may become more suitable for the multiplication of shoots in vitro, which has been tested by other researchers on related plant species [27]. Thus, for example, for growing *Hyssopus officinalis*, an optimal nutrient medium for the induction of callusogenesis was a medium with indolyl-3-butyric acid—1 mg/L, and a medium with a-naphthylacetic acid—0.5 mg/L, kinetin—0.1 mg/Land 6 -benzylaminopurine—1 mg/L, and for the cultivation of plants, H. officinalis in vitro—a medium containing indolyl-3-acetic acid IAA (2 mg/L), and kinetin (0.2 mg/L) [27].

The experimental results demonstrate that the WPM medium may be more suitable for the effective multiplication of *H. cretaceous* shoots *in vitro*. A reduced concentration of nitrogen-containing salts may remain very important for this plant species (patent by Kritskaya and colleagues (2015)).

4.2.3. Matthiola Fragrans

The results obtained on the effect of BAP on the multiplication of *M. fragrans* differ from the information found for *M. incana* [43]. Visually, when using *m*T, the leaves of regenerating plants increased, and with KT, the shoots were slightly stretched upwards. The effects of using KT are confirmed by the literature information on the example of *M. incana* [44].

When using *m*T, the multiplication coefficient did not fall below 1.95 $\pm$ 0.22 per plant, which is the highest minimum value compared to those for other phytohormones.

Based on a comparative analysis of the average values of the multiplication coefficient of *M. fragrans* shoots, the use of *m*T and KT turns out to be more effective compared to the use of BAP.

The analysis of variance showed interesting data (Appendix A Table A3). Based on the results of the analysis, the mineral base B5 has a significant effect on the "work" of growth regulators. When using it, BAP and *m*T have relatively the same strength of action at all the concentrations studied. KT behaves similarly at concentrations of 1.5 and 2.0 mg/L. Of the studied factors affecting the multiplication of *M. fragrans* shoots, the type of growth regulator is the most significant. The effects of *m*T, BAP, and KT at different concentrations and different mineral bases turned out to be very different. This is especially noticeable in the graphs describing the behavior of BAP relative to other growth regulators (Figure 5). The concentration of phytohormone also affects the degree of multiplication. Based on the graphs in Figure 5, the effectiveness of using a particular hormone varies greatly from the chosen concentration.

The effectiveness of the use of *m*T on BAP was shown for many valuable and threatened species of plants in the plant world, such as *Spathifphyllum floribundum* [45], *Uniola paniculate* [46], species and varieties of *Pelargonium* [47], and *Musa* [48], *Aloe polyphylla* [12], *Actinidia chinensis*, *Coccoloba uvifera* [49], *Beta vulgaris* [50], *Cannabis sativa* [51], *Manihot esculenta* [52], *Sesamum indicum* [53], and species and varieties of *Prunus* [54].

Based on the results of statistical analysis, the use of *m*T and KT is more effective for stimulating the multiplication of *M. fragrans* in vitro compared to BAP. The MS environment is less suitable for this task compared to B5 and WPM (Figure 4). Low concentrations of phytohormones (0.5 and 1.0 mg/L) stimulate the multiplication of *M. fragrans* in vitro more effectively than elevated ones.

Different plant growth regulators belonging to the same class of phytohormones have different physiological activity for different plants. Thus, in our study, three different species from three families (*Fabaceae, Lamiaceae, Brassicaceae*) were studied. As can be seen in Figure 5 and Tables 2–4, the influence of the same growth regulator has a different effect for different species. For example, on medium B5 under the action of *m*T at a concentration of 1 mg/L, the reproduction coefficient for *H. grandiflorum* is 2.9, for *H. cretaceus*—2.05, and for *M. fragrans*—1.95. Thus, there is no universal nutrient medium for the most rare and endangered plant species, and only thanks to the mathematical planning of nutrient media, it is possible to obtain an effective scheme for the selection of a mineral base, such as a growth regulator and its concentration.

5. Conclusions

The method of seed sterilization used during the study made it possible to achieve high sterility rates for all three objects of study. A mixture of ethanol and hydrogen peroxide effectively cleanses the outer covers of diaspores and firing in a flame allows for additional removal of hard-seeding. In the case of sterilization of *M. fragrans* seeds, it may be necessary to reduce the exposure time in sterilizing agents to further increase sterility.

According to the results of laboratory experiments and statistical analyses, as part of a study on the selection of nutrient media and effective concentrations of phytohormones at the stage of multiplication of shoots for *H. grandiflorum*, it is recommended to use *m*T at concentrations of 0.5–2.0 mg/L on a mineral base B5; the use of BAP proved detrimental to the multiplication of *H. cretaceus* shoots *in vitro*, while the use of low (0.5–0.1 mg/L) concentrations of mT and KT on a B5 and WPM medium proved effective; also, low concentrations of *m*T and KT together with nutrient media B5 and WPM are more effective for microcloning of *M. fragrans* in vitro in comparison with BAP, as well as MS nutrient medium.

For the studied rare and endangered plant species of the Rostov region that have needs for specific substrates, the MS nutrient medium rich in ammonium nitrogen adversely affects the growth and development of regenerating plants *in vitro*. A reduced concentration of ammonium in the medium or its complete absence can play a big role in the success of the selected species in vitro animation.

Author Contributions: Conceptualization, V.A.C., S.D.B. and O.Y.E.; methodology, S.D.B. and V.V.S.; validation, B.L.K., P.A.D. and T.V.V.; formal analysis, I.V.K.; data curation, S.D.B.; writing—original draft preparation, V.A.C. and S.D.B.; writing—review and editing, V.A.C., A.A.B. and V.D.R.; supervision, V.A.C. and T.V.V.; project administration, V.A.C. All authors have read and agreed to the published version of the manuscript.

Funding: The research was financially supported by the Ministry of Science and Higher Education of the Russian Federation within the framework of the state task in the field of scientific activity (no. 0852-2020-0029).

Institutional Review Board Statement: Not applicable.

Informed Consent Statement: Not applicable.

Data Availability Statement: Not applicable.

Acknowledgments: Using the equipment of the laboratory of Cellular and Genomic Technologies of Plants of the Botanical Garden of the SFedU, center of "Biotechnology, Biomedicine and Environmental Monitoring", and the center of "High Technologies".

Conflicts of Interest: The authors declare no conflict of interest.

Appendix A

Table A1. Results of multivariate analysis of variance values of *H. grandiflorum* shoot multiplication coefficient *in vitro*.

Effect	One-Dimensional Significance Criterion for the Number of Shoots Per Plant, pcs. (Data Table 2) Sigma-Limited Parametrization Decomposition of the Hypothesis				
	SS	Degrees	MS	F	p
Medium type	37.580 *	2 *	18.790 *	42.277 *	0 *
Hormone	2.327	2 *	1.163	2.617	0.073570
Hormone concentration, mg/L	18.733 *	4 *	4.683 *	10.538 *	0 *
Medium type * Hormone	48.953 *	4 *	12.238 *	27.536 *	0 *
Medium type * Hormone concentration, mg/L	59.453 *	8 *	7.432 *	16.721 *	0 *
Hormone * Hormone concentration, mg/L	9.240 *	8 *	1.155 *	2.599 *	0.008249 *
Medium type * Hormone * Hormone concentration, mg/L	26.713 *	16 *	1.670 *	3.757 *	0.000001 *
Error	380.000	855	0.444		

* The values with reliability are underlined. $p = 0.05$.

Table A2. Results of multivariate analysis of variance values of *H. cretaceus* shoot multiplication coefficient *in vitro*.

Effect	One-Dimensional Significance Criterion for the Number of Shoots Per Plant. pcs. (Data Table 3) Sigma-Limited Parametrization Decomposition of the Hypothesis				
	SS	Degrees	MS	F	p
Medium type	43.469 *	2 *	21.734 *	19.254 *	0 *
Hormone	18.729 *	2 *	9.364 *	8.296 *	0.000270 *
Hormone concentration, mg/L	86.196 *	4 *	21.549 *	19.090 *	0 *
Medium type * Hormone	90.884 *	4 *	22.721 *	20.128 *	0 *
Medium type * Hormone concentration, mg/L	61.231 *	8 *	7.654 *	6.780 *	0 *
Hormone * Hormone concentration, mg/L	30.338 *	8 *	3.792 *	3.359 *	0.000844 *
Medium type * Hormone * Hormone concentration, mg/L	114.816 *	16 *	7.176 *	6.357 *	0 *
Error	965.150	855	1.129		

* The values with reliability are underlined. $p = 0.05$.

Table A3. Results of multivariate analysis of variance values of *M. fragrans* shoot multiplication coefficient *in vitro*.

Effect	One-Dimensional Significance Criterion for the Number of Shoots Per Plant, pcs. (Data Table 4) Sigma-Limited Parametrization Decomposition of the Hypothesis				
	SS	Degrees	MS	F	p
Medium type	64.642 *	2 *	32.321 *	22.327 *	0 *
Hormone	88.169 *	2 *	44.084 *	30.453 *	0 *
Hormone concentration, mg/L	31.473 *	4 *	7.868 *	5.435 *	0.000253 *
Medium type * Hormone	60.551 *	4 *	15.138 *	10.457 *	0 *
Medium type * Hormone concentration, mg/L	26.180 *	8 *	3.272 *	2.261 *	0.021535 *
Hormone * Hormone concentration, mg/L	46.553 *	8 *	5.819 *	4.020	0.000106 *
Medium type * Hormone * Hormone concentration, mg/L	81.993 *	16 *	5.125 *	3.540	0.000003 *
Error	1237.700	855	1.448		

* The values with reliability are underlined. $p = 0.05$.

References

1. Cardoso, J.C.; Sheng Gerald, L.T.; Teixeira da Silva, J.A. Micropropagation in the twenty-first century. In *Plant Cell Culture Protocols. Methods in Molecular Biology*; Loyola-Vargas, V., Ochoa-Alejo, N., Eds.; Humana Press: New York, NY, USA, 2018; Volume 1815. [CrossRef]
2. Molkanova, O.; Shirnina, I.; Mitrofanova, I. Conservation and micropropagation of rare and endemic species in genepool collections of the Russian Federation. *J. Biotechnol.* **2018**, *280*, S83–S88. [CrossRef]
3. Chokheli, V.A.; Dmitriev, P.A.; Rajput, V.D.; Bakulin, S.D.; Azarov, A.S.; Varduni, T.V.; Stepanenko, V.V.; Tarigholizadeh, S.; Singh, R.K.; Verma, K.K.; et al. Recent Development in Micropropagation Techniques for Rare Plant Species. *Plants* **2020**, *9*, 1733. [CrossRef] [PubMed]
4. Kamelin, R.V.; Gizatulin, R.R.; Mitvol, O.L.; Amirkhanov, A.M.; Bardunov L.V.; Novikov, V.S.; Orlov, V.A.; Stepanitsky, V.B.; Belanovich, D.M.; Varlygina, T.I.; et al. *Red book of the Rostov Region. Ministry of Natural Resources and Ecology of the Rostov Region*, 2nd ed.; Association of Scientific Publications KMK: Rostov-on-Don, Russia, 2008. (In Russian)

5. Erst, A.A.; Zvyagina, N.S.; Novikova, T.I.; Dorogina, O.V. Clonal Micropropagation of a Rare Species *Hedysarum theinum* Krasnob. (Fabaceae) and Assessment of the Genetic Stability of Regenerated Plants Using ISSR Markers. *Russ. J. Genet.* **2015**, *51*, 158–162. [CrossRef]
6. Akhmetova, A.S.; Zaripova, A.A. In vitro propagation of some species of the genus *Hedysarum* L. In *Biology, Biochemistry and Genetics*; News of the Ufa Scientific Center of the Russian Academy of Sciences: Ufa, Russia, 2017; pp. 28–33.
7. Kaviani, B. Kaviani Behzad Micropropagation of ten weeks (*Matthiola incana*) and lisianthus (*Eustoma grandiflorum*) (two ornamental plants) by using kinetin (KIN), naphthalene acetic acid (NAA) and 2,4-dichlorophenoxyacetic acid (2,4-D). *Acta Sci. Pol. Hortorum Cultus* **2014**, *13*, 141–154.
8. Mohajjel, S.H.; Kharrati, S.H. Effects of different hormonal treatments on growth parameters and secondary metabolite production in organ culture of *Hyssopus officinalis* L. *J. Biotechnol. Comput. Biol. Bionanotechnol.* **2021**, *102*, 33–41.
9. Hossein, B.; Morteza, A.; Hassani, A.; Morad, J.; Rahimi, A. High-Frequency *in Vitro* Direct Shoot Regeneration from Nodal Explants of Hyssop Plant (*Hyssopus officinalis* L.). *J. Med. Plants By-Prod.* **2016**, *2*, 187–193.
10. Prakash, J. Micropropagation of ornamental perennials: Progress and problems. *Acta Hortic.* **2009**, *812*, 289–294. [CrossRef]
11. Werbrouck, S.P.O.; van der Jeugt, B.; Dewitte, W.; Prinsen, E.; Van Onckelen, H.A.; Debergh, P.C. The metabolism of benzyladenine in Spathiphyllum floribundum "Schott Petite" in relation to acclimatisation problems. *Plant Cell Rep.* **1995**, *14*, 662–665. [CrossRef] [PubMed]
12. Bairu, M.W.; Stirk, W.A.; Dolezal, K.; Van Staden, J. Optimizing the micropropagation protocol for the endangered Aloe polyphylla: Can meta-topolin and its derivatives serve as replacement for benzyladenine and zeatin? *Plant Cell Tissue Organ Cult.* **2007**, *90*, 15–23. [CrossRef]
13. Strnad, M.; Hanus, J.; Vanek, T.; Kaminek, M.; Ballantine, J.A.; Fussell, B.; Hanke, D.E. Meta-topolin, a highly active aromatic cytokinin from poplar leaves (Populus × canadensis Moench., CV. Robusta. *Phytochemistry* **1997**, *45*, 213–218. [CrossRef]
14. Zlenko, V.A.; Kotikov, I.V.; Volynkin, V.A.; Troshin, L.P. Optimization of nutrient medium composition by the mathematical design of experiment for shoot tip development in four grapevine genotypes. *Sci. J. KubSAU* **2010**, *55*, 237–254. Available online: http://ej.kubagro.ru/2010/01/pdf/05.pdf (accessed on 14 September 2022). (In Russian).
15. Imachueva, D.R.; Serebryanaya, F.K. Use of Capillary Electrophoresis Method in Determination of Quantitative Content of Mangiferin in Grass of Species of Genus *Hedysarum* (*Hedysarum Caucasicum* M.Bieb., *Hedysarum Grandiflorum* Pall., *Hedysarum Daghestanicum* Rupr. ex Boiss) of flora of the North Caucasus. *Dev. Regist. Med. (Razrabotka i registraciya lekarstvenny'x sredstv)* **2021**, *10*, 90–96. (In Russian)
16. Shevchuk, O.M.; Korotkov, O.I.; Malaeva, E.V.; Feskov, S.A. Component composition of essential oil in *Hyssopus cretaceus* Dubj. and *Hyssopus officinalis* L. *Ind. Bot. (Promy'shlennaya botanika)* **2019**, *19*, 49–54. (In Russian)
17. Kamelin, R.V.; Gizatulin, R.R.; Mitvol, O.L.; Amirkhanov, A.M.; Bardunov, L.V.; Novikov, V.S.; Orlov, V.A.; Stepanitsky, V.B.; Belanovich, D.M.; Varlygina, T.I.; et al. *Red Book of the Russian Federation (Plants and Fungi)*; Association of Scientific Publications KMK: Moscow, Russia, 2008. (In Russian)
18. Murashige, T.; Skoog, F. A revised medium for rapid growth and bioassay with tobacco tissue cultures. *Plant Physiol.* **1962**, *15*, 437–497. [CrossRef]
19. McCown, B.H.; Lloyd, G. Woody Plant Medium (WPM)—A Mineral Nutrient Formulation for Microculture of Woody Plant Species. *HortScience* **1981**, *15*, 437–497.
20. Lakin, G.F. *Biometriya*, 4th ed.; Revised and additional; Higher School: Moscow, Russia, 1990; 351p. (In Russian)
21. Bakulin, S.D. Cultivation of the Red Book plant species of the Rostov region *Hedysarum cretaceum* Fisch. In *Vitro Culture. Materials of the International Youth Scientific Forum "LOMONOSOV-2020"*; Aleshkovsky, I.A., Andriyanov, A.V., Antipov, E.A., Eds.; MAKS Press: Moscow, Russia, 2020. Available online: https://lomonosov-msu.ru/archive/Lomonosov_2020/index.htm (accessed on 5 October 2022). (In Russian)
22. Bliudneva, E.A.; Kritckaia, T.A.; Kashin, A.S.; Kirillova, I.M. Conservation of Plant Species and Cultivars in Botanical Garden Saratov State University in vitro Collection. *Proc. Saratov Univ. Nov. Ser. Ser. Chem. Biol. Ecol.* **2014**, *14*, 48–53. [CrossRef]
23. Duque, A.S.; Araújo, S.S.; Fevereiro, P.; Barradas, A.; Godinho, B.; Silva, A.R.; Crespo, J.P. Development of protocols for micropropagation of elite genotype forage allogamous legume species. *Acta Hortic.* **2015**, *1083*, 409–413. [CrossRef]
24. Murakami, Y.; Omoto, T.; Asai, I.; Shimomura, K.; Yoshihira, K.; Ishimaru, K. Rosmarinic acid and related phenolics in transformed root cultures of Hyssopus officinalis. *Plant Cell Tissue Organ Cult.* **1998**, *53*, 75–78. [CrossRef]
25. Zayova, E.; Geneva, M.; Stancheva, I.; Dimitrova, L.; Petrova, M.; Hristozkova, M.; Salamon, I. Evaluation of the antioxidant potential of *in vitro* propagated hyssop (*Hyssopus officinalis* L.) with different plant growth regulators. *Med. Plants* **2018**, *10*, 295–304. [CrossRef]
26. Zybkina, D.R. *Introduction to the Culture of Some Rare and Medicinal Plants of the Lamiaceae Family (Labiaceae): Final Qualifying Work*; Belgorod State University: Belgorod, Russia, 2017; 68p.
27. Maslova, E.; Gulya, N.; Perelugina, T.; Semykina, V.; Kalashnikova, E. Introduction of *Hyssopus officinalis* L. into *in vitro* culture to optimize the conditions for obtaining callus tissues and microclonal propagation as a promising metod of innovative agrobiotechnologies. *BIO Web Conf.* **2021**, *30*, 05006. [CrossRef]
28. Golub, N.O.; Cherednichenko, M.Y. In vitro Introduction of Two Matthiola Species. In Proceedings of the 3rd International Symposium on EuroAsian Biodiversity, Minsk, Belarus, 5–8 July 2017.
29. Akhmetova, A.S. Morphogenesis of *Hedysarum argyrophyllum* Ledeb. in *in vitro* culture. *Agrochemistry. Plant Growth Regul.* **2013**, *9*, 55–58. (In Russian)

30. Gamburg, K.Z. A possibility of saving an endangered endemic of the Lake Baikal shore, *Hedysarum zundukii* Peschkova (Fabaceae Lindl.) using clonal micropropagation. *Nat. Sci.* **2013**, *5*, 1289–1297.
31. Erst, A.A.; Nuzhdina, N.S. In vitro propagation of endemic species Hedysarum chaiyrakanicum (Tuva Republic, Russia) and its widespread congener, H. gmelini (Fabaceae). *BIO Web Conf. EDP Sci.* **2020**, *24*, 00021. [CrossRef]
32. Alieva, Z.M. Reproduction of rare plants of dagestan *in vitro*. Botany in the modern world. In Proceedings of the XIV Congress of the Russian Botanical Society and the Conference "Botany in the Modern World", Makhachkala, Russia, 18–23 June 2018; pp. 233–234. (In Russian).
33. Erst, A.A.; Zheleznichenko, T.V.; Novikova, T.I.; Dorogina, O.V.; Banaev, E.V. Ecological and geographic variability of *Hedysarum theinum* and features of its propagation in vitro. *Contemp. Probl. Ecol.* **2014**, *7*, 67–71. [CrossRef]
34. Konurbaeva, R.U.; Aldayrbek kyzy, G.; Umralina, A.R. Introduction to in vitro culture and maintenance of in vitro collections of Hedysarum genus species of Kyrgyzstan. *News Univ.* **2015**, 126–131. (In Russian)
35. Gamborg, O.L.; Miller, R.A.; Ojima, K. Nutrient requirements of suspension cultures of soyabean root cells. *Exp. Cell Res.* **1968**, *50*, 151–158. [CrossRef]
36. Bennett, I.J.; McDavid, D.A.J.; McComb, J.A. The influence of ammonium nitrate, pH and indole butyric acid on root induction and survival in soil of micropropagated Eucalyptus globulus. *Biol. Plant.* **2003**, *47*, 355–360. [CrossRef]
37. Rahman, M.H.; Haider, S.A.; Hossain, M.; Islam, R. Effect of potassium and ammonium nitrate media on in vitro growth response of potato (*Solanum tuberosun* L.). *Int. J. Biosci. (IJB)* **2011**, *1*, 54–57.
38. Luciani, G.; Marinangeli, P.A.; Curvetto, N.R. Increasing nitrate/ammonium ratio for improvement of garlic micropropagation. *Sci. Hortic.* **2001**, *87*, 11–20. [CrossRef]
39. Croser, J.S.; Lülsdorf, M.M.; Davies, P.A.; Clarke, H.J.; Bayliss, K.L.; Mallikarjuna, N.; SiddiqueToward, K.H.M. Toward doubled haploid production in the Fabaceae: Progress, constraints, and opportunities. *Crit. Rev. Plant Sci.* **2006**, *25*, 139–157. [CrossRef]
40. Mongomaké, K.; Hilaire, K.T.; Daouda, K.; Michel, Z.; Justin, K.Y.; Sergio, J. Ochatt In vitro plantlets regeneration in Bambara groundnut [*Vigna subterranea* (L.) Verdc. (Fabaceae)] through direct shoot bud differentiation on hypocotyl and epicotyl cuttings). *Afr. J. Biotechnol.* **2009**, *8*, 1466–1473.
41. Nanova, Z.; Slavova, Y.; Nenkova, D.; Ivanova, I. Microclonal propagation of hyssop (*Hyssopus officinalis* L.). *Bulg. J. Agric.* **2007**, *13*, 213–219.
42. Rolli, E.; Ricci, A.; Bianchi, A.; Bruni, R. Optimisation ofin vitropropagation of *Hyssopus officinalis* L. using two-node explants and N-phenyl-N′-benzothiazol-6-yl-urea (PBU), a new urea-type cytokinin. *J. Hortic. Sci. Biotechnol.* **2011**, *86*, 141–145. [CrossRef]
43. Kaviani, B. Micropropagation of *Matthiola incana* using BA and IBA. *Iran. J. Plant Physiol.* **2014**, *4*, 1071–1078.
44. Kaviani, B.; Hesar, A.A.; Kharabian-Masouleh, A. In vitro propagation of Matthiola incana (Brassicaceae)-an ornamental plant. *Plant Omics J.* **2011**, *4*, 435–440.
45. Werbrouck, S.P.O.; Strnad, M.; Van Onckelen, H.A.; Debergh, P.C. Meta-topolin, an alternative to benzyladenine in tissue culture? *Physiol. Plant.* **2008**, *98*, 291–297. [CrossRef]
46. Valero-Aracama, C.; Kane, M.E.; Wilson, S.B.; Philman, N.L. Substitution of benzyladenine with meta-topolin during shoot multiplication increases acclimatization of difficult- and easy-to-acclimatize sea oats (*Uniola paniculata* L.) genotypes. *Plant Growth Regul.* **2009**, *60*, 43–49. [CrossRef]
47. Wojtania, A. Effect of meta-topolin on in vitro propagation of Pelargonium hortorum and Pelargonium hederaefolium cultivars. *Acta Soc. Bot. Pol.* **2010**, *79*, 101–106. [CrossRef]
48. Escalona, M.; Cejas, I.; Gonzalez-Olmedo, J.; Capote, I.; Roles, S.; Cañal, M.J.; Rodríguez, R.; Sandoval, J.; Debergh, P. The effect of metatopolin on plantain propagation using a Temporary Immersion Bioreactor. *Infomusa* **2003**, *12*, 28–30.
49. Podwyszyńska, M.; Wojtania, A.; Rojek, A.; Kowalczyk, J. Mikrorozmnażanie perukowca podolskiego [Micropropagation of smoke tree]. In Proceedings of the Polish Conf. "New Technology in Nursery Production", Poznań-Kórnik, Poland, 8–9 September 2000; pp. 92–95.
50. Kubaláková, M.; Strnad, M. The effect of aromatic cytokinins on micropropagation and regeneration of sugarbeet *in vitro*. *Biol. Plant* **1992**, *34*, 578–579.
51. Hemant, L.; Suman, C.; Natascha, T.; Khana, I.A.; ElSohlya, M.A. *In vitro* mass propagation of *Cannabis sativa* L.: A protocol refinement using novel aromatic cytokinin meta-topolin and the assessment of eco-physiological, biochemical and genetic fidelity of micropropagated plants. *J. Appl. Res. Med. Aromat. Plants* **2016**, *3*, 18–26. [CrossRef]
52. Chauhan, R.D.; Taylor, N.J. Meta-topolin stimulates de novo shoot organogenesis and plant regeneration in cassava. *Plant Cell Tissue Organ Cult.* **2018**, *132*, 219–224. [CrossRef] [PubMed]
53. Elayaraja, D.; Subramanyam, K.; Vasudevan, V.; Sathish, S.; Kasthurirengan, S.; Ganapathi, A.; Manickavasagam, M. Meta-Topolin (mT) enhances the in vitro regeneration frequency of *Sesamum indicum* (L.). *Biocatal. Agric. Biotechnol.* **2019**, *21*, 101320. [CrossRef]
54. Gentile, A.; Ja'quez Gutie´rrez, M.; Martinez, J.; Frattarelli, A.; Nota, P.; Caboni, E. Effect of meta-Topolin on micropropagation and adventitious shoot regeneration in *Prunus* rootstocks. *Plant Cell Tissue Organ Cult.* **2014**, *118*, 373–381. [CrossRef]

horticulturae

MDPI

Article

Potential of *Suaeda nudiflora* and *Suaeda fruticosa* to Adapt to High Salinity Conditions

Abhishek Joshi [1], Vishnu D. Rajput [2,*], Krishan K. Verma [3], Tatiana Minkina [2], Karen Ghazaryan [4] and Jaya Arora [1,*]

1 Laboratory of Biomolecular Technology, Department of Botany, Mohanlal Sukhadia University, Udaipur 313001, India
2 Academy of Biology and Biotechnology, Southern Federal University, 344090 Rostov-on-Don, Russia
3 Key Laboratory of Sugarcane Biotechnology and Genetic Improvement (Guangxi), Ministry of Agriculture and Rural Affairs/Guangxi Key Laboratory of Sugarcane Genetic improvement/Sugarcane Research Institute, Guangxi Academy of Agricultural Sciences, Nanning 530007, China
4 Faculty of Biology, Yerevan State University, Yerevan 0025, Armenia
* Correspondence: rajput.vishnu@gmail.com (V.D.R.); jaya890@gmail.com (J.A.)

Abstract: The deposition of salts in soil seems likely to become a significant barrier for plant development and growth. Halophytes that flourish in naturally saline habitats may sustain extreme salt levels by adopting different acclimatory traits. Insight into such acclimatory features can be useful for devising salt-resilient crops and the reclamation of saline soil. Therefore, salinity-induced responses were studied in two halophytes, i.e., *Suaeda nudiflora* and *Suaeda fruticosa*, at a high soil salinity level (ECe 65) to explore their possible tolerance mechanisms in their natural habitat. Samples of different tissues were collected from both *Suaeda* species for the determination of physio-biochemical attributes, i.e., ionic (Na^{+}, K^{+}, Ca^{2+}, Cl^{-}) content, osmo-protective compounds (proline, soluble sugars, soluble proteins), total phenolic content, and antioxidant components. Heavy metal composition and accumulation in soil and plant samples were also assessed, respectively. Fourier transform infrared spectroscopy (FTIR) analysis was conducted to explore cellular metabolite pools with respect to high salinity. The results showed that both species considerably adjusted the above-mentioned physio-biochemical attributes to resist high salinity, demonstrated by quantitative differences in their above-ground tissues. The FTIR profiles confirmed the plants' differential responses in terms of variability in lipids, proteins, carbohydrates, and cell wall constituents. The high capacity for Na^{+} and Cl^{-} accumulation and considerable bioaccumulation factor (BAF) values for metals, mainly Fe and Zn, validate the importance of both *Suaeda* species as phytodesalination plants and their potential use in the phytoremediation of salt- and metal-polluted soils.

Keywords: antioxidants; adaptive mechanism; halophytes; phytoremediation; soil salinity

Citation: Joshi, A.; Rajput, V.D.; Verma, K.K.; Minkina, T.; Ghazaryan, K.; Arora, J. Potential of *Suaeda nudiflora* and *Suaeda fruticosa* to Adapt to High Salinity Conditions. *Horticulturae* **2023**, *9*, 74. https://doi.org/10.3390/horticulturae9010074

Academic Editors: Adalberto Benavides-Mendoza, Yolanda González-García, Fabián Pérez Labrada and Susana González-Morales

Received: 5 December 2022
Revised: 29 December 2022
Accepted: 4 January 2023
Published: 6 January 2023

1. Introduction

Soil or land salinization is a severe ecological problem worldwide that consistently increases nearly 10% annually. Nowadays, more than 3% (424 mha) of global topsoil (0–30 cm) and more than 6% (833 mha) of global sub-soil (30–100 cm) are affected by salinity or sodicity. Remarkably, more than two-thirds of global salt-contaminated soils are established in arid and semi-arid climatic zones, of which 64% are located in arid deserts and steppes [1]. Individually, Europe has a maximum share of saline land, which accounts for nearly 3.3% of the world's total saline land. In the rest of the world, including Asia, Africa, America, and Australia, the majority of cultivated land has been salinized and become uncultivable [2]. Salt accumulation in the soil constrains agricultural production and the global economy. It has been estimated that the per hectare cost of salinity-induced land degradation is approximately USD 441, which is further responsible for the loss of USD 27 billion per annum [3]. In India, soil salinization is highly worrisome in arid and semi-arid areas as it obstructs plant growth and can ultimately limit the distribution of plant communities [4,5].

Salinity imposes deleterious impacts on plants' basic physiology and metabolism in the form of ionic toxicity and osmotic and oxidative stress [6]. Halophytes are plants that grow throughout salty environments and are able to not only sustain but also proliferate by implementing distinct adaptive mechanisms, such as (a) accumulation, exclusion, or compartmentalization of toxic ions at the cellular or whole-plant level, (b) synthesis and accumulation of osmo-protective compounds, i.e., proline, glycine betaine, etc., (c) activation of enzymatic and non-enzymatic antioxidant activities, and (d) modulation of various metabolic cascades, such as the photosynthetic pathway, plant hormones, and signaling molecules [7–9]. In addition, salinity tolerance varies among species of taxonomically identical or related taxa, which should be attributed to plant habitat, growth form, and some specialized structures such as the salt gland, salt bladder, and Kranz anatomy [10,11]. Several studies have reported that halophytes of the same habitat, even the same taxa, respond differently to salinity levels through quantitative and qualitative differences in their response mechanisms [12–14].

The present study explored two species of *Suaeda* (Figure 1), namely *Suaeda nudiflora* (Willd.) Moq., and *Suaeda fruticosa* (L.) Forssk., that dominate a hypersaline region close to the Thar Desert. Despite being genetically identical, these species differ slightly in their aerial morphology, which has been reported as an adaptation to saline conditions [15]. *Suaeda nudiflora* is a perennial under-shrub with smooth stems, elliptic-oblong or linear-obovate glabrous leaves, like spike inflorescences, and black seeds with curved embryos. *Suaeda fruticosa* is a perennial shrub with an erect glabrous stem, usually about 3 m tall, with fleshy and subsessile oblong or elliptic leaves and black seeds. They are considered cash crops because of their medicinal, nutritional, and economic value, as well as their potential use in phytoremediation [16–18].

(a) **(b)**

Figure 1. Appearance of the plants in their native environment, (**a**) = *Suaeda nudiflora*, (**b**) = *Suaeda fruticosa*.

Several investigations have described salinity-induced modulations, particularly in *Suaeda* species [19,20]. A few studies have examined these two species under controlled laboratory conditions [21–23]. However, there is scant information about their responses in habitats with high salt exposure and which acclimatory mechanisms help the plants complete their life cycle under high salinity circumstances. Studies on the differential behavior of halophytes in their natural habitats can help explain species-specific salt tolerance and provide a framework for the development of salt-resilient crops and a restoration strategy for saline soils [24–26].

The current study aimed to assess the influence of soil salinity on the differential physiological traits of both *Suaeda* species. The hypothesis expressed here is that the successful adaptation of different *Suaeda* species to high salinity conditions is determined by both the magnitude of salinity in their rhizosphere soil as well as their individual salt tolerance evolutionary strategies. Furthermore, species-specific and common physiological responses may operate within the species. In order to gain insight into possible common and species-specific tolerance mechanisms in these species, a comparative study of their physiological responses to the physicochemical attributes of rhizospheric soil would be a useful approach. This study is designed specifically to (i) identify key biochemical indicators and cellular metabolites of *Suaeda nudiflora* and *Suaeda fruticoca* relevant to their tolerance mechanism towards high salinity, and (ii) determine the bioaccumulation capacity of salts and heavy metals by both species for phytoremediation purposes in the future.

2. Materials and Methods

2.1. Description of Sampling Site

The sampling site, Sambhar Salt Lake (26°58′0″ N to 75°5′0″ E), which is recognized as India's largest inland salt lake, was chosen to represent high soil salinity levels. The lake is elliptically shaped, with a length of approximately 36 km, a breadth varying 3–11 km, and located in the Rajasthan state of India (Figure 2). It is an extensive saline wetland, receiving water from six rivers, including the Medtha, Samaod, Mantha, Rupangarh, Khari, and Khandela. The eastern area of the lake is accompanied by numerous salt reservoirs, canals, salt pans, and halophyte vegetation [27]. It is estimated that silt from the Aravalli hills, which is generally encrusted in schists (medium-grade metamorphic rock) and gneisses (high grade regional metamorphic rock), is the major source of the salt composites. The sodium composites in the silt dissolve in rainwater and enter the lake via rivers, and the salt remains in the lake after the rainwater evaporates [28]. Extremely hot summers and mild winters are features of the temperate-continental environment. The average annual temperature was 25.1 °C with a multiannual minimum of 5 °C and maximum of 50 °C in May/June.

2.2. Collection and Analysis of the Samples

In the month of May 2019, soil and plant samples (sampled area; 10 m × 15 m, sample weight; 100 g) were collected in order to ascertain the physicochemical properties. The soil samples were drawn from a depth of 20–25 cm, carefully packed inside polybags, and sent to the laboratory for further examination and analysis. The experiments were performed in triplicate. At the same location, 12 plants of each species were carefully taken out of the soil and maintained at 4 °C to prevent the destruction of the sample's constituents until analysis.

2.3. Soil Analysis

Standardized protocols described in the USDA Handbook were applied to calculate the pH, electric conductivity (ECe), and organic carbon (OC) content of the soil [29]. The accessibility of phosphorous (available form P_2O_5) and potassium content (available form K_2O) were determined by the methods described by Olsen [30] and Merwin and Peech [31], respectively. According to the method described by Prakash and Prathapasenan [32], the soluble salt content was determined after extraction with distilled water (soil: water ratio,

1:5) using a flame photometer (Eppendorf; Na^+ and K^+), atomic absorption spectrometer (Perkin Elmer, Analyst 200, Rodgau, Germany; Ca^{2+}), and chloridometer (Buchler-Cotlove; Cl). With the use of an atomic absorption spectrometer (Perkin Elmer, Analyst 200 Germany), the availability of iron (Fe), zinc (Zn), manganese (Mn), and copper (Cu) was assessed following digestion with a di-acid mixture (HCl/HNO_3 mixture and concentrated $HClO_4$) according to Tüzen's technique [33].

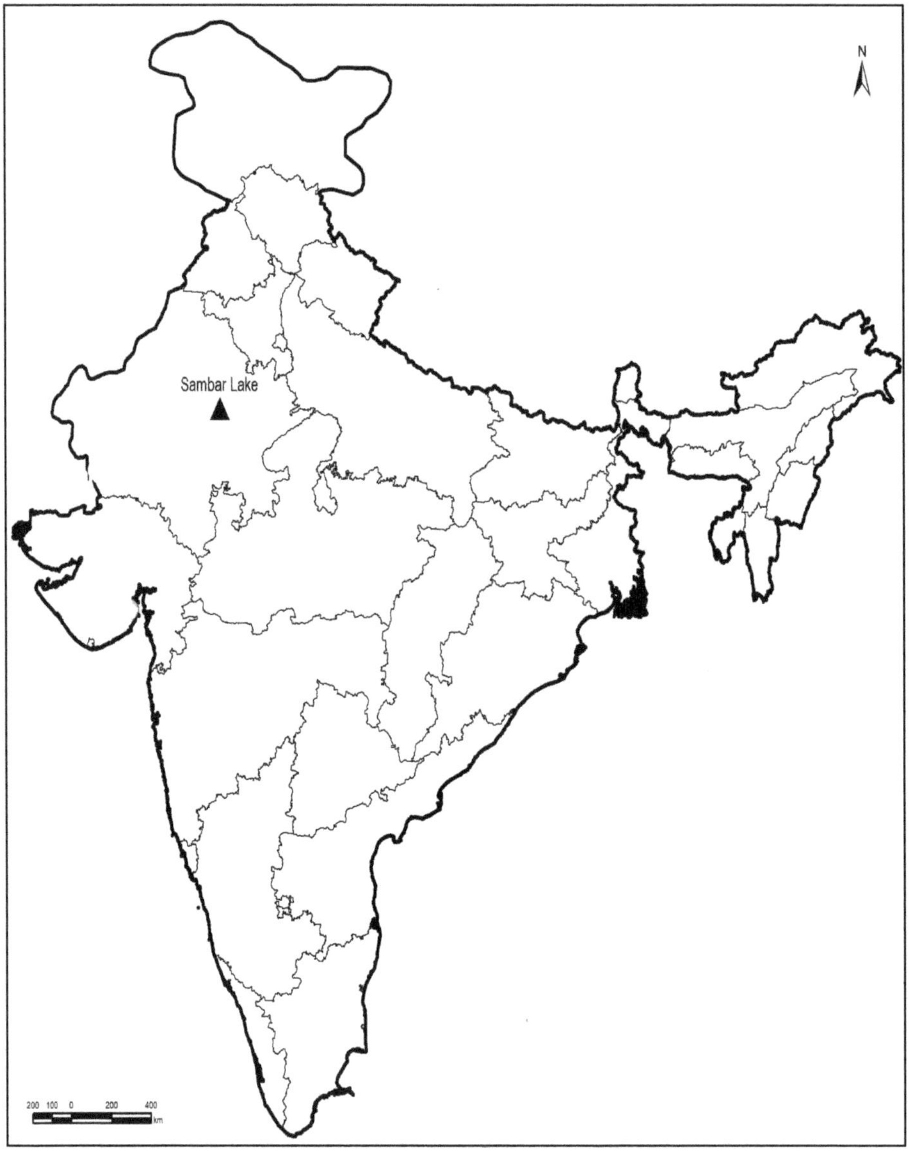

Figure 2. GIS map of the sampling site.

2.4. Determination of Soluble Ions in Plants

Plant parts such as the leaves and stem were cleaned and dried inside an oven at 60 °C for 72 h. The completely dried leaves and stem were ground into a fine powder using a mortar and pestle. The extracts were prepared by digestion with HNO_3, according to Prakash and Prathapasenan [32]. Similar to the soil above, the concentrations of Na^+, K^+, Ca^{2+}, and Cl^- in the plant extract were determined using a flame photometer, atomic absorption spectrometer, and chloridometer, respectively.

2.5. Determination of Osmo-Protective Compounds

The ninhydrin technique was utilized to quantify the proline content [34]. Soluble proteins were assessed according to the Bradford method [35] using bovine serum albumin as the standard. A modification of the phenol–sulfuric acid method was used to determine the total soluble sugar content [36].

2.6. Determination of Total Phenolic Content (TPC) and Antioxidant Activity

A 250 mg sample of dried powdered plant material was extracted with 5 mL of 70% methanol and dried in test tube rotator at room temperature for 12 h. The total phenolic content (TPC) and antioxidant activity of the extract were determined in triplicate. Phenolic compound analysis was performed using Foline-Ciocalteu reagent with the Farkas and Kiraly method [37]. TPC was calculated using a calibration curve for gallic acid at 650 nm and represented as mg gallic acid equivalents (GAE g^{-1} DW). The method described by Hatano et al. [38] was used to measure the 1.1-diphenyl-2-picrylhydrazyl (DPPH) radical scavenging activity. DPPH scavenging activity (%) was calculated as (%) = 100(A B)/A, where A and B are the control and corrected absorption of the sample reaction mixture at 517 nm, respectively. The Benzie and Strain technique [39] was applied to calculate the ferric reducing power (FRAP).

2.7. Determination of Metal Contents and Bioaccumulation Factor in Plants

After separating the plant parts (leaves and stem), they were completely cleaned with distilled water and then dried in an oven at 65 °C. The dried plant organs (0.5 g) were heated in a muffle furnace at 550 °C for 12 h. Subsequently, an extract was made in accordance with the method described by Tuzen [33], and an atomic absorption spectrometer was employed to measure the amount of metals in the plant sample extract (Perkin Elmer, Analyst 200 Germany). The ratio of the concentrations of metal in various parts of the plant is known as the bioaccumulation factor (BAF). BAF refers to the ability of plants to take up, transport, and store metals in its above-ground tissue [40], and it was determined as follows:

(BAF) Leaves: [Metal in leaves tissue]/[Metal in soils]
(BAF) Stem: [Metal in stem tissue]/[Metal in soils]

2.8. Fourier Transform Infrared Spectroscopy (FTIR) Analysis

FTIR analysis was used to identify functional groups in the leaves and stem parts of the plants. Three different samples of leaves and stem were taken from both species for FTIR analysis. Thereafter, pelleted samples of leaves and stem were scanned in the mid-infrared region (4000–400 cm^{-1}) using an FTIR spectrometer (Bruker, Model OPUS 7.5.18). These samples contributed to the generation of three unique FTIR spectra. Analysis software was used for the identification of functional groups in the leaves and stem samples.

2.9. Experimental Design and Statistical Analysis

The study was carried out using a complete randomized block design (CRBD), which was performed twice. In every test, soil and plant samples were replicated at least three times (n = 3). Duncan's multiple range test as used to determine whether there was a statistically significant difference ($p < 0.05$) between the means of the different species. The results are presented as the mean ± SD of three separate trials, which were then analyzed using one-way analysis of variance (ANOVA). Statistical analysis was performed using SPSS version 17 software (SPSS Inc., Chicago, IL, USA).

3. Results

3.1. Soil Physicochemical Properties

The physicochemical properties of the soil at the sampling sites are summarized in Table 1. The soil in the study area had an alkaline nature with high values for pH (9.89) and ECe (65 dS/m^{-1}). The average available organic carbon content and phosphorous and potassium concentrations were 0.19%, 8.49 kg/ha, and 84.13 kg/ha, respectively. Among the soluble cations, the concentration of Na^+ was relatively high (1485 mg/100 g) followed by K^+ (41.23 mg/100 g) and Ca^{2+} (19.51 mg/100 g). The average Cl^- concentration was 1.02 mg/100 g. Heavy metal analysis of the soil sample revealed Fe as the major metal ion with the highest concentration (4185 mg/kg), followed by Zn (38.64 mg/kg), Mn (131.9 mg/kg), and Cu (6.47 mg/kg).

Table 1. Physiochemical properties of the soil at the collection sites (mean ± SD, n = 3).

S. No.	Parameters	Values
1	Soil pH	9.89 ± 0.6
2	ECe (dS/m^{-1})	65 ± 0.7
3	Organic carbon (%)	0.19 ± 0.04
4	P_2O_5 (kg/ha)	8.49 ± 0.32
5	K_2O (kg/ha)	84.13 ± 0.79
6	Na^+ (mg/100 g dry soil)	1485 ± 16.11
7	K^+ (mg/100 g dry soil)	41.23 ± 2.4
8	Ca^{2+} (mg/100 g dry soil)	19.51 ± 1.8
9	Cl^- (mg/100 g dry soil)	1.02 ± 0.74
10	Iron (mg/kg)	4185 ± 70.4
11	Zinc (mg/kg)	38.64 ± 1.85
12	Manganese (mg/kg)	131.9 ± 4.6
13	Copper (mg/kg)	6.47 ± 0.97

3.2. Accumulation of Soluble Ions in Plants

The amount of solubilized ions in the different tissues of both species was affected by the salinity, as indicated in Table 2. Measurement of cation (Na^+, K^+, Ca^{2+}) and anion (Cl^-) concentrations in the plants revealed clear differences between the species, with relatively high concentrations in *S. fruticosa*. The corresponding value was higher in the leaves than in the stem and was higher in *S. fruticosa* leaves. Except for Ca^{2+}, the mean concentrations of soluble ions were approximately 1.5-fold higher in the *S. fruticosa* leaves (Na^+ 71.01 mg/g; K^+ 19.54 mg/g; Cl^- 14.02 mg/g) than in the *S. nudiflora* leaves (Na^+ 45.56 mg/g; K^+ 12.65 mg/g; Cl^- 11.68 mg/g).

Table 2. Deposition of various ions in leaves and stem parts of the plants.

Species	Plant Parts	Na^+	K^+	Ca^{2+}	Cl^-	Na^+/K^+ Ratio
S. nudiflora	Leaves	45.56 ± 1.34 ^{b}	12.65 ± 1.05 ^{b}	11.72 ± 1.12 ^{b}	11.68 ± 1.21 ^{b}	3.60
	Stem	21.32 ± 1.08 ^{d}	7.63 ± 0.98 ^{d}	6.89 ± 0.65 ^{d}	5.56 ± 0.99 ^{d}	2.79
S. fruticosa	Leaves	71.01 ± 1.71 ^{a}	19.54 ± 1.45 ^{a}	11.54 ± 0.91 ^{a}	14.02 ± 1.01 ^{a}	3.63
	Stem	23.84 ± 1.19 ^{c}	8.55 ± 0.77 ^{c}	7.90 ± 0.84 ^{c}	7.11 ± 1.18 ^{c}	2.78

Note: Statistically significant differences ($p \leq 0.05$) between plant parts are marked with superscripts a, b, c, and d.

In the stem, the concentrations of soluble ions were approximately 1.2-fold higher in *S. fruticosa* (Na^+ 23.84 mg/g; K^+ 8.55 mg/g; Ca^{2+} 7.90 mg/g; Cl^- 7.11 mg/g) than in *S. nudiflora* (Na^+ 21.32 mg/g; K^+ 7.63 mg/g; Ca^{2+} 6.89 mg/g; Cl^-; 5.56 mg/g). Given the Na^+ and K^+ accumulation patterns in the different tissue, the Na^+/K^+ ratio was relatively high in the leaves (3.6 on average) compared to that in the stem tissue (2.8) of both species.

3.3. *Accumulation of Osmo-Protective Compounds*

Measurements of common osmo-protective compounds in plants, including proline, total soluble sugar (TSS), and total soluble proteins (TSP), are presented in Table 3. Proline and TSS were probably the dominant osmo-protective compounds in both species. The absolute concentrations of accumulated proline in the stem were significantly greater than in the leaves, by nearly 2-fold, and were highest in the *S. nudiflora* stem (22.41 μmoles/g). The corresponding values of TSS were higher in the leaves than in the stem, with minor quantitative differences between species.

Table 3. Deposition of osmo-protective chemicals in leaves and stem parts of halophytes.

Species	Plant Part	Proline Content (μmoles/g)	Soluble Sugar Content (mg/g)	Soluble Protein Content (mg/g)
S. nudiflora	Leaves	7.99 ± 0.20 [d]	9.87 ± 0.36 [a]	7.23 ± 0.91 [a]
	Stem	22.41 ± 0.30 [a]	6.35 ± 0.38 [b]	3.24 ± 0.13 [c]
S. fruticosa	Leaves	8.82 ± 0.14 [c]	9.43 ± 0.24 [d]	4.37 ± 0.41 [b]
	Stem	18.57 ± 0.90 [b]	6.02 ± 0.26 [d]	1.98 ± s0.12 [d]

Note: Statistically significant differences ($p \leq 0.05$) between plant parts are marked with superscripts a, b, c, and d.

The highest value of TSS was in the *S. nudiflora* leaves (9.87 mg/g), followed by that in *S. fruticosa* leaves (9.43 mg/g). The absolute concentration of accumulated TSP in the leaves was higher than in the stem, by approximately 2.2-fold, and was highest in the *S. nudiflora* leaves (7.23 mg/g), followed by that in the *S. fruticosa* leaves (4.37 mg/g).

3.4. *Estimation of TPC and Antioxidant Activity*

The variations in TPC and antioxidant potential of the studied species are illustrated in Figure 3.

The mean values of TPC were significantly greater in the leaves than in the stem and were 1.7-fold higher in the *S. nudiflora* leaves than in the *S. fruticosa* leaves (Figure 3a). In the case of the stem, its corresponding value was greater in *S. fruticosa* than in *S. nudiflora* by 2-fold.

Remarkably, both species displayed differential trends to scavenge the free ferric and DPPH radical ions. In *S. nudiflora*, the free ferric radical ion scavenging potential was significantly greater in the leaves than in the stem, while in *S. fruticosa*, it was greater in the stem than in the leaves, by 1.7- and 1.45-fold, respectively (Figure 3b). Similarly, the DPPH radical scavenging potential was significantly greater in the leaves of *S. nudiflora*, while it was greater in the stem in the case of *S. fruticosa* by 1.03-fold (Figure 3c).

3.5. *Estimation of Metal Concentrations in Leaves and Stem Parts of Plants*

The concentrations of several metals, such as Zn, Mn, Fe and Cu, in the leaves and stem parts of the two species of plant with similar habitat are shown in Figure 4. The metal analysis of the plants revealed Fe as the major metal ion, followed by Zn, Mn, and Cu.

The mean Fe concentration ranged from 59.91 to 61.37 mg/kg and was highest in the *S. nudiflora* leaves. Nevertheless, its corresponding BAF value was similar in both species and remained very low (0.014). The Zn concentration ranged from 15.61 to 17.59 mg/kg and was highest in the *S. nudiflora* stem. The bioaccumulation factor (BAF) values for Zn ranged from 0.403 to 0.455 and were highest in the *S. nudiflora* stem (Table 4). The mean Mn concentration varied from 11.23 to 12.22 mg/kg and was highest in the *S. fruticosa* leaves. The BAF value for Mn was low, ranging from 0.085 to 0.094. Similarly, the mean Cu concentration varied from 2.13 to 2.89 mg/kg and was highest in the *S. fruticosa* leaves. The corresponding BAF value for Cu ranged from 0.329 to 0.446 and was highest in the *S. fruticosa* leaves.

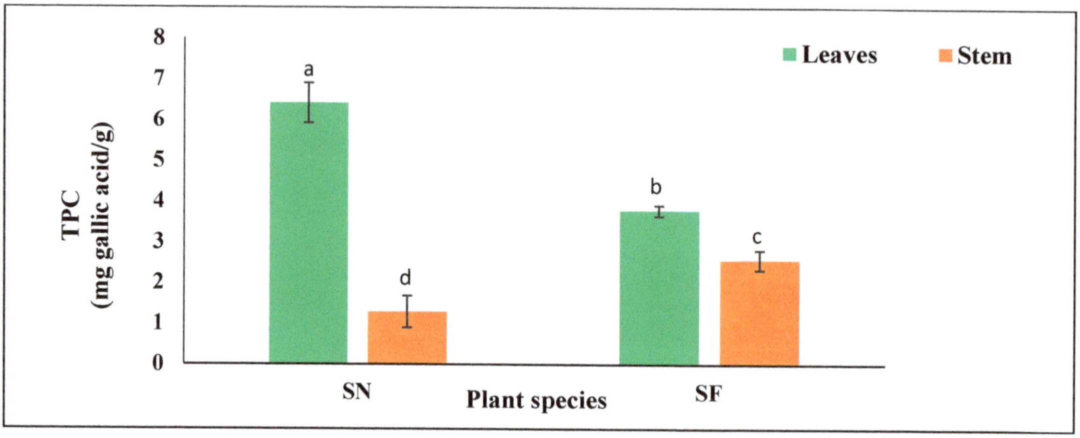

(**a**)

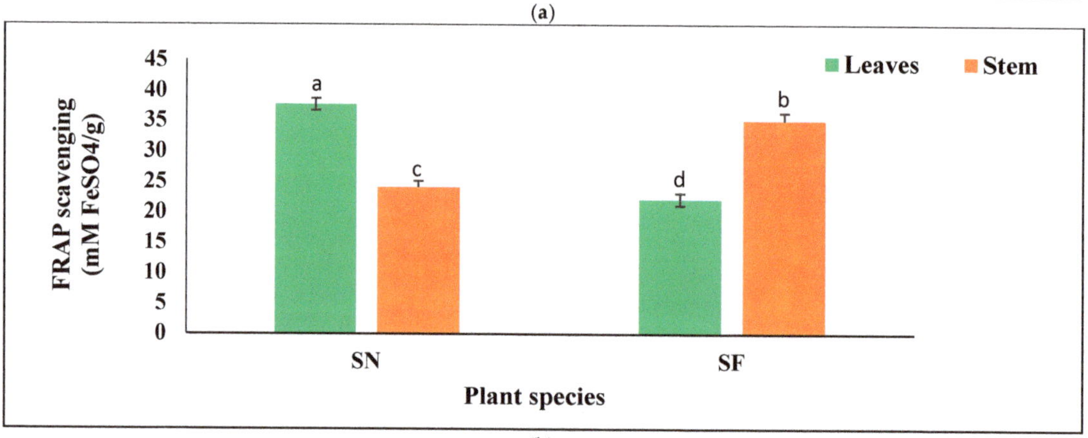

(**b**)

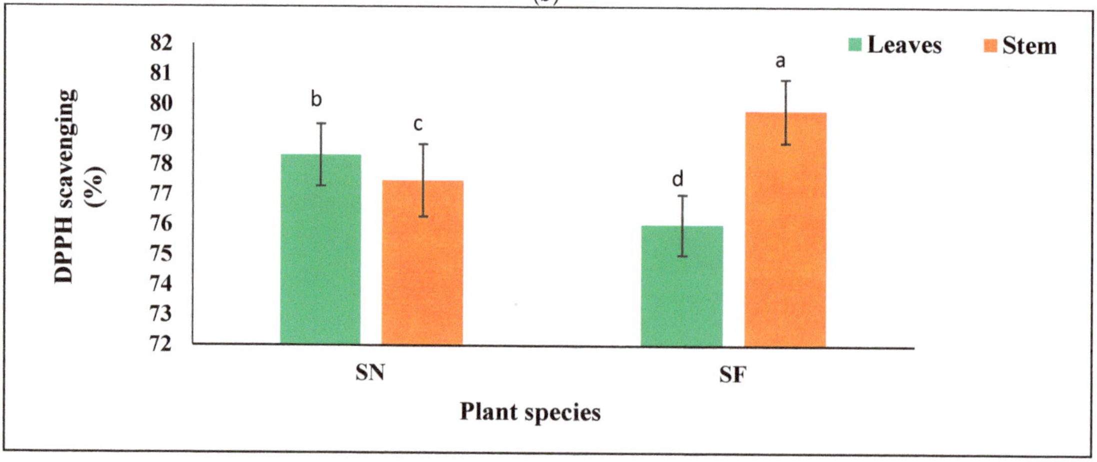

(**c**)

Figure 3. TPC and antioxidant (FRAP and DPPH) activity in leaves and stem parts of the plants (DMRT), (**a**) = TPC, (**b**) = FRAP, (**c**) = DPPH, SN = *S. nudiflora*, SF = *S. fruticosa*. Note: Statistically significant differences ($p \leq 0.05$) between plant parts are marked with superscripts a, b, c, and d.

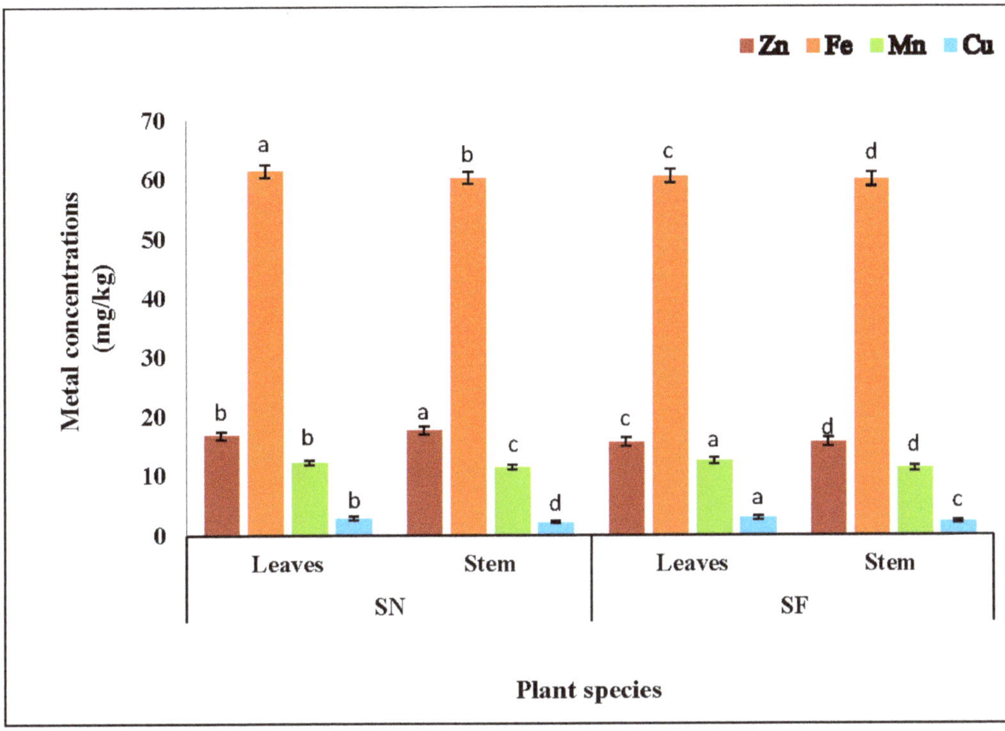

Figure 4. Concentrations of various metal ions in leaves and stem parts of the plants (DMRT), SN = *S. nudiflora*, SF = *S. fruticosa*. Note: Statistically significant differences ($p \leq 0.05$) between plant parts are marked with superscripts a, b, c, and d.

Table 4. Bioaccumulation factor (BAF) values for metals in leaves and stem of plants species.

Species	Plant Parts	Metals			
		Zn	Fe	Mn	Cu
S. nudiflora	Leaves	0.435 ± 0.17	0.014 ± 0.01	0.092 ± 0.05	0.438 ± 0.21
	Stem	0.455 ± 0.23	0.014 ± 0.01	0.086 ± 0.03	0.329 ± 0.14
S. fruticosa	Leaves	0.404 ± 0.15	0.014 ± 0.01	0.094 ± 0.07	0.446 ± 0.20
	Stem	0.403 ± 0.11	0.014 ± 0.01	0.085 ± 0.03	0.341 ± 0.16

Note: Data are presented as the mean ± SD of three separate trials.

3.6. FTIR Analysis

The major FTIR spectra peaks and possible functional groups present in the leaves and stem parts of both plant species are shown in Table 5 and Figure 5. The wavenumber region 3000–2000 cm^{-1} was assigned to lipids. In this region, both species showed a similar FTIR profile, except for a peak at 2921 cm^{-1}, which was not present in the leaves of *S. nudiflora* (Figure 5a). This peak indicated the presence of O-H stretch (Alcohols), S, O–H stretch (carboxylic acids), and =C–H (benzene, alkynes, alkenes). Additionally, the peak at 2854 cm^{-1} was not present in the stem of both species (Figure 5b,d). This peak indicated the presence of C–H stretch (alkenes) and H–C=O:C–H stretch (aldehydes).

Table 5. FTIR spectra illustrating various peaks of functional groups in leaves and stem parts of the plants.

Cellular Metabolites	Wavenumber (cm^{-1})				Probable Functional Group
	S. nudiflora		*S. fruticosa*		
	Leaves	Stem	Leaves	Stem	
Lipids (3000–2000 cm^{-1})	3853.16 ±0.94	3825.61 ±0.91	3853.00 ±1.02	3852.86 ±0.98	O–H stretch (alcohols, phenols)
	3737.46 ±0.72	3738.90 ±0.51	3737.97 ±0.64	3739.05 ±0.85	O–H stretch (alcohols)
	*	2921.35 ±0.63	2921.95 ±0.41	2923.88 ±0.33	O–H stretch (alcohols), S, O–H stretch (carboxylic acids), =C–H (benzene, alkynes, alkenes)
	2853.13 ±0.55	*	2854.78 ±0.61	*	C–H stretch (alkenes), H–C=O:C–H stretch (aldehydes)
	2383.62 ±0.52	2383.79 ±0.49	2384.10 ±0.52	2383.87 ±0.37	P–H (phosphine)
	2312.77 ±0.43	2312.64 ±0.64	2312.28 ±0.50	2312.42 ±0.73	C=C stretch (alkynes)
Proteins (1800–1500 cm^{-1})	1724.21 ±0.39	1728.73 ±0.42	*	*	C=O (esters, carboxylic acids, ketones, aldehydes), C=C (benzenes)
	1638.44 ±0.13	1643.27 ±0.28	1641.08 ±0.21	1642.24 ±0.26	N-H bend (nitro compounds, amides), C–C stretch (amides), C=O stretch (carboxylic acid, ketone), C=C (benzene, alkenes)
	1517.62 ±0.18	1515.17 ±0.25	1517.66 ±0.10	1514.58 ±0.14	N-H bend (nitro compounds), C–O stretch (amides), C=C (benzenes), C=O (ketones)
Carbohydrates (1500– 1000 cm^{-1})	1455.04 ±0.16	1457.20 ±0.23	1460.25 ±0.24	1457.18 ±0.20	C–C stretch (aromatics), C–H bend (alkanes), N–O stretch (nitro compounds), C-O stretch (esters), CO–H bend (aldehydes), O–H bend (alcohols
	*	1377.69 ±0.57	*	1377.27 ±0.13	N=O, CO–H band, O–H band
	1322.34 ±0.27	1322.34	±0.27	1323.60 ±0.63	S(=O)2 stretch (sulfones), N=O stretch (nitro compounds), O–H bend (carboxylic acids, alcohols)
	1240.17 ±0.31	1237.21 ±0.29	1237.34 ±0.25	1233.55 ±0.54	C–N stretch (amines), C–O stretch (esters), C–O stretch (ethers, alcohols), O–H band (carboxylic acids)
	1021.99 ±0.18	1029.64 ±0.23	1030.39 ±0.12	1030.27 ±0.20	S=O stretch (sulfoxides), C–N stretch (amines), C–O stretch (esters, ethers, alcohols), =C–H bend (benzene, alkenes) (cellulose)
Cell wall components (1000– 600 cm^{-1})	*	*	818.93 ±0.11	830.21 ±0.16	C–N stretch (amines), =C–H bend (benzene, alkynes) (xyloglucan)
	777.99 ±0.08	787.99 ±0.15	775.31 ±0.12	782.15 ±0.16	C–N stretch (amines), =C–H bend (benzene), C–C stretch
	*	639.39 ±0.23	*	649.96 ±0.31	C–N stretch (amines), =C–H bend (Bbenzene), C–C stretch (chloride)

Note: Data are presented as the mean ± SD of three separate trials, * = not determined.

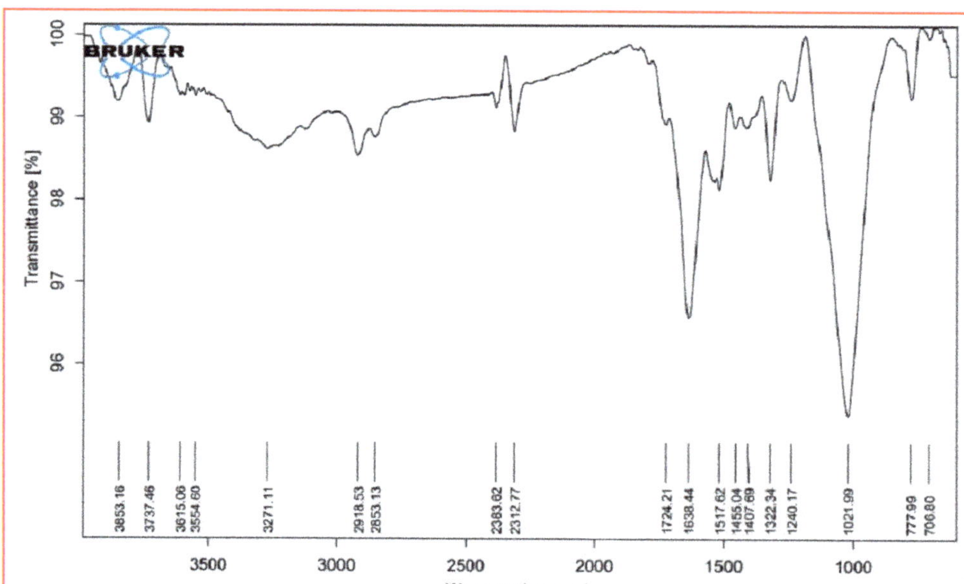

(a)

Figure 5. *Cont.*

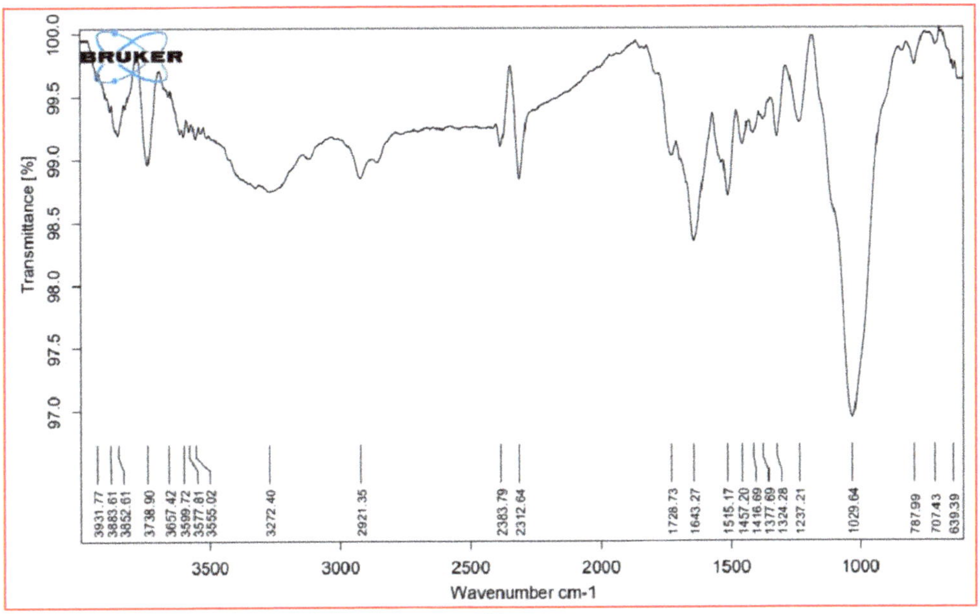

(**b**)

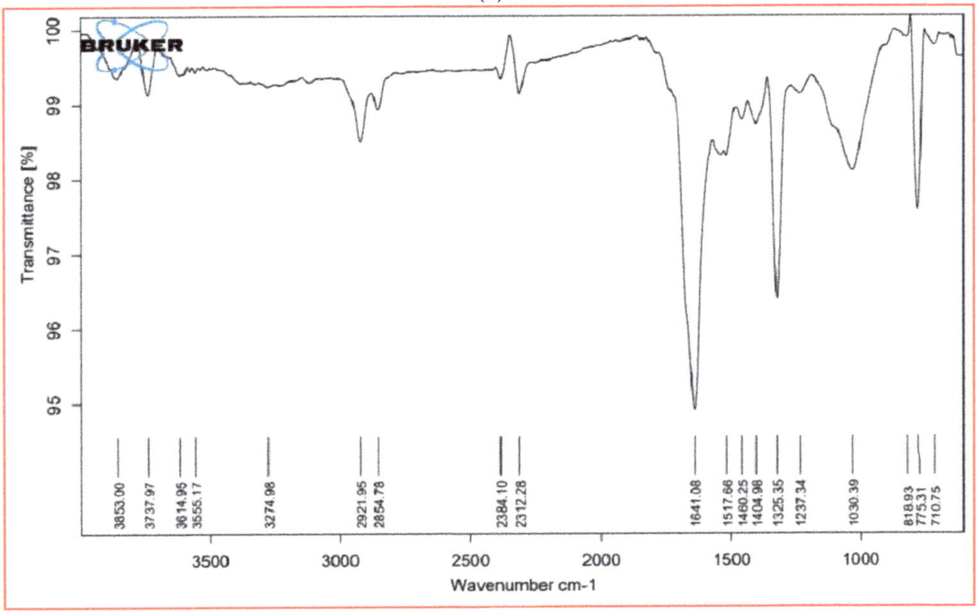

(**c**)

Figure 5. *Cont.*

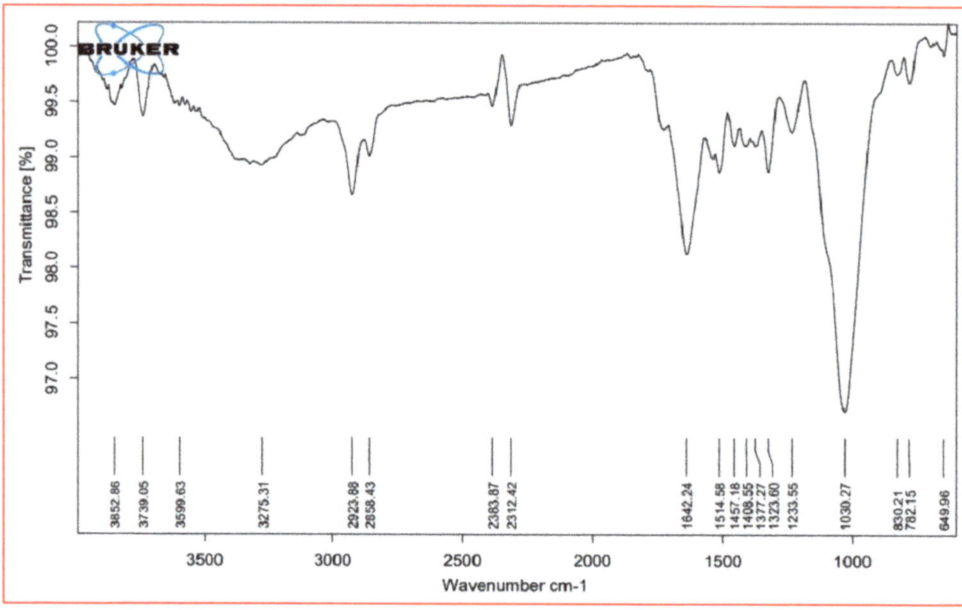

(d)

Figure 5. FTIR profiles of different tissue of the plants, (**a**) = *S. nudiflora* leaves, (**b**) = *S. nudiflora* stem, (**c**) = *S. fruticosa* leaves, (**d**) = *S. fruticosa* stem.

The wavenumber region 1800–1500 cm^{-1} was assigned to proteins. In this region, both species showed a similar FTIR profile, except for a peak at 1724 cm^{-1} that completely disappeared in *S. fruticosa* (Figure 5c,d). These peaks were characteristic of C=O (esters, carboxylic acids, ketones, aldehydes) and C=C (benzenes). The wavenumber region 1500–1000 cm^{-1} represented carbohydrates. Except for the peak at 1377 cm^{-1} (characteristic of N=O, CO–H band, and O–H band), both species displayed similar FTIR profiles. The wavenumber region 1000–600 cm^{-1} was assigned to cell wall components. In both species, a major peak has appeared at 775 cm^{-1}. These peaks were characteristic of C–N stretch (amines), =C–H bend (benzene, alkenes), and C–C (chlorides).

4. Discussion

In arid and semi-arid climates, soil salinization is a vital stress factor that impedes the physiological, biochemical, and molecular attributes of plants and can affect the distribution of plant communities [4,41]. In this study, the sampling site was located near the hyper-arid climate of the Thar Desert, which possesses an alkaline nature with a high ECe value, likely due to the presence of inherent salt sources, a high evaporation rate, and poor surface drainage conditions [15,42]. The ECe value was approximately 17-fold greater than that of cultivated land; therefore, the sampling site could be considered to be a hypersaline habitat. By definition, soil is considered saline when its ECe value is above 4 dS m^{-1} [43]. The higher amount of Na^+ among major cations and Cl^- among anions confirmed the abundance of NaCl salt, which might be responsible for the expansion of hyper salinity at this site. However, its concentration may be subject to seasonal variations due to intricate evaporative mechanisms, particularly during the dry period [28].

pH and ECe values are dominant factors determining the binding and retention capacity of metal ions in soil. They can be used as potential indicators for metal/metalloid pollution [44]. At this site, the corresponding concentrations of metal ions, particularly Fe and Zn, confirmed that metal ions became less mobile and were effectively retained in the soil under high salinity conditions. Although the amounts of all studied metal ions were within the range of threshold values of Indian standards, their corresponding values may threaten soil fertility and productivity [45].

In this study, a series of acclimatory responses of two halophytes, namely *S. nudiflora* and *S. fruticosa*, were investigated to quantify the impact of high salinity on both species in their natural habitat. It was found that both species adjusted several physio-biochemical attributes to survive in the hypersaline habitat. The results also supported the hypothesis that plants of the same taxa respond differently to salinity levels by means of quantitative and qualitative differences in their response mechanisms [25,26]. Adjustment to ionic toxicity is usually achieved through accumulation or compartmentalization of toxic ions in specific plant cells, tissues, or organs, without affecting plant growth and development [46].

In this study, corresponding concentrations of cations (Na^+, K^+, and Ca^{2+}) and anion (Cl^-) were higher in the leaves than in the stem of both species. Additionally, the high Na^+ level was accompanied by elevated levels of K^+ and Ca^{2+}, which might be associated with the conjoint action of anions to confer some degree of halo-tolerance [47]. The facts that K^+ and Ca^{2+} counteract the harmful effects of Na^+ and are essential to sustaining various turgor-driven movements in salt-stressed plants are well clarified and documented [48–50]. A lower Na^+/K^+ ratio in the stem than in the leaves in both species could be part of the constitutive mechanism to maintain ionic homeostasis under high salinity conditions [51]. These findings are consistent with the fact that the internal molecular ratio of Na^+/K^+ in the shoots of dicot halophytes was lower than that of halophyte grasses, indicating some specific features that enabled these dicot species to accumulate, absorb, and compartmentalize Na^+, thus providing inexpensive osmotic particles for adjusting osmotic pressure [52,53]. Thus, these *Suaeda* species may prove to be beneficial for their possible use in the phytoremediation of saline soils since they accumulate high concentrations of Na^+ and Cl^- ions in comparison with several halophytes that have been previously explored for their potential use in reclaiming saline soil [54–56].

Osmo-protective compounds or osmolytes are well known to accumulate in response to a plant's exposure to abiotic stress conditions [57]. Proline accumulates in the cell as a molecular chaperone, playing a vital role in osmotic stress tolerance by protecting cellular structures and metabolic pools [58]. At the same time, sugars directly contribute to osmotic adjustment and can regulate the expression of stress-responsive genes [59]. In the present study, both species accumulated a significantly high amount of proline in the stem, while accumulating TSS and TSP in the leaves. This finding emphasized that such components are efficient osmoprotectants that help *Suaeda* species tolerate high salinity levels. The high proline level in *Suaeda* species could be associated with the plant's ability to reallocate osmolytes in distinct subcellular compartments in order to compensate for water loss and ionic toxicity [60]. The greater accumulation of TSS in the leaves may be important to stabilize protein structures, thereby increasing protein levels when exposed to salt stress conditions [61,62]. Many halophytes have demonstrated similar osmotic adjustment patterns in salty or alkaline habitats [63,64]. There has been evidence that *Suaeda* species collected from sites with high salinity retain higher amounts of soluble protein, sugars, proline, and total organic osmolytes in their aerial tissues than those collected from low salinity sites [65–67].

Under salinity stress, oxidative damage imposed by reactive oxygen species (ROS) is mitigated through enzymatic and non-enzymatic antioxidants machinery [68]. Among several secondary metabolites, polyphenolic bioactive compounds play crucial roles as hydrogen donors, singlet oxygen quenchers, and reducing agents, which makes them one of the most interesting metabolites of antioxidants [69]. In the present study, the TPC of leaves was significantly higher in comparison to the stem, with 1.7-fold higher levels in the *S. nudiflora* leaves than in the *S. fructicosa* leaves. While in the stem, TPC was 2-fold greater in *S. fruticosa* than *S. nudiflora*. The observed differential accumulation of TPC under salinity stress is an indication that TPC is capable of mitigating the effects of oxidative stress in these species. Moreover, its robust accumulation and synthesis could be dependent on the salt sensitivity of the considered species [70]. Contrary to the accumulation of phenolic content, both species showed differential trends in order to scavenge the free ferric (Figure 3b) and DPPH radical ions (Figure 3c). In the case of *S. nudiflora*, FRAP and DPPH scavenging activity were significantly higher in the leaves than in the stem, while in *S. fruticosa*, they were much greater in the stem than in the leaves. These results suggest a differential mechanism for FRAP and DPPH scavenging under high salinity conditions in these species. This could be attributed to the fact that halophytes have developed different strategies to avoid cellular oxidative damage by enhancing the phenolic content and other ROS-detoxifying agents combined with enzymatic antioxidants [68,71]. As these plants contain polyphenolic compounds with high antioxidant activities, they should be considered for cultivation in saline soil in order to achieve a sustainable income for farmers living in arid and semi-arid regions.

Salinity can help to improve the mobility of metals/metalloids in plants, mainly because of the structural complexity and antagonistic actions between metal ions and salt-derived anions or cations for sorption sites [72,73]. In the present study, the plant species accumulated Fe as the major metal ion, followed by Zn, Mn, and Cu, while the accumulation patterns significantly differed between species and tissues. This indicated that high salinity positively influenced the mobility of all four elements, and species-specific translocation mechanisms may lead to their significant accretion in different tissues [74]. Moreover, both species efficiently took up Fe, Zn, Mn, and Cu, but none reached BAF values higher than 1. The BAF is widely used to characterize a plant as phytoremediator, and its value (higher than 1) is a crucial feature to ascertain a feasible hyperaccumulator [75]. In comparison with previous studies investigating the phytoremediation potential of halophytes [76–78], our results demonstrated that the investigated *Suaeda* species cannot be considered as metal accumulators, but they can be used for phytosequestration, especially for Zn and Cu.

FTIR is recognized as a non-destructive technique for exploring structural and chemical changes in plants under salinity stress conditions [79,80]. In this study, both species showed changes in the structural composition and functional groups of primary cellular metabolites, including lipids, proteins, and carbohydrates, detected as variable peaks in the FTIR spectra that are unique to bioactive metabolites found within both leaves and stem parts of plant species. Interestingly, the differential peaks at 1000–600 cm^{-1} revealed the significance of cell wall components in salinity tolerance. It can be argued that the differential responses of cellular metabolites may prevent the adverse impact of high salinity. Previous studies have revealed the application of FTIR-based metabolic analysis to infer salinity-induced responses in halophytes [14,81].

5. Conclusions

The findings of the present study suggested that a high salt content in soil threatens soil fertility and makes it highly susceptible to metal/metalloid corrosion. In the case of *Suaeda* plants, the presented results showed that both species considerably adjusted distinct physiological and biochemical attributes to tolerate high salinity via quantitative differences in their above-ground tissues. The high Na^+ level was accompanied by elevated K^+ and Ca^{2+} levels, which confirmed specific absorption and translocation mechanisms to avoid Na^+ toxicity. Proline acted as an efficient osmo-protective compound in both species to compensate for water loss and ionic toxicity, particularly in the stem. The presence of improved concentrations of soluble sugars and proteins implied a synergistic impact on osmotic adjustment in the leaves of both species. The observed accumulation of TPC was associated with robust antioxidant activity to reduce the oxidative damage caused by free radical ions. The FTIR profiles revealed differential cellular macromolecules that contribute to salinity tolerance. Due to the high capacity of Na^+ and Cl^- accumulation and considerable BAF values for metals, particularly Fe and Zn, *Suaeda* species (*S. nudiflora*, *S. fruticosa*) would be advisable for possible use in the phytoremediation of salt- and metal/metalloid-polluted soils.

Author Contributions: Conceptualization, A.J. and J.A.; Methodology, A.J., V.D.R. and J.A.; Validation, A.J. and J.A.; Formal Analysis, A.J., V.D.R., K.K.V., J.A. and T.M.; investigation, Resources, A.J. and J.A.; Data Curation, A.J. and J.A.; Writing—Original Draft Preparation, A.J. and J.A.; Writing—Review & Editing, A.J., V.D.R.; K.G., K.K.V. and J.A., Visualization, A.J. and J.A.; Supervision, J.A.; Project Administration. J.A.; Funding Acquisition, A.J., V.D.R. and J.A. All authors have read and agreed to the published version of the manuscript.

Funding: Not applicable.

Data Availability Statement: Not applicable.

Acknowledgments: We wish to acknowledge Prabhat Baroliya, MLSU, Udaipur, for providing the FTIR facilities. Abhishek Joshi acknowledges the support of UGC, New Delhi, for awarding the BSR meritorious fellowship [25-1/2014-15(BSR) 7-125/2007(BSR)]. V.D.R. and T.M. would like to acknowledge support from the laboratory of «Soil Health» of the Southern Federal University with the financial support of the Ministry of Science and Higher Education of the Russian Federation, agreement No. 075-15-2022-1122. K.G. acknowledge support by the Science Committee of RA, in the frames of the research project No 21AG-4C075.

Conflicts of Interest: The authors declare no conflict of interest.

References

1. FAO. Global Map of Salt-affected Soils (GSASmap). 2021. Available online: https://www.fao.org/global-soil-partnership/gsasmap/en/ (accessed on 16 January 2022).
2. Hassani, A.; Azapagic, A.; Shokri, N. Global predictions of primary soil salinization under changing climate in the 21st century. *Nat. Commun.* **2021**, *12*, 6663. [CrossRef]
3. Shahid, S.A.; Zaman, M.; Heng, L. Soil salinity: Historical perspectives and a world overview of the problem. In *Guideline for Salinity Assessment, Mitigation and Adaptation Using Nuclear and Related Techniques*, 1st ed.; Zaman, M., Ed.; Springer: Cham, Switzerland, 2018; pp. 43–53.
4. Dagar, J.C. Salinity research in India: An overview. *Bull. Natl. Inst. Ecol.* **2005**, *15*, 69–80.
5. Mandal, A.K.; Reddy, G.O.; Ravisankar, T.; Yadav, R.K. Computerized database of salt-affected soils for coastal region of India. *J. Soil Salin. Water Qual.* **2018**, *10*, 1–13.
6. Flowers, T.J.; Colmer, T.D. Plant salt tolerance: Adaptations in halophytes. *Ann. Bot.* **2015**, *115*, 327–331. [CrossRef]
7. Mishra, A.; Tanna, B. Halophytes: Potential resources for salt stress tolerance genes and promoters. *Front. Plant Sci.* **2017**, *8*, 829. [CrossRef]
8. Zhao, C.; Zhang, H.; Song, C.; Zhu, J.K.; Shabala, S. Mechanisms of plant responses and adaptation to soil salinity. *Innovation* **2020**, *1*, 100017. [CrossRef]
9. Yadav, S.; Mishra, A. Ectopic expression of C4 photosynthetic pathway genes improves carbon assimilation and alleviate stress tolerance for future climate change. *Physiol. Mol. Biol. Plants.* **2020**, *26*, 195–209. [CrossRef]
10. Nikalje, G.C.; Srivastava, A.K.; Pandey, G.K.; Suprasanna, P. Halophytes in biosaline agriculture: Mechanism, utilization, and value addition. *Land Degrad. Dev.* **2018**, *29*, 1081–1095. [CrossRef]
11. Grigore, M.N.; Toma, C. Morphological and anatomical adaptations of halophytes: A review. In *Handbook of Halophytes: From Molecules to Ecosystems towards Biosaline Agriculture*, 1st ed.; Grigore, M.N., Ed.; Springer Nature: Cham, Switzerland, 2021; pp. 1079–1221.
12. Al Hassan, M.; Estrelles, E.; Soriano, P.; López-Gresa, M.P.; Bellés, J.M.; Boscaiu, M.; Vicente, O. Unravelling salt tolerance mechanisms in halophytes: A comparative study on four Mediterranean *Limonium* species with different geographic distribution patterns. *Front. Plant Sci.* **2017**, *8*, 1438. [CrossRef]
13. Podar, D.; Macalik, K.; Réti, K.O.; Martonos, I.; Török, E.; Carpa, R.; Székely, G. Morphological, physiological and biochemical aspects of salt tolerance of halophyte *Petrosimonia triandra* grown in natural habitat. *Physiol. Mol. Biol. Plants.* **2019**, *25*, 1335–1347. [CrossRef]
14. Joshi, A.; Kanthaliya, B.; Rajput, V.; Minkina, T.; Arora, J. Assessment of phytoremediation capacity of three halophytes: *Suaeda monoica, Tamarix indica* and *Cressa critica. Biologia Futr.* **2020**, *71*, 301–312. [CrossRef]
15. Joshi, A.; Kanthaliya, B.; Arora, J. Halophytic Plant Existence in Indian Salt Flats: Biodiversity, Biology, and Uses. In *Handbook of Halophytes: From Molecules to Ecosystems towards Biosaline Agriculture*; Grigore, M.N., Ed.; Springer Nature: Cham, Switzerland, 2020; pp. 1–22.
16. Joshi, A.; Kanthaliya, B.; Arora, J. Halophytes: The Nonconventional Crops as Source of Biofuel Production. In *Handbook of Halophytes: From Molecules to Ecosystems towards Biosaline Agriculture*, 1st ed.; Grigore, M.N., Ed.; Springer Nature: Cham, Switzerland, 2020; pp. 1–28.
17. Joshi, A.; Kanthaliya, B.; Arora, J. Halophytes of Thar Desert: Potential source of nutrition and feedstuff. *Int. J. Bioass.* **2018**, *8*, 5674–5683.

18. Sharma, V.; Joshi, A.; Ramawat, K.G.; Arora, J. Bioethanol production from halophytes of Thar Desert: A "green gold". In *Environment at Crossroads: Challenges, Dynamics and Solutions*; Basu, S.K., Zandi, P., Chalaras, S.K., Eds.; Haghshenass Publishing: Guilan Prov, Iran, 2017; pp. 219–235.
19. Haque, M.I.; Siddiqui, S.A.; Jha, B.; Rathore, M.S. Interactive Effects of Abiotic Stress and Elevated CO2 on Physio-Chemical and Photosynthetic Responses in *Suaeda* Species. *J. Plant Growth Regul.* **2021**, *41*, 2930–2948. [CrossRef]
20. Li, H.; Wang, H.; Wen, W.; Yang, G. The antioxidant system in Suaeda salsa under salt stress. *Plant Signal. Behav.* **2020**, *15*, 1771939. [CrossRef]
21. Joshi, A.; Kanthaliya, B.; Arora, J. Evaluation of growth and antioxidant activity in *Suaeda monoica* and *Suaeda nudiflora* callus cultures under sequential exposure to saline conditions. *Curr. Biotechnol.* **2019**, *8*, 42–52. [CrossRef]
22. Kumar, A.; Kumar, A.; Lata, C.; Kumar, S. Eco-Physiological responses of *Aeluropus lagopoides* (grass halophyte) and *Suaeda nudiflora* (non-grass halophyte) under individual and interactive sodic and salt stress. *S. Afr. J. Bot.* **2016**, *105*, 36–44. [CrossRef]
23. Hameed, A.; Hussain, T.; Gulzar, S.; Aziz, I.; Gul, B.; Khan, M.A. Salt tolerance of a cash crop halophyte *Suaeda fruticosa*: Biochemical responses to salt and exogenous chemical treatments. *Acta Physiol. Plant.* **2012**, *34*, 2331–2340. [CrossRef]
24. Song, J.; Wang, B. Using euhalophytes to understand salt tolerance and to develop saline agriculture: *Suaeda salsa* as a promising model. *Ann. Bot.* **2015**, *115*, 541–553. [CrossRef]
25. Ghanem, A.E.M.F.; Mohamed, E.; Kasem, A.M.; El-Ghamery, A.A. Differential salt tolerance strategies in three halophytes from the same ecological habitat: Augmentation of antioxidant enzymes and compounds. *Plants* **2021**, *10*, 1100. [CrossRef]
26. Ltaeif, H.B.; Sakhraoui, A.; González-Orenga, S.; LandaFaz, A.; Boscaiu, M.; Vicente, O.; Rouz, S. Responses to salinity in four *Plantago* species from Tunisia. *Plants* **2021**, *10*, 1392. [CrossRef]
27. Sinha, R. The Sambhar Lake: The Largest Saline Lake in Northwestern India. In *Landscapes and Landforms of India: World Geomorphological Landscapes*; Kale, V., Ed.; Springer: Dordrecht, The Netherlands, 2014; pp. 239–244.
28. Cherekar, M.N.; Pathak, A.P. Chemical assessment of Sambhar Soda Lake, a Ramsar site in India. *J. Water Chem. Technol.* **2016**, *38*, 244–247. [CrossRef]
29. USDA Handbook. *Diagnosis and Improvement of Saline and Alkali Soils*; Richards, L.A., Ed.; Oxford & IBH Publ. Co. Pvt. Ltd.: New Delhi, India, 1960; p. 60.
30. Olsen, S.R. *Estimation of Available Phosphorus in Soils by Extraction with Sodium Bicarbonate (No. 939)*; US Department of Agriculture: Washington, DC, USA, 1954.
31. Merwin, H.D.; Peech, M. Exchangeability of soil potassium in the sand, silt, and clay fractions as influenced by the nature of the complementary exchangeable cation 1. *Soil Sci. Soc. Am. J.* **1951**, *15*, 125–128. [CrossRef]
32. Prakash, L.; Prathapasenan, G. Efect of NaCl salinity and putrescine on shoot growth, tissue ion concentration and yield of rice (Oryzasativa L. var. GR-3). *J. Agron. Crop Sci.* **1988**, *160*, 325–334. [CrossRef]
33. Tüzen, M. Determination of heavy metals in soil, mushroom and plant samples by atomic absorption spectrometry. *Microchem. J.* **2003**, *74*, 289–297. [CrossRef]
34. Bates, C.J.; Waldren, R.P.; Teare, I.D. Rapid determination of free proline for water-stress studies. *Plant Soil.* **1973**, *39*, 205–207. [CrossRef]
35. Bradford, M.M. A rapid and sensitive method for the quantitation of microgram quantities of protein utilizing the principle of protein-dye binding. *Anal. Biochem.* **1976**, *72*, 248–254. [CrossRef]
36. Dubois, M.; Gilles, K.A.; Hamilton, J.K.; Rebers, P.T.; Smith, F. Colorimetric method for determination of sugars and related substances. *Anal. Chem.* **1956**, *28*, 350–356. [CrossRef]
37. Farkas, G.L.; Kiraly, Z. Role of phenolic compound in the physiology of plant diseases and disease resistance. *Phytopathol Zeitsch.* **1962**, *44*, 105–150. [CrossRef]
38. Hatano, T.; Kagawa, H.; Yasuhara, T.; Okuda, T. Two new flavonoids and other constituents in licorice root: Their relative astringency and radical scavenging effects. *Chem. Pharma. Bulletin.* **1988**, *36*, 2090–2097. [CrossRef]
39. Benzie, I.F.; Strain, J.J. The ferric reducing ability of plasma (FRAP) as a measure of "antioxidant power": The FRAP assay. *Anal. Biochem.* **1996**, *239*, 70–76. [CrossRef]
40. Petelka, J.; Abraham, J.; Bockreis, A.; Deikumah, J.P.; Zerbe, S. Soil heavy metal (loid) pollution and phytoremediation potential of native plants on a former gold mine in Ghana. *Water Air Soil Pollu.* **2019**, *230*, 267. [CrossRef]
41. Hussain, S.; Shaukat, M.; Ashraf, M.; Zhu, C.; Jin, Q.; Zhang, J. Salinity stress in arid and semi-arid climates: Effects and management in field crops. In *Climate Change and Agriculture*; Intech Open: London, UK, 2019; p. 13. [CrossRef]
42. Kar, A. The Thar or the Great Indian Sand Desert. In *Landscapes and Landforms of India*; Kale, V.S., Ed.; Springer: Dordrecht, The Netherlands, 2014; pp. 79–90.
43. Chinnusamy, V.; Jagendorf, A.; Zhu, J.K. Understanding and improving salt tolerance in plants. *Crop Sci.* **2005**, *45*, 437–448. [CrossRef]
44. Ghazaryan, K.A.; Movsesyan, H.S.; Minkina, T.M.; Sushkova, S.N.; Rajput, V.D. The identification of phytoextraction potential of Melilotus officinalis and Amaranthus retroflexus growing on copper-and molybdenum-polluted soils. *Environ. Geochem. Health.* **2021**, *43*, 1327–1335. [CrossRef]
45. Kumar, V.; Sharma, A.; Kaur, P.; Sidhu, G.P.S.; Bali, A.S.; Bhardwaj, R.; Cerda, A. Pollution assessment of heavy metals in soils of India and ecological risk assessment: A state-of-the-art. *Chemosphere* **2019**, *216*, 449–462. [CrossRef]

46. Kumari, A.; Das, P.; Parida, A.K.; Agarwal, P.K. Proteomics, metabolomics, and ionomics perspectives of salinity tolerance in halophytes. *Front. Plant Sci.* **2015**, *6*, 537. [CrossRef]
47. Mangalassery, S.; Dayal, D.; Kumar, A.; Bhatt, K.; Nakar, R.; Kumar, A.; Misra, A.K. Pattern of salt accumulation and its impact on salinity tolerance in two halophyte grasses in extreme saline desert in India. *Indian J. Exp. Biol.* **2017**, *55*, 542–548.
48. Gil, R.; Bautista, I.; Boscaiu, M.; Lidón, A.; Wankhade, S.; Sánchez, H.; Vicente, O. Responses of five Mediterranean halophytes to seasonal changes in environmental conditions. *AoB Plants.* **2014**, *6*, plu049. [CrossRef] [PubMed]
49. Rahman, M.M.; Rahman, M.A.; Miah, M.G.; Saha, S.R.; Karim, M.A.; Mostofa, M.G. Mechanistic insight into salt tolerance of *Acacia auriculiformis*: The importance of ion selectivity, osmoprotection, tissue tolerance, and Na + exclusion. *Front. Plant Sci.* **2017**, *8*, 155. [CrossRef]
50. Ahmed, H.A.I.; Shabala, L.; Shabala, S. Understanding the mechanistic basis of adaptation of perennial *Sarcocornia quinqueflora* species to soil salinity. *Physiol. Plant.* **2021**, *172*, 1997–2010. [CrossRef]
51. Kumar, A.; Mann, A.; Kumar, A.; Kumar, N.; Meena, B.L. Physiological response of diverse halophytes to high salinity through ionic accumulation and ROS scavenging. *Int. J. Phytoremediation* **2021**, *23*, 1041–1051. [CrossRef]
52. Chaudhary, D.R. Ion accumulation pattern of halophytes. In *Halophytes and Climate Change: Adaptive Mechanisms and Potential Uses*; Hasanuzzaman, M., Shabala, S., Fujita, M., Eds.; CAB International Publishing: Wallingford, UK, 2019; pp. 137–151.
53. Rahman, M.M.; Mostofa, M.G.; Keya, S.S.; Siddiqui, M.N.; Ansary, M.M.U.; Das, A.K.; Rahman, M.A.; Tran, L.S.P. Adaptive mechanisms of halophytes and their potential in improving salinity tolerance in plants. *Int. J. Mol. Sci.* **2021**, *22*, 10733. [CrossRef]
54. Ahmadi, F.; Mohammadkhani, N.; Servati, M. Halophytes play important role in phytoremediation of salt-affected soils in the bed of Urmia Lake, Iran. *Sci. Rep.* **2022**, *12*, 12223. [CrossRef] [PubMed]
55. Ravindran, K.C.; Venkatesan, K.; Balakrishnan, V.; Chellappan, K.P.; Balasubramanian, T. Restoration of saline land by halophytes for Indian soils. *Soil Biol. Biochem.* **2007**, *39*, 2661–2664. [CrossRef]
56. Devi, S.; Nandwal, A.S.; Angrish, R.; Arya, S.S.; Kumar, N.; Sharma, S.K. Phytoremediation potential of some halophytic species for soil salinity. *Int. J. Phytoremediation* **2016**, *18*, 693–696. [CrossRef] [PubMed]
57. Deinlein, U.; Stephan, A.B.; Horie, T.; Luo, W.; Xu, G.; Schroeder, J.I. Plant salt-tolerance mechanisms. *Trends Plant Sci.* **2014**, *19*, 371–379. [CrossRef] [PubMed]
58. Pardo-Domènech, L.L.; Tifrea, A.; Grigore, M.N.; Boscaiu, M.; Vicente, O. Proline and glycine betaine accumulation in two succulent halophytes under natural and experimental conditions. *Plant Biosyst.* **2016**, *150*, 904–915. [CrossRef]
59. Ghosh, U.K.; Islam, M.N.; Siddiqui, M.N.; Khan, M.A.R. Understanding the roles of osmolytes for acclimatizing plants to changing environment: A review of potential mechanism. *Plant Signal. Behav.* **2021**, *16*, 1913306. [CrossRef]
60. Slama, I.; Abdelly, C.; Bouchereau, A.; Flowers, T.; Savouré, A. Diversity, distribution and roles of osmoprotective compounds accumulated in halophytes under abiotic stress. *Ann. Bot.* **2015**, *115*, 433–447. [CrossRef]
61. Parida, A.K.; Panda, A.; Rangani, J. Metabolomics-guided elucidation of abiotic stress tolerance mechanisms in plants. In *Plant Metabolites and Regulation under Environmental Stress*; Ahmad, P., Ed.; Academic Press: Cambridge, MA, USA, 2018; pp. 89–131.
62. Saddhe, A.A.; Manuka, R.; Penna, S. Plant sugars: Homeostasis and transport under abiotic stress in plants. *Physiol. Plant.* **2021**, *171*, 739–755. [CrossRef]
63. Marco, P.; Carvajal, M.; del Carmen Martinez-Ballesta, M. Efficient leaf solute partioning in *Salicornia fruticosa* allows growth under salinity. *Environ. Exper. Bot.* **2019**, *157*, 177–186. [CrossRef]
64. Mali, B.S.; Chitale, R.D. Comparision of accumulation of organic and inorganic osmolyte in *Trianthema portulacastrum* L. growing in saline and non-saline habitats. *Eco. Env. Cons.* **2020**, *26*, 155–158.
65. Akcin, A.; Yalcin, E. Effect of salinity stress on chlorophyll, carotenoid content, and proline in *Salicornia prostrata* Pall. and *Suaeda prostrata* Pall. subsp. prostrata (Amaranthaceae). *Braz. J. Bot.* **2016**, *39*, 101–106. [CrossRef]
66. Youssef, A.M. Salt tolerance mechanisms in some halophytes from Saudi Arabia and Egypt. *Res. J. Agricul. Biol. Sci.* **2009**, *5*, 191–206.
67. Ibraheem, F.; Al-Zahrani, A.; Mosa, A. Physiological adaptation of three wild halophytic *Suaeda* species: Salt tolerance strategies and metal accumulation capacity. *Plants* **2022**, *11*, 537. [CrossRef]
68. Rajput, V.D.; Harish Singh, R.K.; Verma, K.K.; Sharma, L.; Quiroz-Figueroa, F.R.; Meena, M.; Mandzhieva, S. Recent developments in enzymatic antioxidant defence mechanism in plants with special reference to abiotic stress. *Biology* **2021**, *10*, 267. [CrossRef] [PubMed]
69. Waśkiewicz, A.; Muzolf-Panek, M.; Goliński, P. Phenolic content changes in plants under salt stress. In *Ecophysiology and responses of plants under Salt Stress*; Ahmad, P., Azooz, M., Prasad, M., Eds.; Springer: New York, NY, USA, 2013; pp. 283–314.
70. Benjamin, J.J.; Lucini, L.; Jothiramshekar, S.; Parida, A. Metabolomic insights into the mechanisms underlying tolerance to salinity in different halophytes. *Plant Physiol. Biochem.* **2019**, *135*, 528–545. [CrossRef]
71. Reginato, M.; Cenzano, A.M.; Arslan, I.; Furlán, A.; Varela, C.; Cavallin, V.; Luna, V. Na2SO4 and NaCl salts differentially modulate the antioxidant systems in the highly stress tolerant halophyte *Prosopis strombulifera*. *Plant Physiol. Biochem.* **2021**, *167*, 748–762. [CrossRef] [PubMed]
72. Lutts, S.; Lefèvre, I. How can we take advantage of halophyte properties to cope with heavy metal toxicity in salt-affected areas? *Ann. Bot.* **2015**, *115*, 509–528. [CrossRef] [PubMed]

73. Joshi, A.; Kanthaliya, B.; Meena, S.; Rajput, V.D.; Minkina, T.; Arora, J. Proteomic and Genomic Approaches to Study Plant Physiological Responses under Heavy Metal Stress. In *Heavy Metal Toxicity in Plants: Physiological and Molecular Adaptations*; Aftab, T., Hakeem, K.R., Eds.; CRC Press: Boca Raton, FL, USA, 2021; pp. 231–247.
74. Sánchez-Gavilán, I.; Rufo, L.; Rodriguez, N.; de la Fuente, V. On the elemental composition of the Mediterranean euhalophyte *Salicornia patula* Duval-Jouve (Chenopodiaceae) from saline habitats in Spain (Huelva, Toledo and Zamora). *Environ. Sci. Pollu. Res.* **2021**, *28*, 2719–2727. [CrossRef]
75. Rajput, V.; Minkina, T.; Semenkov, I.; Klink, G.; Tarigholizadeh, S.; Sushkova, S. Phylogenetic analysis of hyperaccumulator plant species for heavy metals and polycyclic aromatic hydrocarbons. *Environ. Geochem. Health* **2021**, *43*, 1629–1654. [CrossRef]
76. Caparrós, P.G.; Ozturk, M.; Gul, A.; Batool, T.S.; Pirasteh-Anosheh, H.; Unal, B.T.; Altay, V.; Toderich, K.N. Halophytes have potential as heavy metal phytoremediators: A comprehensive review. *Environ. Exper. Bot.* **2022**, *193*, 104666. [CrossRef]
77. Alam, M.R.; Islam, R.; Tran, T.K.A.; Le Van, D.; Rahman, M.M.; Griffin, A.S.; Yu, R.M.K.; MacFarlane, G.R. Global patterns of accumulation and partitioning of metals in halophytic saltmarsh taxa: A phylogenetic comparative approach. *J. Hazard. Mater.* **2021**, *414*, 125515. [CrossRef]
78. Mousavi Kouhi, S.M.; Moudi, M. Assessment of phytoremediation potential of native plant species naturally growing in a heavy metal-polluted saline–sodic soil. *Environ. Sci. Pollu. Res.* **2020**, *27*, 10027–10038. [CrossRef] [PubMed]
79. Afifi, A.A.; Youssef, R.A.; Hussein, M.M. Fourier transform infrared spectometry study on early stage of salt stress in *Jujube* plant. *Life Sci. J.* **2013**, *10*, 1973–1981.
80. Akyuz, S.; Akyuz, T.; Celik, O.; Atak, C. FTIR spectroscopy of protein isolates of salt-tolerant soybean mutants. *J. Appl. Spectroscopy.* **2018**, *84*, 1019–1023. [CrossRef]
81. Nikalje, G.C.; Kumar, J.; Nikam, T.D.; Suprasanna, P. FT-IR profiling reveals differential response of roots and leaves to salt stress in a halophyte *Sesuvium portulacastrum* (L.) L. *Biotechnol. Rep.* **2019**, *23*, e00352. [CrossRef] [PubMed]

horticulturae

Review

The Content of Heavy Metals in Medicinal Plants in Various Environmental Conditions: A Review

Natalya Vinogradova [1], Alexander Glukhov [2], Victor Chaplygin [3], Pradeep Kumar [4], Saglara Mandzhieva [3,*], Tatiana Minkina [3] and Vishnu D. Rajput [3,*]

1 Pharmaceutical and Medical Chemistry Department, State Educational Institution of Higher Professional Education, M. Gorky Donetsk National Medical University, Donetsk 283003, Russia
2 Natural Flora and Conservation Affairs Department, Donetsk Botanical Garden, Donetsk 283059, Russia
3 Academy of Biology and Biotechnology, Southern Federal University, Rostov-on-Don 344090, Russia
4 Department of Botany, MMV, Banaras Hindu University, Varanasi 221005, India
* Correspondence: msaglara@mail.ru (S.M.); rajput.vishnu@gmail.com (V.D.R.)

Abstract: Nowadays people are becoming poisoned through the consumption of herbal remedies that comprise heavy metals (HMs) worldwide. It is possible for HMs to be present in pharmaceutical herb materials coming from anthropogenic activities like agriculture, industrial waste, and natural sources. In various ethnic groups, there is evidence that contaminants were purposefully added in the belief that they had some sort of therapeutic benefit. HM toxicity of medicinal plant products has been linked to a wide range of adverse health effects, causing dysfunction of the liver, kidney, and heart, and even death. Natural plant-based products established around the world have progressed to the point that they now combine a variety of synthetic products for their purported medical benefits. This assessment focuses on the impacts of HMs on plants, sources of HMs, herbal sample collection, and identification techniques, especially in medicinal plant samples. At the same time, it focuses on the sociocultural applications of HMs as well as the dangers associated with their usage in conventional therapies. It is necessary to implement appropriate regulation and monitoring systems for natural supplements due to the prevalence of hazardous HMs.

Keywords: heavy metals (HMs); herbal plants; atomic absorption spectroscopy; X-ray florescence; neutron activation analysis; anodic stripping voltammetry

Citation: Vinogradova, N.; Glukhov, A.; Chaplygin, V.; Kumar, P.; Mandzhieva, S.; Minkina, T.; Rajput, V.D. The Content of Heavy Metals in Medicinal Plants in Various Environmental Conditions: A Review. *Horticulturae* **2023**, 9, 239. https://doi.org/10.3390/horticulturae9020239

Academic Editors: Adalberto Benavides-Mendoza, Yolanda González-García, Fabián Pérez Labrada and Susana González-Morales

Received: 17 January 2023
Revised: 30 January 2023
Accepted: 6 February 2023
Published: 9 February 2023

1. Introduction

Currently, despite significant advances in the field of chemical synthesis, the interest of scientific medicine in herbal medicines remains high. Every third medicinal product on the world market is made of plant-based raw materials [1]. In some pharmaceutical groups, the share of herbal medicinal products is even higher; for example, about 70% of drugs used to treat cardiovascular diseases are plant-based [2].

In the Russian Federation, about 40% of the total number of medicinal products used in practical medicine are herbal medicinal products [3]. Their advantages over synthetic drugs are a wide spectrum of action, relatively low cost, high bioavailability, low toxicity when used rationally, lower likelihood of side effects, and the possibility of long-term use [1,4]. At present, interest in the study of medicinal herbs continues to increase. This is reflected in the number of relevant publications, which more than tripled from 2008 to 2018 (from 4686 to 14,884 publications) [5].

According to the World Health Organization (WHO), the share of herbal medicines will reach 60% in the near future [4]. An important problem that limits the possibilities of using medicinal herbs in medicine is the continuing reduction of territories that are not exposed to anthropogenic pressure. In this regard, it is not always possible to harvest medicinal plant materials only in environmentally clean areas. Therefore, one of the urgent issues in modern pharmacy is to analyze the possibility of using plants growing under technogenic pressure for medicinal purposes. This problem is of particular significance in Russia, as a major share of medicinal plant materials is harvested in the European part of the country, which is populated and industrialized [6].

Medicinal plant materials harvested under such conditions can be a source of various toxicants entering the human body, primarily heavy metals (HMs), as well as pesticides, nitrates, and other xenobiotics, and damage human health [7,8]. A good example of this is the study by Buettner et al. [9] that found a 10% increase in lead concentration in the blood of females who took herbal dietary supplements for a month compared the control group. Lead concentration in the blood of females who used Ayurvedic plants and plants of traditional Chinese medicine was more than 24% higher than that in the control group [9]. Similar results were obtained in epidemiological studies conducted in Taiwan [10]. Therefore, the present study aimed at systematizing and analyzing information on the content of HMs in medicinal plant materials and herbal medicinal products.

2. Heavy Metals, Medicinal Herbs, and Regulatory Documents

The common HMs in the environment are lead, cadmium, and mercury, and their major sources are vehicles, industrial and thermal power plants, waste incinerators, and agricultural production [11–14]. Plants, especially trees, act as a barrier to the spread of HMs. A comparative analysis of approaches to regulation of HMs in medicinal plant materials and herbal medicinal products adopted in Russia, Europe, the United States, and Asian countries were carried out. For this purpose, the current editions of the Russian, European, American, and Asian pharmacopoeias, as well as international standards, have been studied (Table 1).

Table 1. Permissible concentrations of heavy metals according to regulatory documents in different countries.

Food Items and Medicinal Plant Materials	Regulatory Document	Permissible Concentrations (mg/kg)			
		Pb	Cd	Hg	As
Medicinal plant materials and herbal medicinal products	Russian pharmacopoeia [15]	6.0	1.0	0.1	0.5 (Laminaria 90)
Herbal medicines, medicinal herbs	European pharmacopoeia [16]	5.0	1.0	0.1	There are no general regulations (Laminaria 90)
Herbal medicines	United States pharmacopeia [17]	5.0	0.5	1.0 (total) methyl mercury 0.2	Non-organic 2.0
Medicinal plant materials and herbal medicinal products	Eurasian Economic Union pharmacopeia [18,19]	6.0	1.0	0.1	0.5
Medicinal plant materials (underground organs)	Pharmacopoeia of the People's Republic of China [20]	5.0	0.3	0.2	2.0
Herbs consumed by humans	World Health Organization [21]	10.0	0.3	–	1.0
Traditional Chinese herbal medicines	ISO international standards [7]	10.0	2.0	3.0	4.0
Medicinal herbal preparations	Ayurvedic pharmacopoeia [22]	10.0	0.3	1.0	3.0
Medicinal herbal preparations	Thai pharmacopoeia [7]	10.0	0.3	–	4.0
Medicinal plant materials	Korean pharmacopoeia [7]	5.0	0.3	0.2	3.0
Traditional medicine products	Singapore Health Sciences Authority [23]	20.0	–	0.5	5.0

According to the current *State Pharmacopoeia of the Russian Federation* (14th edition), the content of lead in medicinal plant materials and herbal medicinal products should not exceed 6.0 mg/kg; for cadmium—1.0 mg/kg; for mercury—0.1 mg/kg; and for arsenic—0.5 mg/kg [15]. The content limits of HMs in medicinal plant materials are similar to those for dry herbal dietary supplements and are less stringent than similar requirements for fruits, berries, and drinks (Table 1). A comparative analysis of the regulatory documentation in various countries has found that the requirements for the environmental safety of medicinal plant materials and herbal medicinal products differ significantly.

The Russian standard for lead is lower than that recommended by World Health Organization and the International Organization for Standardization (ISO) (10 mg/kg). Nevertheless, it is slightly higher than the corresponding standard set by the European, American, and Chinese pharmacopoeias (5 mg/kg). For cadmium, the standard established by the Russian pharmacopoeia is similar to the European one and is more than three times higher than those recommended by WHO and many Asian countries (0.3 mg/kg). Interestingly, in the recent edition (2020) of the Chinese pharmacopoeia, the maximum permissible concentration (MPC) for cadmium was changed from 0.3 mg/kg to 1.0 mg/kg [7,23].

This is an illustrative example of a trade-off between safety for human health and the need to use medicinal herbs that grow in a particular region.

The WHO, Russia, and a number of countries have similar requirements for mercury concentration; however, in the Chinese and Indian pharmacopoeias, the MPC is 10 times higher, while the American pharmacopoeia separately regulates the content of methylmercury. The Russian pharmacopoeia imposes the most stringent requirements for arsenic concentrations; the regulatory documents of other countries set standards that exceed the Russian ones by 2–8 times. When comparing the methods for analyzing medicinal plant materials for HMs, it has been revealed that they do not always correspond to each other, which may explain the difference in maximum permissible concentration (MPC) values. Thus, in foreign pharmacopoeias, the arsenic concentration in medicinal plant materials and herbal medicinal products is determined by decomposition in closed vessels, which eliminates the loss of the element at the stage of sample preparation.

There are a number of imperfections in the methods for determining HMs given in the current *State Pharmacopoeia of the Russian Federation* [15]. The incorrectness of using standard samples to analyze medicinal plant materials and herbal medicinal products, in which HMs are found the form of inorganic salts and are not associated with organic compounds, was noted [24]. According to the authors, it is advisable to use standard samples of plant materials certified for the content of HMs, as the organic matrix has a significant influence on measurement results. In addition to the listed metals, the concentration of copper is regulated in China and Singapore, and the content of nickel is regulated in European countries. The introduction of MPCs for these metals, as well as zinc, iron, and manganese, possibly taking into account the regional characteristics of industrial activities, is also a pressing issue in Russian Federation. Another aspect to consider is the part of the plant used in pharmacy. The *State Pharmacopoeia of the Russian Federation* [15] imposes uniform requirements on the content of toxicants for all types of plant materials. In the Chinese pharmacopoeia [20], the standards for HM contents in underground organs are set separately, which is logical, because when growing on polluted soil, many species are able to limit the supply of xenobiotics to the aerial part, especially to the generative organs.

There is large number of publications that analyzes the regional features of HMs' accumulation in medicinal herbs [6,24–32]. For example, the environmental purity of *Cichorium intybus* L. in the Trans-Ural region of the Republic of Bashkortostan (Russia) has been assessed according to the MPCs of chemical elements in feed for farm animals and feed additives (5 mg/kg Pb, 0.3 mg/kg Cd) [31]. The authors have concluded that it is inappropriate to harvest the raw material under study as a medicinal plant material due to the excess of cadmium concentration by 3.0–6.5 times. However, by the current MPC for medicinal plant materials safety assessment, the results would have been different. Although the content of HMs in plants depends on the intensity of contamination in the harvesting area, a general conclusion about the unfavorable environmental situation is not enough to assess the possibility of growing medicinal herbs in a particular area. The safety of 51 samples of *Tanacetum vulgare* L. flowers, harvested in urban and agroecosystems of the Voronezh region, was established with respect to the content of HMs and arsenic [6]. The roots of *Taraxacum officinale* F.H. Wigg. plants growing in the Voronezh region near roads and railways were also found to be safe, while in the samples collected near a thermal power plant and a chemical enterprise, an excess of arsenic was recorded [6]. The aboveground parts of *Artemisia frigida* Willd. and *Artemisia jacutica* Drob. growing in the Republic of

Buryatia, as well as *A. frigida* growing in Mongolia, had lead and cadmium concentrations within the normal range (except for the year in which forest fires occurred) [33]. It was found that *Plantago major* L. leaves harvested in the park area of the central part of the city of Kursk (Russia) were environmentally safe, while in the industrial area the concentration of lead in these medicinal plant materials was 20.5 times the MPC [34]. The content of HMs in medicinal plant materials growing in the Grodno region (in the Republic of Belarus) did not exceed the MPC [29]. This applied to both wild (*Vaccinium vitis-idaea* L., *Vaccinium myrtillus* L., *Elytrigia repens* (L.) Nevski, *Artemisia absinthium* L., *Hypericum maculatum* Crantz, *Angelica sylvestris* L.) and cultivated (*Calendula officinalis* L., *Chamomilla recutita* (L.) Rauschert, *Aesculus hippocastanum* L., *Paeonia anomala* L.) plants.

In Russia, wild-growing medicinal herbs are mainly used in pharmacy, in contrast to Europe and Asia. A significant number of studies were carried out in China. This can be attributed to the popularity of traditional medicine in the country: medicinal herbs were used to treat as many as 85.2% of COVID-19 (60,107 cases), yielding positive results at all stages of the disease [35]. In 2021, the results of a large-scale study on the analysis of HM contents in plant raw materials purchased at the major medicinal plant markets in China were published (1773 medicinal plant material samples were analyzed). In 541 samples, the content of HMs exceeded the MPC, of which 75 samples demonstrated excess content of two metals, 24 samples demonstrated excess content of three HMs, and 9 samples demonstrated excess content of four HMs. The highest concentration of copper was found in *Schisandra chinensis* (Turcz.) Baill. plant material (34.01 mg/kg), the highest arsenic concentration in *Plantago asiatica* L. (14.53 mg/kg), the highest cadmium concentration in *Curcuma longa* L. (6.20 mg/kg), the highest mercury concentration in *Chrysanthemum indicum* L. (8.69 mg/kg), and the highest lead concentration in *Tetradium ruticarpum* (A.Juss.) T.G.Hartley (50.11 mg/kg) [35].

A similar study was conducted in the United Arab Emirates [36]. A total of 81 medicinal plant samples (*Petroselinum crispum*, *Ocimum basilicum*, *Salvia officinalis*, *Origanum vulgare*, *Mentha spicata*, *Thymus vulgaris* and *Matricaria chamomilla*) purchased from Dubai markets were analyzed. The authors claim that 29% of the samples exceeded the MPC for cadmium compared to the WHO standard (0.3 mg/kg). If the assessment had been carried out according to the Russian standards (1.0 mg/kg), the concentration of cadmium would have exceeded the MPC only in the *Ocimum basilicum* plant. The results for lead were similar: its concentration in the analyzed samples varied from less than 1.0 to 23.52 mg/kg, and in 64% of the samples it exceeded the MPC established by the WHO (10 mg/kg), while in Russia a more stringent lead concentration standard of 6.0 mg/kg was adopted.

A number of studies have focused on the analysis of HM contents in pharmacy medicinal plant materials. The average contents of lead, cadmium, zinc, copper, and nickel in *Hypericum perforatum* L. and *Achillea millefolium* L. harvested in the vicinity of Dubna turned out to be significantly lower than those for samples purchased in the pharmacy network in the same city [37]. Similar results were obtained in India for various species of *Berberis* L.: market samples were found to be more contaminated with HMs (at a level exceeding the MPCs set by the WHO) [38]. The study of 26 samples of *Ginkgo biloba* L.-based products purchased in pharmacies in the capital of Mexico found that the contents of lead, arsenic, and cadmium in dietary supplements and medicinal plant materials were higher than in herbal preparations, and in some cases the concentrations of lead and arsenic exceeded the recommended daily allowances (10 μg/day) [39].

When assessing the environmental safety of medicinal plant materials, much attention was given to cadmium and lead determination, while mercury and arsenic were much less studied, which was due to the labor-intensive nature of the methods for their determination. However, even if a medicinal plant material contains metals in quantities not exceeding the MPCs, their presence can pose a potential threat to health due to their ability to accumulate, which can be observed in cases of long-term use of herbal decoctions, infusions, and teas in order to prevent diseases. This reality necessitates long-term studies to track the risk. Numerous works showed the content of HMs in leaves under conditions of pollution, whereas other types of plant materials (flowers, fruits, underground organs, bark) are often used to manufacture medicines, and different organs of the same plants are known to respond unequally to the effects of toxicants [14]. This makes it expedient to conduct a comparative analysis of the metal content in different organs of each type of HM.

The Russian pharmacopeia provides a single value on the content of metals in medicinal plant materials and herbal medicinal products that does not take into account the type of extractant, the dosage form, or production technology peculiarities. The standards for the total content of HMs in tinctures (not more than 0.001%) and extracts (not more than 0.01%) are set separately [15]. The colorimetric method is used, i.e., the whole range of HMs, including essential ones, is determined. In Russia, there are many fewer studies of HM concentrations in various dosage forms and the consequences of their impact on the human body than studies of their contents in plants. The studies of pharmaceutical herbal medicines do not provide information on the concentrations of toxicants in the raw materials the herbal medicinal products are made of. This does not allow tracing metals' migration from the plant to the medicinal product.

When studying the safety of medicinal herbs, the transfer of HMs from the plant material to various dosage forms is of great importance; the literature data on this issue are contradictory. In the case of wild *Rosa* spp., *Vaccinium subg. Oxycoccus, Glycyrrhiza glabra, Zingiber officinale*, and *Aronia melanocarpa*, it was shown that 15–96% of the HMs was transferred to the extract, the maximum transition being recorded for mercury (from 60 to 96%), and the minimum for arsenic (from 24 to 37%) [3]. Potassium, magnesium, sodium, strontium, cadmium, and nickel were found (the leaves of *Plantago major* L. were taken as an example) to transfer to decoctions to a large extent; copper, lead, and iron were characterized by a low degree of extraction [40]. When analyzing raw plant materials representing 16 species of various morphological groups and their herbal preparations, the transition of HMs into tinctures and liquid extracts (alcohol extracts) was found to be significantly lower compared to infusions and decoctions (water extracts) [41]. In 197 samples of herbal medicinal products obtained from various morphological parts of plants (powders, collections, liquid, and solid extracts) lead, cadmium, mercury, and arsenic showed low degrees of extraction from plant material when obtaining extracts and tinctures [24]. *Hypericum perforatum* L. is a super-concentrator of cadmium; however, the maximum extraction into decoctions did not exceed 23%, and 5% of its content in the herb into tinctures, which was due to the formation of physiologically inactive complexes, with lead extraction being 45–70% in decoctions and 30–60% in tinctures [26]. The minimum extractability of HMs, regardless of the extractant, was observed in the dustiest plants (with a higher content of total ash and residue insoluble in 10% HCl); apparently, some HMs in strongly bound forms were found in fine soil particles deposited on the surface of plants [26].

When assessing environmental safety, the dosage and duration of particular drugs' use are important. To this end, international publications calculated indicators such as health risk index, target hazard quotient, hazard index, cancer risk, non-carcinogenic risk assessment, and average daily intake [27,28,32,42]. Russian researchers were most often limited to comparing the HMs' content with the requirements of regulatory documentation. The environmental safety and pharmaceutical value of MHs growing in Donbass under various levels of anthropogenic pressure were analyzed [30]. It was revealed that the concentrations of cadmium, lead, and mercury in the medicinal raw materials of *Sorbus aucuparia* L., *Sorbus intermedia* (Ehrh.) Pers, *Rosa lupulina* Dubovik, *Crataegus fallacina* Klokov, and *Sambucus nigra* L., growing in the urban environment of Donbass, did not exceed the MPCs for medicinal plant materials, even for plants of first-row planting along urban highways with heavy traffic, and the contents of active substances met the requirements of regulatory documentation.

In places where plant material was harvested, soil samples were also collected, the mobile (the most accessible to plants) forms of HMs were determined, and the coefficients of biological accumulation (CBA) equal to the ratio of the toxicant content in the plant to that in the soil were calculated [11]. This made it possible to reveal disproportions between HM contents and CBAs, with the barrier function of the plant root system and the species specificity of plants in relation to specific metals being of major importance. It was found that the CBAs of cadmium and lead for most of the studied fruits and flowers under pollution conditions decreased, indicating the ability of plants to limit metal intake into the generative organs. The content of mercury in all analyzed soil samples was minimal; however, higher concentrations of mercury were found in plants, indicating its aerotechnogenic intake into aboveground organs. Nevertheless, when studying the leaves of *Cotinus coggygria* Scop. harvested in Donbass under the same conditions, completely different results were obtained [30]. The content of lead in the leaves of *C. coggygria* from the control (conditionally ecologically clean) territory was minimal, while in the urbanized environment it was 54 mg/kg, which was nine times higher than the MPC. It should be noted that CBA for lead in *C. coggygria* leaves under conditions of severe technogenic pollution was 19.3, which indicated its vigorous accumulation. The increase in lead concentration in the leaves of *C. coggygria* was more pronounced than the increase in its content in the soil. Thus, based on the results obtained in the study of several plant species, it is impossible to make general conclusions about the suitability of a particular area for harvesting medicinal plant materials. It is necessary to continue studying the characteristics of each individual type of MH in the conditions of a particular urbanized region.

HMs affect the quality of medicinal plant materials, thus indirectly causing a change in the contents of active substances. The concentration of biologically active substances in plants varies species-specifically, largely depending on how much one or another group of substances contributed to the antioxidant system of a particular species. Under conditions of technogenic pressure in most of the studied species, increases in the concentrations of hydroxycinnamic acids, anthocyanins, flavonoids, and carotenoids in contrast to decreases in the levels of tannins, free organic acids, and sugars has been revealed.

In other words, there is a decrease in the content of primary metabolites (which may indicate inhibition of their biosynthesis and/or intensive consumption) and an increase in the concentration of secondary metabolites known for high antioxidant activity and the ability to chelate HMs. The concentrations of those groups of phenolic compounds that are most effective for adapting a particular species to the technogenic environment (and probably make a significant contribution to its antioxidant system) increase in contrast to decreases in the contents of other phenolic compounds, which is due to the common way of their biosynthesis. In reference to the foregoing, when studying medicinal herbs growing in an anthropogenically transformed environment, it is necessary to study their pharmaceutical value along with the assessment of environmental safety.

3. Impact of Heavy Metal on Herbal Plants

Soils have become a significant source of HM pollution and possess a high conversion of ion capability. Certain important HMs become crucial elements which are required in extremely low quantities for proper development of plant [43]. Such HMs play a lead role in physiochemical processes in plants. Plant roots absorbs HMs from soil by the phenomenon of diffusion [44]. These HMs dissolve into their complex structures around the surface of root tissues and are taken up through the apoplast and symplast mechanisms [45]. In contrast, Fe is a biological molecule that is rapidly reduced and oxidized in a wide range of biological reactions.

It is also a crucial mediator for metabolic catalysts that participate in respiration, photosynthesis, and nutrient absorption [46]. At the cellular level, several unique metal transporters like IRT1(Fe, Zn, Mn, and Cd transport), ZIP (Zn transport), and NRAMP (Ni transport) are found on the biological membrane. Several HMs like Co, Cu, Fe, Mn, Ni, and Zn in small amounts are vital for plants. These HMs are necessary for the induction of morphological and metabolic process, the regulation of photosynthesis, the synthesis of chlorophyll, a high rate of the production of bioactive compounds, transpiration, protection of DNA, distraction of ROS, and enhancing nitrogen fixation in plants [47]. However, a high amount of HMs causes lipid peroxidation, ROS generation, and DNA damage. Several studies have found that higher concentrations of Zn have a negative impact on plant development and metabolism [48]. Additionally, according to Arora et al. [49], an increase in plant Fe^{2+} levels triggers the formation of free radicals that destroy protein molecules, DNA, and membranes (Figure 1).

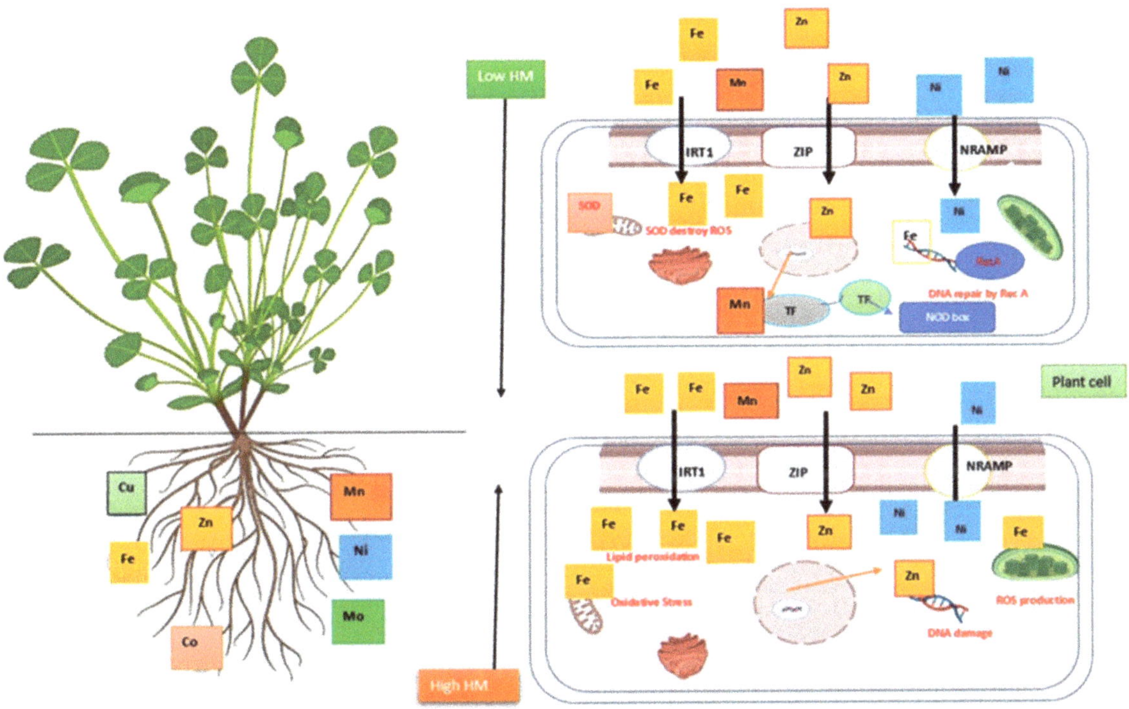

Figure 1. Impact of heavy metals' contribution to the creation of proteins, nucleic acids, photosynthesis pigments, and cell membrane function and structure at low concentrations [50]. Mn enhances antioxidant capacity [51], while Fe increases N_2-fixation and DNA repair [52]. At high concentrations, it causes conversion of numerous significant functional groups, lipid peroxidation (LPO), mitochondrial dysfunction, ROS production, and biochemical disruption via changing enzymatic activity [53,54].

The major causes of HM pollution are anthropogenic activities like resource extraction, agricultural production, construction, industrial activities, inadequate garbage management processes, and excessive use of agrochemicals (Figure 2). HMs enter the environment through these activities and accumulate in living systems. These HMs are toxic in nature and cause several chronic illnesses like weakened immunity, cardiac instability, neonatal disorders, psychological disorders, and sensorimotor behavior issues [55]. Several HMs like Pb, As, Hg, and Cd are not required by plants or the human body, and they also cause a variety of health complications associated with the brain, liver, lungs, heart, kidneys, and nervous system, including hypertension, abdominal pain, rashes, intestinal ulcers, and other symptoms linked to various cancer types [56,57]. Although copper is a major element in several enzymatic reactions, consuming excess amounts of it can lead to internal organ injury but also induce skin infections, elevated lung tissue inflammation, abdominal discomfort, nausea, vomiting, and diarrhea [58].

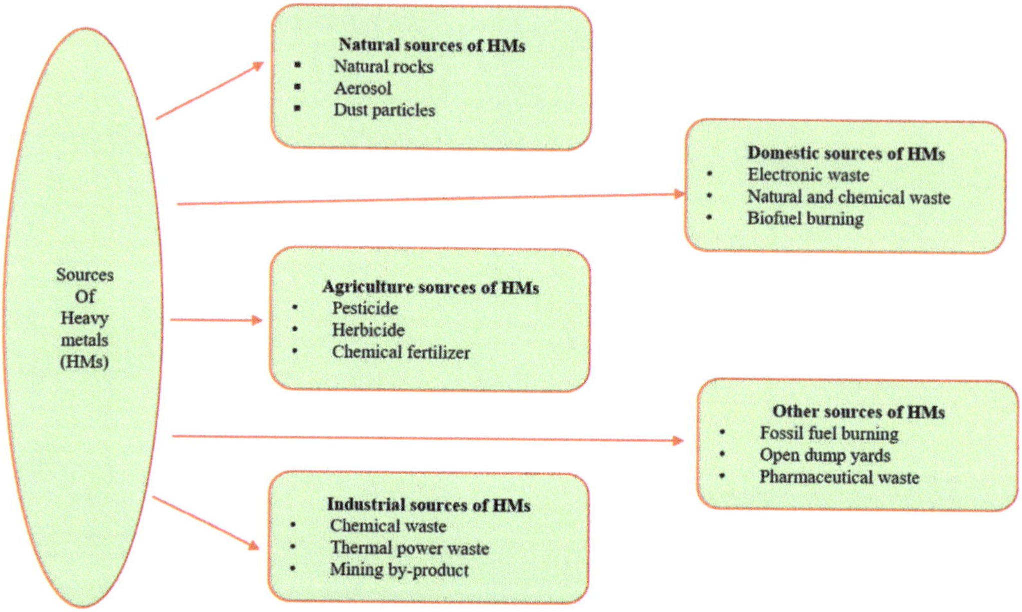

Figure 2. Various sources of heavy metals.

As they are known to cause harm, it is critical and urgent to conduct a thorough investigation into the dangers of metal contamination in medicinal herbs. As a result of the survey's exact methodology, it is now clear that more screening and dosing frequency guidance for herbal medicines is required. This research explored pollutant levels and their risks, particularly when they are present in herbal remedies. The results serve as a groundwork for future research into preventative methods, uniform guidelines, and exogenous contaminants. Research studies conducted have led to suggestions that can rapidly lower or completely remove HM contents in active pharmaceutical ingredients.

4. Herbal Sample Preparation

The processing of samples is an essential step throughout the examination of HM components. The choice of an appropriate technique for sample processing is one of the most critical factors in determining the precision of the measurement. The perfect method for preparing samples is one that does not produce any pollutants, does have a high response time, is easy to use, offers enhanced exactness, and works quickly. Even during the process of sample pre-treatment, there is an assortment of variables that need to be carefully considered. Some of these variables include the physical and chemical features of the matrix, the number of samples, the state of the samples, the composition of the compound that needs to be examined, as well as the estimated accuracy. There are three basic methods for sample processing: wet digestion, the ashing method, and direct injection (Figure 3) [59].

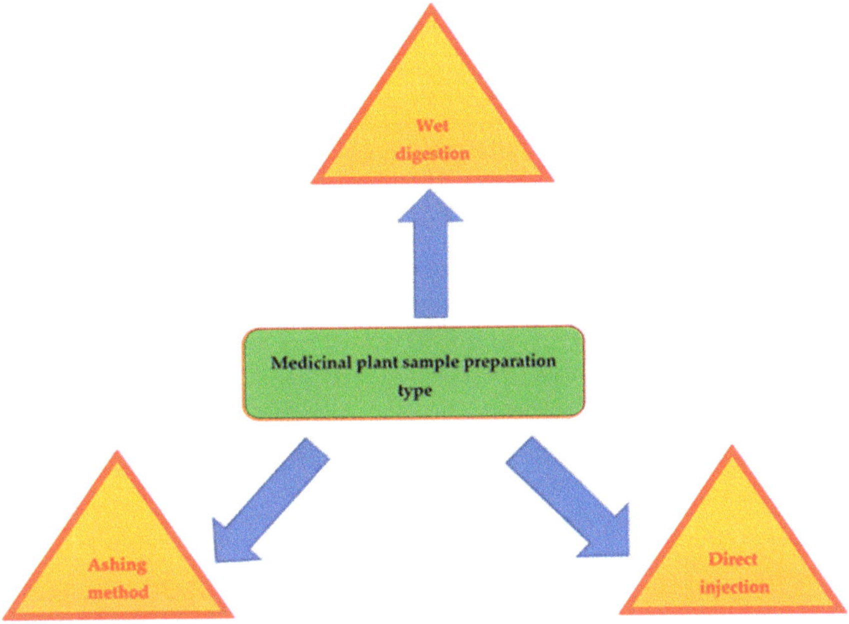

Figure 3. Methods of herbal sample preparation for heavy metal detection.

5. Detection Method of the Heavy Metals in Medicinal Plants

There are several techniques like inductively coupled plasma mass spectrometry (ICPMS), atomic emission spectroscopy (AES), X-ray fluorescence (XRF), neutron activation analysis (NAA), anodic stripping voltammetry (ASV), thermolysis-coupled atomic absorption spectroscopy (TCAS), atomic absorption spectrometry (AAS), and graphite furnace atomic absorption spectrometry (GFAAS) which are used for the quantification of the HMs in herbal plant samples (Figure 4).

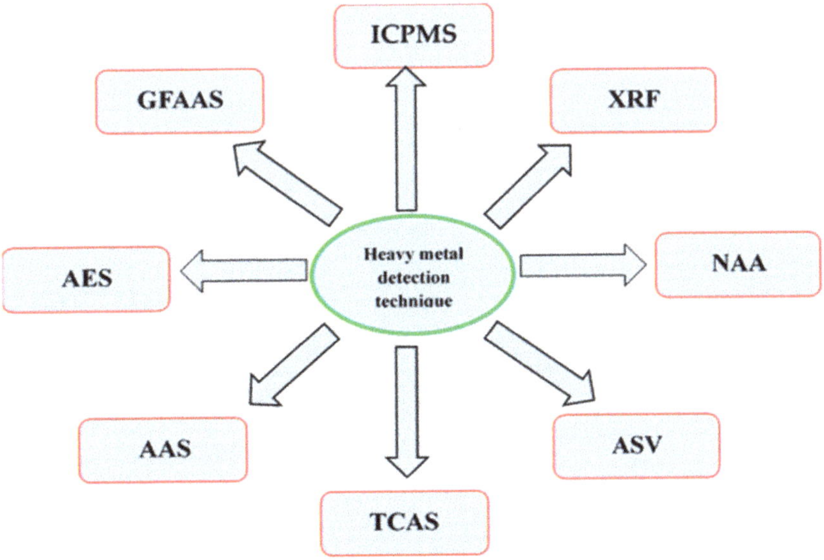

Figure 4. Techniques of heavy metal detection in herbal samples.

5.1. ICPMS

ICPMS detects the HMs on the basis of m/z ratio measurement. As an atomization source for atomic spectroscopy, it is significantly more difficult to use than a graphite furnace due to its high degree of atomization in argon plasma at 7000 K. This method possesses a maximum output capacity of multi-element detection ability in a broad range. ICPMS and AES can identify numerous metallic pollutants in an extra-specific way and can quantify an impurity with a significantly greater degree of sensitivity. The typical detection limits in a solvent are between 0.01 and 1 ppm. The mass of the study sample can be as low as 10 mg. ICPMS assessments are currently being performed and have a solid foundation. It is cost-effective while also producing superior outcomes [60].

5.2. AES

The majority of the time, this method is combined with optical emission spectroscopy. The material is made to become stimulated through the absorption of either a thermal or electrical charge, and then the emission that the agitated material gives off is investigated. Additionally, this method is connected to a solid; however, fluids are often the samples that are examined. It is able to analyze over 70 different elements at dosages as low as one part per million (ppm) [61].

5.3. XRF

When a specimen comes into contact with high-energy X-rays or gamma rays, it emits "secondary" (or fluorescent) X-rays with a unique spectrum. This technique is used in geochemical analyses, toxicology, as well as paleontology. It is also adopted in the analysis of elements and chemicals, specifically in the study of metallic materials, glass, ceramic materials, and construction materials [62].

5.4. NAA

The NAA method is used for determination of the heavy metals present in an herbal material by analyzing the energy and intensity of rays (mostly gamma rays) generated through immediate radiation or radionuclide decay. This method makes use of neutron bombardment [63]. NAA need not require material processing beforehand. Because it is an excitation research methodology, it involves bombarding atoms as well as neutrons produced by reactors, accelerators, or isotope neutron sources; as a result, it is ideal for the qualitative and quantitative measurement of the compositions and occurrences of various elements [63]. In addition to this, NAA is sensitive to a wide variety of elements, and this sensitivity enables it to precisely estimate the levels of trace elements present in a wide variety of samples, including MPs. The use of NAA is important for the rapid identification of the various components of MPs.

5.5. ASV

By using differential pulse anodic stripping, it is a straightforward procedure to simultaneously determine the contents of heavy metals in pharmaceutical herbs. The specimen is prepared through the dry ashing approach, in which 1 g of fine powdered plant matter is kept at 5000 °C for 2.5 h. A hanging mercury electrode and platinum wire are employed as the functioning and countering electrodes in this straightforward voltametric device. These potentials are assessed in relation to a reference electrode made of Ag/Ag Cl and KCl. Prior to the analysis, pure N_2 is bubbled throughout the specimen for 400 s. The ASV approach is more sensitive. This technique yields accurate, repeatable outcomes. It has several drawbacks, like fouling, and it is time-consuming. Its use is constrained, as electrodes (E + 0.4 V) of As and Hg are readily oxidized [64].

5.6. TCAS

According to this procedure, plant materials or herbal products are heated. When the herbal sample is heated to a high temperature, the atomic absorption (AA) detectors are used to analyze HM vapors which have been thermally liberated from molecular HMs of the herbal sample. The procedure does not require any pre-processing of herbal samples. This procedure requires less time than other ones (four minutes every cycle). There is no requirement for calibrating, allowing for continuous testing to be performed. It has a few limitations, such as it only being able to identify Hg. HM level assessments cannot to be performed simultaneously [65].

5.7. AAS

AAS is a common technique of spectrum analysis that is used for both quantitative and qualitative analyses of the HMs present in MPs. Specific analysis of a wide spectrum of elements like Cd, Ni, Pb, and Zn in either solid or liquid specimens is possible with the help of such a technique [66]. In AAS, components such as a beam of light, atomizer, splitter, and detection system are implemented [67]. Although significant strides have been made in the advancement of technology for the identification of heavy metals, AAS is still widely used.

5.8. GFAAS

In GFAAS, an herbal sample is first introduced directly to the graphite tubes in the equipment known as a graphite furnaces. Here only a small portion of the nebulized vapor actually makes it to the flame after the atomized sample has quickly passed through a straightforward surface. Therefore, in order to boost the analytic sensitivity, advanced models employing graphite furnaces for electrical vaporization were utilized. The residual atomic samples of herbs are again employed by GFAAS, which detects the presence of lead and cadmium and the concentrations of copper, arsenic, and mercury in the herbal materials after the solvent and matrix materials have been removed using heat [68].

6. Conclusions

Herbal medicines are an important part of the modern pharmaceutical market worldwide. In order to expand the possible harvesting areas for medicinal plant materials, it must be taken into account that contamination of medicinal plant materials and herbal preparations can occur during storage and manipulation, so safety must be monitored throughout the entire manufacturing process, from harvesting to market. Studies by scientists from different countries show the potential use of plants growing under conditions of technogenic pressure for pharmaceutical purposes. However, it is necessary to continue fundamental and applied research of medicinal plants growing under varying intensities of anthropogenic pressure, taking into account regional characteristics. To date, the problem of setting standards for HM contents in medicinal plant materials in the territory of Russia has not been completely resolved. It is necessary to develop standards for the contents of toxicants in various medicinal products, taking into account the peculiarities of the manufacturing technology, as well as the permissible concentrations of other metals such as zinc, copper, nickel, manganese, and iron in medicinal plant materials and herbal medicinal products. The use of indicators that take into account the dose and frequency of use in assessing the environmental purity of plant raw materials or plant-based preparations will reduce the risk of excessive intake of HMs into the human body.

Author Contributions: N.V., conceptualization; A.G. and V.C., data curation; S.M., writing—original draft preparation; T.M. and P.K., writing—review and editing; V.D.R., supervision. All authors have read and agreed to the published version of the manuscript.

Funding: This research received no external funding.

Data Availability Statement: Not applicable.

Acknowledgments: The study was supported by the Russian Science Foundation (project No. 22-77-10097) at the Southern Federal University.

Conflicts of Interest: The authors declare no conflict of interest.

References

1. Mafimisebi, T.E.; Oguntade, A.E.; Ajibefun, I.A.; Mafimisebi, O.E.; Ikuemonisan, E.S. The expanding market for herbal, medicinal and aromatic plants in Nigeria and the international scene. *Med. Aromat. Plants.* **2013**, *144*, 2167–2412.
2. Brinckmann, J.A. Geographical indications for medicinal plants: Globalization, climate change, quality and market implications for geo-authentic botanicals. *World J. Tradit. Chin. Med.* **2015**, *1*, 16–23. [CrossRef]
3. Nikulin, A.; Potanina, O.; Alyussef, M.; Vasil'ev, V.; Abramovich, R.; Novikov, O.; Boyko, N.; Khromov, A.; Platonov, E. Development of a technique for determining cadmium, lead, arsenic with the etaas method in medicinal plant raw materials. *Farmacia* **2021**, *69*, 566–575. [CrossRef]
4. Petrukhina, I.K.; Yagudina, R.I.; Ryazanova, T.K.; Kurkin, V.A.; Pervushkin, S.V.; Egorova, A.V.; Loginova, L.V.; Khusainova, A.I.; Blinkoval, P.R. Analysis of the implementation of the federal assurance program of supporting beneficiaries with indispensable medicinal preparations in the subjects of the Russian Federation. *Farmatsiya Farmakol.* **2021**, *8*, 273–284. [CrossRef]
5. Fitzgerald, M.; Heinrich, M.; Booker, A. Medicinal plant analysis: A historical and regional discussion of emergent complex techniques. *Front. Pharmacol.* **2020**, *10*, 1–14. [CrossRef]
6. D'yakova, N.A.; Samylina, I.A.; Slivkin, A.I.; Gaponov, S.P.; Myndra, A.A. Estimated heavy-metal and Arsenic contents in medicinal plant raw materials of the Voronezh region. *Pharm. Chem. J.* **2018**, *52*, 220–223. [CrossRef]
7. Chen, Y.G.; Huang, J.H.; Luo, R.; Ge, H.Z.; Wołowicz, A.; Wawrzkiewicz, M.; Gładysz-Płaska, A.; Li, B.; Yu, Q.X.; Kołodyńska, D.; et al. Impacts of heavy metals and medicinal crops on ecological systems, environmental pollution, cultivation, and production processes in China. *Ecotoxicol. Environ. Safet.* **2021**, *219*, 17. [CrossRef]
8. Carrubba, A.; Scalenghe, R. The scent of *Mare Nostrum*: Medicinal and aromatic plants in Mediterranean soils. *J. Sci. Food Agric.* **2012**, *92*, 1150–1170. [CrossRef]
9. Buettner, C.; Mukamal, K.J.; Gardiner, P.; Davis, R.B.; Phillips, R.S.; Mittleman, M.A. Herbal supplement use and blood lead levels of United States adults. *J. Gen. Intern. Med.* **2009**, *24*, 1175–1182. [CrossRef]
10. Ernst, E. Risks of herbal medicinal products. *Pharmacoepidemol. Drug Saf.* **2004**, *13*, 767–771. [CrossRef]
11. Liu, X.; Ju, Y.; Mandzhieva, S.; Pinskii, D.; Minkina, T.; Rajput, V.D.; Roane, T.; Huang, S.; Li, Y.; Ma, L.Q.; et al. Sporadic Pb accumulation by plants: Influence of soil biogeochemistry, microbial community and physiological mechanisms. *J. Hazard. Mater* **2023**, *444*, 130391. [CrossRef]
12. Chaplygin, V.; Dudnikova, T.; Chernikova, N.; Fedorenko, A.; Mandzhieva, S.; Fedorenko, G.; Sushkova, S.; Nevidomskaya, D.; Minkina, T.; Sathishkumar, P.; et al. *Phragmites australis* cav. As a bioindicator of hydromorphic soils pollution with heavy metals and polyaromatic hydrocarbons. *Chemosphere* **2022**, *308*, 136409. [CrossRef]
13. Bezuglova, O.S.; Gorbov, S.N.; Tischenko, S.A.; Aleksikova, A.S.; Tagiverdiev, S.S.; Sherstnev, A.K.; Dubinina, M.N. Accumulation and migration of heavy metals in soils of the Rostov region, south of Russia. *J. Soils Sediments* **2016**, *16*, 1203–1213. [CrossRef]
14. Erofeeva, E.A. Hormesis in plants: Its common occurrence across stresses. *Curr. Opin. Toxicol.* **2022**, *30*, 100333. [CrossRef]
15. *State Pharmacopoeia of the Russian Federation*, XIV ed.; Ministry of Health of the Russian Federation: Moscow, Russia, 2018; Volume II, p. 1449.
16. *European Directorate for the Quality of Medicines & HealthCare (EDQM)*; Council of Europe: Strasbourg, France, 2019; Volume 7.
17. *USP44–NF39*; 561 Articles of Botanical Origin. United States Pharmacopeia: Rockville, MD, USA, 2020; p. 15.
18. *European Pharmacopoeia*, 10th ed.; Council of Europe: Strasbourg, France, 2019; Volume 1, p. 4370.
19. *Pharmacopoeia of the Eurasian Economic Union*; Ministry of Health of the Russian Federation: Moscow, Russia, 2021; 568p.
20. *Pharmacopoeia of the People's Republic of China*; China Food and Drug Administration: Beijing, China, 2015; Volume 1, 2266p.
21. World Health Organization (WHO). *Quality Control Methods for Medicinal Plant Materials*; World Health Organization: Geneva, Switzerland, 2005.
22. Debnath, M.; Paul, N.; Bhattacharya, S.; Biswas, M.; Haldar, P.K. Formulation and assessment of microbial and heavy metal contents of Vidangadilouham: A classical Ayurvedic formulation. *Int. J. Herb. Med.* **2020**, *8*, 101–102.
23. Vyas, P.; Vohora, D. Pharmaceutical regulations for complementary medicine. In *Pharmaceutical Medicine and Translational Clinical Research*; Vohora, D., Singh, G., Eds.; Academic Press: Cambridge, MA, USA, 2018; Chapter 13; pp. 233–264. [CrossRef]
24. Shchukin, V.M.; Kuzmina, N.E.; Erina, A.A.; Yashkir, V.A.; Merkulov, V.A. Comparative analysis of the heavy metal, Aluminum, and Arsenic contents in brown algae of various origins. *Pharm. Chem. J.* **2018**, *52*, 627–634. [CrossRef]
25. Chizzola, R.; Michitsch, H.; Franz, C. Monitoring of metallic micronutrients and heavy metals in herbs, spices and medicinal plants from Austria. *Eur. Food Res. Technol.* **2003**, *216*, 407–411. [CrossRef]
26. Siromlya, T.I. Influence of traffic pollution on ecological state of *Plantago major* L. *Contemp. Probl. Ecol.* **2011**, *4*, 499–507. [CrossRef]
27. Parveen, R.; Abbasi, A.M.; Shaheen, N.; Shah, M.H. Accumulation of selected metals in the fruits of medicinal plants grown in urban environment of Islamabad, Pakistan. *Arab. J. Chem.* **2017**, *13*, 308–317. [CrossRef]
28. Li, J.; Wang, Y.; Yang, H.; Tang, Y. Five heavy metals accumulation and health risk in a traditional Chinese medicine cortex Moutan collected from different sites in China. *Hum. Ecol. Risk Assess.* **2018**, *24*, 2288–2298. [CrossRef]
29. Kuzovkova, A.A.; Drebenkova, I.V.; Velentei, Y.N.; Pleshkova, A.A.; Bychok, G.E.; Chernik, D.V.; Maskalevich, N.V. Heavy metal contamination of wild and cultivated medicinal plants in the Republic of Belarus. *Occup. Med. Human Ecol.* **2020**, *4*, 112–117.
30. Vinogradova, N.A.; Glukhov, A.Z. Ecological and phytochemical features of *Crataegus fallacina* Klokov under conditions of technogenic pollution. *Contemp. Probl. Ecol.* **2021**, *14*, 90–97. [CrossRef]

31. Buskunova, G.G.; Khasanova, R.F.; Semenova, I.N.; Ilbulova, G.R. The heavy metals in the system "soil as a wild-growing medicinal plant" on the example of *Tanacetum vulgare* L. *Ecol. Ind. Russ.* **2020**, *24*, 37–41. [CrossRef]
32. Zhang, Z.; Song, J.; Zhang, H.; Zheng, Z.; Li, T.; Wu, S.; He, B.; Mao, B.; Yu, Y.; Fang, H. Analysis method development and health risk assessment of pesticide and heavy metal residues in *Dendrobium Candidum*. *RSC Adv.* **2022**, *1*, 6869–6875. [CrossRef]
33. Dylenova, E.P.; Zhigzhitzhapova, S.V.; Randalova, T.E.; Radnaeva, L.D.; Shiretorova, V.G.; Pavlov, I.A. Biophile elements and heavy metals in *Artemisia frigida* willd. and *Artemisia jacutica* drob. *Khimiya Rastitel'nogo Syr'ya* **2019**, *4*, 199–205. [CrossRef]
34. Babkina, L.A.; Lukyanchikov, D.S.; Lukyanchikova, O.V. Features of the accumulation of heavy metals by plantain leaves. *Samara Sci. Bull.* **2018**, *7*, 19–24. [CrossRef]
35. Luo, L.; Wang, B.; Jiang, J.; Fitzgerald, M.; Huang, Q.; Yu, Z.; Li, H.; Zhang, J.; Wei, J.; Yang, C.; et al. Heavy metal contaminations in herbal medicines: Determination, comprehensive risk assessments, and solutions. *Front. Pharmacol.* **2021**, *11*, 595335. [CrossRef]
36. Dghaim, R.; Al-Khatib, S.; Rasool, H.; Ali Khan, M. Determination of heavy metals concentration in traditional herbs commonly consumed in the United Arab Emirates. *J. Environ. Public Health* **2015**, 973878. [CrossRef]
37. Kamanina, I.Z.; Kaplina, S.P.; Salikhova, F.S. The content of heavy metals in medicinal plants Scientific review. *Biol. Sci.* **2019**, *1*, 29–34.
38. Srivastava, S.K.; Rai, V.; Srivastava, M.; Rawat, A.K.S.; Mehrotra, S. Estimation of heavy metals in different berberis species and Its market samples. *Environ. Monit. Assess.* **2006**, *116*, 315–320. [CrossRef]
39. Rojas, P.; Ruiz-Sánchez, E.; Ríos, C.; Ruiz-Chow, A.; Resendiz-Albor, A.A. A health risk assessment of lead and other metals in pharmaceutical herbal products and dietary supplements containing *Ginkgo biloba* in the Mexico city metropolitan area. *Int. J. Environ. Res. Public Health* **2021**, *18*, 8285–8304. [CrossRef]
40. Siromlya, T.I.; Zagurskaya, Y.V.; Bayandina, I.I. The elemental composition of *Hypericum Perforatum* plants sampled in environmentally different habitats by the example of West Siberia. *Bot. Pac.* **2020**, *9*, 127–132. [CrossRef]
41. Shikov, A.N.; Shikova, V.A.; Whaley, A.O.; Burakova, M.A.; Flisyuk, E.V.; Whaley, A.K.; Terninko, I.I.; Generalova, Y.E.; Gravel, I.V.; Pozharitskaya, O.N. The ability of acid-based natural deep eutectic solvents to co-extract elements from the roots of *Glycyrrhiza glabra* L. and associated health risks. *Molecules* **2022**, *27*, 7690. [CrossRef]
42. Yang, C.M.; Chien, M.Y.; Chao, P.C.; Huang, C.M.; Chen, C.H. Investigation of toxic heavy metals content and estimation of potential health risks in Chinese herbal medicine. *J. Hazard Mater* **2021**, *412*, 125142. [CrossRef]
43. Alloway, B.J. Heavy metals and metalloids as micronutrients for plants and animals. In *Heavy Metals in Soils*; Springer: Dordrecht, The Netherlands, 2013; pp. 195–209.
44. Peralta-Videa, J.R.; Lopez, M.L.; Narayan, M.; Saupe, G.; Gardea-Torresdey, J. The biochemistry of environmental heavy metal uptake by plants: Implications for the food chain. *Int. J. Biochem. Cell Biol.* **2009**, *41*, 1665–1677. [CrossRef]
45. Hossain, M.A.; Piyatida, P.; da Silva, J.A.T.; Fujita, M. Molecular mechanism of heavy metal toxicity and tolerance in plants: Central role of glutathione in detoxification of reactive oxygen species and methylglyoxal and in heavy metal chelation. *J. Bot.* **2012**, 37. [CrossRef]
46. Hell, R.; Stephan, U.W. Iron uptake, trafficking and homeostasis in plants. *Planta* **2003**, *216*, 541–551. [CrossRef]
47. Blaylock, M.J.; Huang, J.W. Phytoextraction of metals. In *Phytoremediation of Toxic Metals: Using Plants to Clean Up the Environment*; Wiley: New York, NY, USA, 2000; pp. 53–70.
48. Rout, G.R.; Das, P. Effect of metal toxicity on plant growth and metabolism: I. Zinc. In *Sustainable agriculture*; Lichtfouse, E., Navarrete, M., Debaeke, P., Véronique, S., Alberola, C., Eds.; Springer: Dordrecht, The Netherlands, 2009; pp. 873–884. [CrossRef]
49. Arora, A.; Sairam, R.K.; Srivastava, G.C. Oxidative stress and antioxidative system in plants. *Curr. Sci.* **2002**, *82*, 1227–1238.
50. Oves, M.; Khan, S.; Qari, H.; Felemban, N.; Almeelbi, T. Heavy Metals: Biological importance and detoxification strategies. *J. Bioremed. Biodegrad.* **2016**, *7*, 334.
51. Shenker, M.; Plessner, O.E.; Tel-Or, E. Manganese nutrition effects on tomato growth, chlorophyll concentration, and superoxide dismutase activity. *J. Plant Physiol.* **2004**, *161*, 197–202. [CrossRef]
52. Moller, I.M.; Jensen, P.E.; Hansson, A. Oxidative modifications to cellular components in plants. *Annu. Rev. Plant Biol.* **2007**, *58*, 459–481. [CrossRef] [PubMed]
53. Anjum, N.A.; Duarte, A.C.; Pereira, E.; Ahmad, I. Plant-beneficial elements status assessment in soil-plant system in the vicinity of a chemical industry complex: Shedding light on forage grass safety issues. *Environ. Sci. Pollut. Res. Int.* **2015**, *22*, 2239–2246. [CrossRef] [PubMed]
54. de Oliveira Jucoski, G.; Cambraia, J.; Ribeiro, C.; de Oliveira, J.A.; de Paula, S.O.; Oliva, M.A. Impact of iron toxicity on oxidative metabolism in young *Eugenia uniflora* L. plants. *Acta Physiol. Plant* **2013**, *35*, 1645–1657. [CrossRef]
55. Dehno, M.A.; Harami, S.R.M.; Noora, M.R. Environmental geochemistry of heavy metals in coral reefs and sediments of Chabahar Bay. *Results Eng.* **2022**, *13*, 100346. [CrossRef]
56. Bharti, R.; Sharma, R. Effect of heavy metals: An overview. *Mater. Today Proc.* **2022**, *51*, 880–885.
57. Rahman, M.M.; Hossain, M.K.F.B.; Afrin, S.; Saito, T.; Kurasaki, M. *Effects of Metals on Human Health and Ecosystem*; Springer: Berlin/Heidelberg, Germany, 2022; pp. 1–39.
58. Pirhadi, M.; Shariatifar, N.; Bahmani, M.; Manouchehri, A. Heavy metals in wheat grain and its impact on human health: A mini-review. *J. Chem. Health Risks* **2022**, *12*, 421–426.
59. Guo, C.; Lv, L.; Liu, Y.; Ji, M.; Zang, E.; Liu, Q.; Zhang, M.; Li, M. Applied analytical methods for detecting heavy metals in medicinal plants. *Crit. Rev. Anal. Chem.* **2023**, *53*, 339–359. [CrossRef]

60. Wilschefski, S.C.; Baxter, M.R. Inductively coupled plasma mass spectrometry: Introduction to analytical aspects. *Clin. Biochem. Rev.* **2019**, *40*, 15. [CrossRef]
61. Rehan, I.; Gondal, M.A.; Aldakheel, R.K.; Almessiere, M.A.; Rehan, K.; Khan, S.; Sultana, S.; Khan, M.Z. Determination of nutritional and toxic metals in black tea leaves using calibration free LIBS and ICP: AES technique. *Arab. J. Sci. Eng.* **2022**, *47*, 7531–7539. [CrossRef]
62. Nguyen, H.M.; Huynh, N.T.K.; Nguyen, N.T.Y.; Ha, L.T.; Pham, T.T. Evaluating the content of some metal elements in soil and their effects on the total phenolic and flavonoid contents of some medicinal plants using X-ray fluorescence (XRF) Method. *Res. Sq.* **2022**, *1*, 1–24. [CrossRef]
63. Garg, A.N.; Singh, R.; Maharia, R.S.; Dutta, R.K.; Datta, A. Quantification of minor, trace and toxic elements in stems of *Santalum album* (L.), *Mangiferra indica* (L.) and *Tinospora cordifolia* by instrumental neutron activation analysis. *J. Plant Sci. Phytopathol.* **2022**, *6*, 8–14. [CrossRef]
64. Khamcharoen, W.; Duchda, P.; Songsrirote, K.; Ratanawimarnwong, N.; Limchoowong, N.; Jittangprasert, P.; Mantim, T.; Siangproh, W. An application of miniaturized electrochemical sensing for determination of arsenic in herbal medicines. *Anal. Methods* **2022**, *14*, 3087–3093. [CrossRef]
65. Papadopoulos, A.; Assimomytis, N.; Varvaresou, A. Sample preparation of cosmetic products for the determination of heavy Metals. *Cosmetics* **2022**, *9*, 21. [CrossRef]
66. Lawi, D.J.; Abdulwhaab, W.S.; Abojassim, A.A. Health risk study of heavy metals from consumption of drugs (solid and liquid) samples derived from medicinal plants in Iraq. *Biol. Trace Elem. Res.* **2022**, *9*, 1–13. [CrossRef]
67. Hyder, Z.; Rizwani, G.H.; Ahmed, I.; Shareef, H.; Azhar, I.; Aqeel, E. Determination of Heavy metals content, Lead (Pb), Mercury (Hg), Cadmium (Cd), Nickel (Ni), and Copper (Cu) with risk assessment to human consumption as a food and medicine in herbal species through Atomic Absorption Spectroscopy. *Res. Sq.* **2022**, *1*, 1–16. [CrossRef]
68. Alinia-Ahandani, E.; Nazem, H.; Malekirad, A.A.; Fazilati, M. The safety evaluation of toxic elements in medicinal plants: A Systematic Review. *J. Hum. Environ. Health Promot.* **2022**, *8*, 62–68. [CrossRef]

horticulturae

MDPI

Article

Effects of Citric Acid and Humic-like Substances on Yield, Enzyme Activities, and Expression of Genes Involved in Iron Uptake in Tomato Plants

Fabián Pérez-Labrada [1], Adalberto Benavides-Mendoza [2], Antonio Juárez-Maldonado [1], Susana Solís-Gaona [3] and Susana González-Morales [4,*]

[1] Departamento de Botánica, Universidad Autónoma Agraria Antonio Narro, Saltillo 25315, Mexico; fabian.perezl@uaaan.edu.mx (F.P.-L.); juma841025@gmail.com (A.J.-M.)
[2] Departamento de Horticultura, Universidad Autónoma Agraria Antonio Narro, Saltillo 25315, Mexico; abenmen@gmail.com
[3] United Phosphorus Limited, UPL, Saltillo 25290, Mexico; susana.solis@upl-ltd.com
[4] Departamento de Horticultura, CONAHCyT-Universidad Autónoma Agraria Antonio Narro, Saltillo 25315, Mexico
* Correspondence: sgonzalezmo@conahcyt.mx or qfb_sgm@hotmail.com Tel.: +52-844-1222471

Abstract: Iron (Fe) deficiency is a common abiotic stress on plants growing in calcareous soils where low organic matter content, high carbonate–bicarbonate concentration, and high pH precipitate Fe in unavailable forms. Enzymatic activity is a mechanism for plants to access soil nutrients; enzymes such as H^+-ATPase, phosphoenolpyruvate carboxylase (PEPC), and the intracellular enzyme ferric reduction oxidase (FRO) are involved in Fe absorption. The effects of the application of citric acid (CA) and humic-like substances (HLS) on the yield, H^+-ATPase, PEPC, and FRO enzyme activity, and expression of *LeHA1*, *LePEPC1*, and *LeFRO1* genes in tomato plants grown under calcareous soil were studied. CA and HLS improved the SPAD units and increased the number of harvested fruits and yield per plant. Temporary alterations in enzyme activity, which reduced PEPC and FRO activity in roots, were documented. In leaf tissue, CA resulted in lower expression of *LeHA1* and *LePEPC1* and the induction of *LeFRO1* expression, whereas HLS application resulted in higher expression of *LePEPC1* and *LeFRO1*. In roots, *LeHA1* expression increased with HLS, whereas *LePEPC1* and *LeFRO1* showed lower expression with CA and HLS, respectively. The application of CA and HLS through a nutrient solution in combination with Fe-chelate can improve Fe nutrition in tomato plants potted in calcareous soil by inducing temporal alterations in PEPC and FRO enzyme activity and *LeFRO1* and *LeHA1* gene expression.

Keywords: ferric chelate reductase; H^+-ATPase; humic acid; phosphoenolpyruvate carboxylase

Citation: Pérez-Labrada, F.; Benavides-Mendoza, A.; Juárez-Maldonado, A.; Solís-Gaona, S.; González-Morales, S. Effects of Citric Acid and Humic-like Substances on Yield, Enzyme Activities, and Expression of Genes Involved in Iron Uptake in Tomato Plants. *Horticulturae* **2023**, *9*, 630. https://doi.org/10.3390/horticulturae9060630

Academic Editor: Zhengguo Li

Received: 21 April 2023
Revised: 24 May 2023
Accepted: 25 May 2023
Published: 27 May 2023

1. Introduction

Iron (Fe) plays a crucial role in photosynthesis and mitochondrial respiration, and, as a cofactor of enzymes such as superoxide dismutase, acotinase, lipoxygenases, nitrogenase, carotenoid cleavage oxygenases/dioxygenase, lutein cleavage dioxygenase, and cupins, among others, it is generally found in high proportions in chloroplasts—up to 80% [1]. Therefore, plants require an average of ~10^{-9}–10^{-4} M Fe ions for optimal development [2]. Fe is taken up from the soil; however, despite its abundance in soil, a high percentage of Fe exists as Fe^{3+}, which is highly insoluble and has very low availability [3]. This subsequently generates abiotic stress in the form of Fe deficiency. An Fe-deficient environment is generated in calcareous soils, where Fe precipitates in forms unavailable to plants due to high pH (>7.5), high content of carbonates ($CaCO_3$), and low content of organic matter [4]. Similarly, the high bicarbonate (HCO_3^-) concentration (in conjunction with the high pH) decreases Fe mobility within the plant, triggering iron deficiency chlorosis (IDC), which is characterized by the loss of chlorophyll, interveinal chlorosis of young leaves, deformation of leaves, reduction of photosynthesis and growth, and decrease in plant production [1].

Because most of the Fe in the soil is present as Fe^{3+}, plants use enzymatic mechanisms to access it, employing a reduction-based strategy (I) or a chelation-based strategy (II) [5]. Strategy II plants (with *Oryza sativa* as a model) secrete phytosiderophores using transporter of mugineic acid 1 (TOM1), which chelates Fe^{3+} [6]. The phytosiderophores-Fe^{3+} complex is subsequently absorbed by yellow stripe 1 (YS1) or yellow stripe-like 1 (YSL) [7]; Fe^{2+} uptake by iron-regulated transporter (IRT) 1/2 may also occur [8].

In contrast, strategy I plants (with *Arabidopsis thaliana* L. as a model) initially induce rhizospheric acidification through a membrane-associated proton pump, H^+-ATPase [9], which releases Fe^{3+} from the soil, and through ferric reduction oxidase (FRO), which reduces Fe^{3+} to Fe^{2+}. Fe^{2+} is captured and finally absorbed and transported into the plant by an iron-regulated transporter-like protein (ZIP) family transporter, IRT1, and natural resistance-associated macrophage protein (NRAMP1) [6,10]. The *FRO*, *IRT1*, and *NRAMP1* genes are regulated transcriptionally by the central basic helix–loop–helix (bHLH) FER-like Fe deficiency-induced transcription factor (FIT) [11]. Similarly, phenolic compounds such as coumarins [12,13] and organic acids [6] are exuded by roots and are involved in the Fe-making machinery by facilitating the solubilization and reduction of unavailable Fe in the soil.

Under conditions of Fe deprivation, plants regulate the enzymatic activity associated with Fe access at the genic level. For example, strategy I plants, such as tomato (*Solanum lycopersicum* L.), substantially increase FRO enzymatic activity in the roots [14]. In pea plants (*Pisum sativum* L.), in addition to higher FRO activity, overexpression of *FRO1*, *IRT1*, and *HA1* was observed in the roots [15]. Similarly, in cucumber (*Cucumis sativus* L.) an increase in *CsHA1* transcripts was reported in a higher proportion in the roots than in the leaves [16], whereas overexpression of *HA6*, which allowed for the outflow of H^+ from the root, was reported in a citrus species [17]. A key regulator in the activation of these mechanisms is FIT, which facilitates Fe uptake [18].

The application of synthetic chelates is commonly recommended for the prevention and reduction of Fe deficiencies in plants grown in calcareous soils. However, these products can be lost by leaching or adsorbing on the soil particles because of the high pH values and high carbonate content of the soil and because of their high affinity with calcium or magnesium [19]. In this sense, citric acid (2-hydroxy-propane-1,2,3-tricarboxylic acid) and leonardite-derived substances (humic, fulvic acid, and humin) improve Fe nutrition in calcareous soils by increasing Fe availability and optimizing synthetic chelate efficiency [20,21].

The exogenous application of citric acid in nutrient solution was found to increase leaf tissue Fe content in tomatoes potted in calcareous soils [21]. Citric acid improves Fe bioavailability by converting Fe to its plant-available form [22]. Similarly, the application of leonardite plus ferrous sulfate heptahydrate was found to improve the fertility of calcareous vineyard soils by increasing their Fe content [23]. The application of Fe complexed with extractable humic substances in water and Fe-citrate led to the overexpression of *FRO*, *IRT1*, and *NRAMP* in cucumber [24] as well as a modulation of the transcripts of the gene encoding H^+-ATPase in tomato [25]. Moreover, this strategy also provides a long-term supply of Fe to plants grown in calcareous soils [26]. The aim of this study was to determine the yield, H^+-ATPase, phosphoenolpyruvate carboxylase (PEPC), and FRO enzyme activity, and expression of *LeHA1*, *LePEPC1*, and *LeFRO1* in tomato plants potted in calcareous soil under the application of citric acid (CA) and humic-like substances (HLS).

2. Materials and Methods

2.1. Plant Growth Conditions

Tomato (*Solanum lycopersicum* L. cv. 'Río Grande') (Crow Seed) plants were grown in a greenhouse with a polyethylene cover and 70% natural irradiance (with recorded averages of 50–60% relative humidity, 32 °C ambient temperature, 736.5 $\mu M\ m^{-2}\ s^{-1}$ photosynthetically active radiation, and 430 ppm CO_2 concentration). The tomato seeds were grown for 35 days in a tray containing a germination substrate, and then the seedlings were transplanted into pots containing 9 L of calcareous soil (pH 8.5, 0.2% organic matter, and 5 mg kg^{-1} Fe content). The plants were fertilized from transplantation to the end of the experiment with Steiner solution [27] at 100% concentration, which contained 4.5 mmol $Ca(NO_3)_2 \cdot 4H_2O$, 2.0 mmol $MgSO_4$, 0.7 mmol KNO_3, 2.0 mmol K_2SO_4, and 1.6 mmol KH_2PO_4. Micronutrients were applied from a stock solution containing H_3BO_3 (2.86 g L^{-1}), $MnSO_4 \cdot H_2O$ (2.15 g L^{-1}, $ZnSO_4 \cdot 7H_2O$ (0.39 g L^{-1}), and $CuSO_4 \cdot 5H_2O$ (0.078 g L^{-1}), and $Na_2MoO_4 \cdot 5H_2O$ (0.09 mg L^{-1}). Finally, Fe was applied in the form of Fe gluconate-ethylenediaminetetraacetic acid (EDTA)-type chelate (at 3 mg L^{-1}), except in the Fe-deprivation treatment, where it was not added. The pH of the nutrient solution was maintained at 6.3 with an electrical conductivity of 2.0 dS cm^{-1}.

The treatments studied were CA (0.1 mM + Fe gluconate-EDTA), HLS (400 μL L^{-1} + Fe gluconate-EDTA), "without organic amendment" (WOA, only Fe gluconate-EDTA), and an Fe-deprivation control (ID, no organic acid and Fe in solution). The concentrations used were selected according to previous results cited by Pérez-Labrada et al. [21]. The citric acid used was food-grade (99.9% purity). The HLS (Arysta LifeScience) contained 10 g L^{-1} humic acid carbon, 90 g L^{-1} fulvic acid carbon, 6 g L^{-1} total nitrogen, 6 g L^{-1} urea nitrogen, and 38 g L^{-1} water soluble K_2O with a density of 1.15 g mL^{-1} and at pH 9.2. The treatments CA and WOA were applied daily through localized irrigation during plant development. The HLS treatment was only applied weekly with the corresponding irrigation volume (15 total applications during crop development). The control (ID) was irrigated daily through localized irrigation during plant development. All treatment compositions were added to the nutrient solution. The nutrient solution was provided to the plants through localized irrigation from transplanting until the end of the experiment.

Considering the phenology of the tomato plant, the sampling dates of leaf-root tissue were chosen according to these stages: vegetative growth (27 days after transplantation, DAT), flowering-anthesis (49 DAT), and physiological maturity of fruits of the 2nd bunch of fruit (89 DAT). We collected the three youngest leaves that had completely expanded, placed them in aluminum bags, immediately froze them with liquid N_2, and stored them at –80 °C. Root tissue was collected on the same leaf-sampling date; 5 cm of secondary roots (obtained at 10 cm depth and 5 cm stem distance) were collected, briefly washed with deionized water, placed in aluminum bags, immediately frozen with liquid N_2, and stored at –80 °C. Plant growth parameters were determined at 27, 49, and 84 DAT. The stem diameter was measured with a sliding caliper 150 mm between the first and second pair of true leaves from the base. Plant length was measured with a flexometer from the base of the stem (soil surface) to the distal growing apex. On fully expanded young leaves, SPAD-unit readings were taken using a Minolta SPAD-502 chlorophyll meter (Konica Minolta, Inc., Osaka, Japan). SPAD values were taken as an average of three readings taken at different locations from the base to the apex of each leaf. Finally, the yield per plant was calculated as the sum of the total number of harvested fruits.

2.2. Enzymatic Activity

Briefly, the enzyme extract was obtained by homogenizing the tissue in liquid N_2 and adding buffer at 4 °C (50 mM Tris-HCL, pH 7.5, 10% glycerol, 20% PVPP, 10 mM $MgCl_2$, 1 mM EDTA, 14 mM β-mercaptoethanol, 1 mM PMSF, and 10 μg mL^{-1} leupeptin; Sigma-Aldrich, St. Louis, MO, USA). The extract was then filtered through a nylon micropore (0.45 μm pore size) and centrifuged at 10,000 rpm for 15 min. The supernatant was centrifuged again at 10,000 rpm for 30 min, and the precipitate was resuspended in the same buffer and stored at –80 °C [28] until use in the determination of H^+-ATPase and PEPC activity. H^+-ATPase activity (EC 7.1.2.1) was determined using a spectrophotometric method [29], coupling the hydrolysis of ATP to the oxidation of NADH [28]; 100 μL of extract was taken, and 25 mM MOPS-BTP buffer (pH 6.5) was added, containing 250 mM sucrose, 50 mM KCl, 1 mM ATP, 1 mM PEP, 0. 25 mM NADH, 15 μg mL^{-1} lactate dehydrogenase (EC 1.1.1.27), 30 μg mL^{-1} pyruvate kinase (EC 2.7.1.40), and 0.015% Brij® 58 (Sigma-Aldrich). Changes in absorbance were measured in a spectrophotometer at 340 nm. The H^+-ATPase activity was calculated using the NADH standard curve. Phosphoenolpyruvate carboxylase activity (PEPC, EC 4.1.1.31) was determined by coupling its activity to the malate dehydrogenase catalyzed by the oxidation of NADH [30]; to initiate the reaction, 100 μL of extract was taken, and standard buffer containing Tris-HCl (100 mM, pH 8.0), $MgCl_2$ (5 mM), PEP (2.5 mM), NADH (0.2 mM), $NaHCO_3$ (10 mM), and MDH (15 μg mL^{-1}) was added. Changes in absorbance were quantified at 340 nm. PEPC activity was calculated using the NADH standard curve.

Chelate ferric reductase activity (FRO, EC 1.16.1.7) was quantified spectrophotometrically by Fe (II)-BPDS concentration according to Romera et al. [31]; 20 mg of tissue was placed in an Eppendorf tube, and 2 mL of $CaSO_4$-$7H_2O$ (0.2 mM) was added and allowed to stand for 5 min. Then, the sample was centrifuged (5 min at 10,000 rpm at 4 °C), and the precipitate was recovered and placed in a new tube, where 10 mL of fresh nutrient solution (without Fe) supplemented with 0.3 mM BPDS (Sigma-Aldrich) and 100 μM Fe(III)-EDTA (Sigma-Aldrich) was added. The pH of the solution was adjusted to 5.5 with 5 mM MES-NaOH (Sigma-Aldrich). It was incubated for 1 h in the dark at room temperature, and finally, the absorbance was measured at 535 nm. BPDS forms a red water-soluble complex with Fe^{2+} and only a weak complex with Fe^{3+}. The amount of reduced iron was calculated by the concentration of the Fe^{2+}-BPDS complex using an extinction coefficient of 22.14 mM cm^{-1}. The protein content of the microsomal fraction membranes was determined using 5 μL of the enzyme extract and 250 μL of Bradford reagent, which were incubated at room temperature for 5 min. Afterward, the absorbance was measured at 630 nm in an ELISA plate reader (BioTek, ELx808 model, Winooski, VT, USA) with BSA as a protein standard [32].

2.3. *Real-Time Reverse-Transcriptase PCR*

The RNA of leaves and roots was extracted using TRIzol reagent [33]; the tissue was ground in liquid N_2, 100 mg was taken and placed in an Eppendorf tube, and immediately after, 1 mL of TRI Reagent® (MRC, TR 118) was added; the mixture was homogenized gently and incubated for 5 min at room temperature. Subsequently, 200 μL of chloroform was added, shaken vigorously, and incubated for 15 min. Then, the samples were centrifuged at 12,000 rpm (15 min at 4 °C), and the supernatant was recovered and placed in a new Eppendorf tube. Then, 500 μL of isopropanol (4 °C) was added, mixed gently, and incubated for 10 min at room temperature. The sample was centrifuged for 10 min at 12,000 rpm and 4 °C, and the supernatant was removed by removing excess isopropanol from the formed RNA pellet. The RNA pellet was washed with 500 μL of 70% ethanol (4 °C), and excess was removed and allowed to dry. The pellet was suspended in 50 μL of water (dissolved at 60 °C). Finally, the RNA solution was treated with DNase I (Sigma Aldrich) and stored at 4 °C. The RNA quantity and quality were determined using a UV-Vis spectrophotometer (260/280 nm ratio) and via denaturing electrophoresis, respectively. An ImProm-II™ Reverse Transcription System Kit (Promega, Madison, WI, USA) was used for the synthesis of cDNA following the manufacturer's instructions; cDNA was synthesized from 1 μg of RNA sample. The primers corresponded to the endogenous internal control gene (*ACT*) and study genes *LeFRO1*, *LePEPC1*, and *LeHA1* (Table 1). For *LeFRO1* and *LePEPC1*, the sequences cited by Paolacci et al. [34] and Diamantopoulous et al. [35], respectively, were considered. The primers were designed using AmplifiX 17.0 (CNRS by Nicolas Jullien, Marseille, France), OligoAnalyzer 3.1 (Integrated DNA Technologies IDT, Coralville, IA, USA), and Primer-BLAST (National Center for Biotechnology Information, Bethesda, MD, USA).

Table 1. Sequences of primers used for gene analysis.

Name Gene	Nomenclature	Forward Primer 5′-3′	Reverse Primer 5′-3′	Tm (°C)
Actin	*ACTIN*	CCCAGGCACACAGGTGTTAT	CAGGAGCAACTCGAAGCTCA	60
H^+-ATPase	*LeHA1*	GAACCCTTCATGGGCTCCAA	GCAACTCACGTAGCCTAGCA	60
PEPC	*LePEPC1*	TGCTGCATTGTTCGACAAGC	CAAAAGTTCGCCGAAAGACAAC	60
FRO	*LeFRO1*	GCGGTGTTGAATATGCTAATC	AAACTTTCCATCTCCCTATCG	60

Real-time PCR was performed in a final volume of 20 μL. For *ACT*, 10 μL of SYBR® Select Master Mix (Applied Biosystems, Foster City, CA, USA), 0.10 μL of forward primer (72 nM), 0.08 μL of reverse primer (60 nM), 2 μL of cDNA, and 7.82 μL of nuclease-free water were added. For *LeHA1*, 10 μL of SYBR® Select Master Mix (Applied Biosystems), 0.13 μL of forward primer (100 nM), 0.13 μL of reverse primer (100 nM), 2 μL of cDNA diluted 1:5, and 7.74 μL of nuclease-free water were added. For *LePEPC1*, 10 μL of SYBR® Select Master Mix (Applied Biosystems), 0.13 μL of forward primer (100 nM), 0.27 μL of reverse primer (200 nM), 2 μL of cDNA, and 7.60 μL of nuclease-free water were added. For *LeFRO1*, 10 μL of SYBR® Select Master Mix (Applied Biosystems), 0.13 μL of forward primer (100 nM), 0.13 μL of reverse primer (100 nM), 2 μL of cDNA, and 7.74 μL of nuclease-free water were added. The *ACT* gene was used to normalize the expression ratio of each gene, and changes in expression were calculated using the standard relative curve method [36].

2.4. Data Analysis

A completely randomized design was used for the experimental development, with 15 biological replicates per treatment. The experimental unit was an individual plant in a pot. One-way analysis of variance (ANOVA) and Fisher's least significant difference test ($p < 0.05$) were applied to the plant growth parameters. Data of enzymatic activity were analyzed using a two-way ANOVA and a multiple means comparison using Fisher's least significant difference test ($p < 0.05$), in which one factor was the organ evaluated (root and leaf), and the other factor was the treatment applied. The gene expression was normalized compared to the internal reference gene (*ACT*). A one-way Kruskal–Wallis nonparametric test by ranks was applied to the gene expression data ($p < 0.05$). All statistical analyses were performed using IBM SPSS v. 19.0.

3. Results

3.1. Plant Growth

The application of CA and HLS resulted in statistically significant differences ($p < 0.05$) in the growth parameters evaluated (Table 2) compared to the control (ID). For example, the addition of CA to the fertilizer solution increased stem diameter (14% at 84 DAT), plant height (39% at 27 DAT), and leaf number (43% at 49 DAT). The weekly application of HLS increased SPAD units by 110% and 223% at 49 and 84 DAT, respectively, compared to ID. Similarly, the total fruit harvested and maximum production increased by 93% and 265%, respectively, in HLS-treated plants. In addition, the use of AC and HLS stimulated stem diameter at different sampling times up to 15% with respect to WOA. Plant height was lower under HLS at 27 and 49 DAT compared to CA and WOA. Plants treated with HLS and CA showed an improvement in leaf number at 27 and 84 DAT compared to WOA. The CA and WOA treatments increased this variable by 18% over HLS at 49 DAT. SPAD units were improved with CA and HLS with respect to WOA by up to 14.8%. Finally, HLS showed a higher number of fruits harvested and a higher yield per plant than the WOA and CA treatments.

Table 2. Plant growth parameters of tomato plants treated with CA and HLS.

Sampling	Treatment	Stem Diameter (mm)	Plant Height (cm)	Number of Leaves	SPAD-Unit	Total of Fruit Harvest	Production per Plant (kg)
27 DAT	CA	10.80 ± 0.52 a †	45.70 ± 3.38 a	13.20 ± 0.45 a	54.19 ± 2.78 a	-	-
	HLS	10.24 ± 0.96 a	39.40 ± 4.39 b	13.40 ± 0.89 a	53.65 ± 2.38 a	-	-
	WOA	10.00 ± 0.29 a	45.50 ± 3.87 a	12.60 ± 0.89 a	53.85 ± 2.95 a	-	-
	ID	8.22 ± 1.90 b	32.80 ± 5.54 c	10.80 ± 2.17 b	33.55 ± 5.64 b	-	-
49 DAT	CA	11.66 ± 1.21 a	71.20 ± 3.63 a	18.00 ± 2.24 a	55.43 ± 3.42 a	-	-
	HLS	12.54 ± 1.03 a	68.20 ± 8.32 a	15.20 ± 2.17 b	57.91 ± 3.51 a	-	-
	WOA	12.02 ± 0.45 a	73.40 ± 5.94 a	18.00 ± 0.71 a	54.85 ± 2.17 a	-	-
	ID	9.74 ± 0.56 b	53.40 ± 2.30 b	12.60 ± 1.67 c	27.51 ± 5.13 b	-	-
84 DAT	CA	12.98 ± 1.28 ab	101.60 ± 11.04 a	23.00 ± 2.35 a	54.21 ± 4.12 a	57.75 ± 6.44 ab	2.76 ± 0.28 a
	HLS	13.84 ± 1.23 a	102.60 ± 13.45 a	23.80 ± 1.92 a	56.03 ± 4.30 a	61.95 ± 6.73 a	2.82 ± 0.27 a
	WOA	11.98 ± 0.67 bc	103.40 ± 4.10 a	21.20 ± 1.48 a	48.77 ± 5.91 a	53.40 ± 4.38 b	2.67 ± 0.08 a
	ID	11.40 ± 1.18 c	64.60 ± 9.42 b	17.00 ± 3.46 b	17.34 ± 16.92 b	32.05 ± 3.09 c	0.77 ± 0.15 b

DAT = days after transplant. CA = citric acid. HLS = humic-like substances. WOA = without organic amendment. ID = Fe deprivation. Values are means ± standard deviations, $n = 5$. † Within a row, values not sharing a letter are significantly different ($p < 0.05$).

3.2. *Enzymatic Activity*

Regardless of treatment, there was higher H^+-ATPase and PEPC activity in leaf tissue, whereas the FRO activity was higher in root tissue. Except for H^+-ATPase activity at 84 DAT, there were significant differences ($p < 0.001$) among the plant tissues studied.

3.2.1. H^+-ATPase Activity

The highest H^+-ATPase activity was documented at 27 DAT in leaf tissue. In this sampling, the CA and HLS treatments resulted in 61% and 51% reductions, respectively, compared to ID. At 49 DAT, the activity increased in root tissue, whereas it was reduced in leaf tissue compared to the first sampling; higher activity was documented under CA. During the fruit harvest stage (84 DAT), the activity of this enzyme increased in the leaf tissue, with 46% more activity in CA than in ID (Table 3).

3.2.2. PEPC Activity

In leaf tissue at 27 DAT, 84% higher PEPC activity was found in WOA, and 22% lower PEPC activity was found in plants receiving the HLS treatment relative to ID. On the other hand, at 49 DAT, the HLS and CA treatments promoted 55% and 50% increases in PEPC activity, respectively. Finally, at 84 DAT, plants treated with HLS showed a 150% increase in PEPC activity compared to ID. In root tissue, at 27 DAT, there was a PEPC-activity reduction of 39% and 67% in the HLS and CA treatments, respectively, whereas at 84 DAT, the CA and WOA treatments showed the lowest PEPC activity compared to ID (Table 3).

3.2.3. FRO Activity

In leaf tissue at 27 DAT, FRO activity was reduced by 12.5% in all treatments relative to ID, whereas at 49 and 84 DAT, the activity increased by 35.8% and 56.70% in the WOA treatment. On the other hand, similar FRO activity was found between treatments in root tissue, except at 27 DAT, where the CA treatment minimally reduced FRO activity compared to ID (Table 3).

Table 3. Enzymatic activity in tomato tissue treated with CA and HLS.

Tissue	Sampling	Treatment	H^+-ATPase	PEPC	FRO
Leaf	27 DAT	CA	1.46 ± 0.18 b †	482.3 ± 64.0 ab	7.2 ± 0.2 b
		HLS	1.84 ± 0.07 ab	267.1 ± 65.7 cd	7.4 ± 0.1 b
		WOA	1.57 ± 0.05 ab	633.6 ± 154.2 a	7.2 ± 0.3 b
		ID	3.76 ± 2.22 a	343.5 ± 56.9 bc	9.6 ± 0.3 b
	49 DAT	CA	0.92 ± 0.11 b	834.2 ± 151.4 a	7.7 ± 0.2 b
		HLS	0.89 ± 0.04 b	856.5 ± 140.6 a	7.5 ± 0.3 b
		WOA	0.89 ± 0.05 b	784.9 ± 194.0 a	9.1 ± 0.4 b
		ID	0.65 ± 0.05 c	552.7 ± 113.0 a	6.7 ± 0.3 b
	84 DAT	CA	2.65 ± 1.33 a	465.6 ± 221.8 b	10.7 ± 1.2 b
		HLS	1.58 ± 0.06 a	759.6 ± 136.5 a	8.2 ± 0.2 bc
		WOA	1.25 ± 0.13 a	141.0 ± 27.0 cd	10.5 ± 1.1 b
		ID	1.81 ± 0.63 a	304.0 ± 99.5 bc	6.7 ± 1.1 c
Root	27 DAT	CA	0.016 ± 0.0005 b	51.5 ± 29.7 e	15.8 ± 2.1 a
		HLS	0.017 ± 0.0007 b	95.2 ± 43.1 de	17.2 ± 1.5 a
		WOA	0.017 ± 0.0004 b	191.8 ± 16.0 cde	17.3 ± 2.4 a
		ID	0.017 ± 0.0009 b	156.9 ± 48.3 cde	14.5 ± 1.3 a
	49 DAT	CA	1.35 ± 0.05 a	26.2 ± 21.5 b	22.0 ± 1.2 a
		HLS	1.28 ± 0.02 a	11.6 ± 7.0 b	23.2 ± 0.7 a
		WOA	1.26 ± 0.13 a	16.6 ± 12.2 b	22.2 ± 2.9 a
		ID	1.32 ± 0.05 a	4.6 ± 0.1 b	25.3 ± 1.5 a
	84 DAT	CA	1.50 ± 0.05 a	4.5 ± 0.0 d	21.6 ± 0.9 a
		HLS	1.27 ± 0.11 a	205.6 ± 21.1 bcd	20.1 ± 1.7 a
		WOA	1.29 ± 0.03 a	11.8 ± 7.6 d	21.4 ± 1.1 a
		ID	1.48 ± 0.04 a	261.9 ± 16.5 bcd	20.1 ± 1.8 a

DAT = days after transplant. CA = citric acid. HLS = humic-like substances. WOA = without organic amendment. ID = Fe deprivation. H^+-ATPase activity expressed as: μmol NADH min^{-1} mg^{-1} prot. PEPC activity expressed as: nmol NADH min^{-1} mg^{-1} prot. FRO activity expressed as: nmol Fe^{2+} reduced (g FW tissue) $^{-1}$ h^{-1}. Values are means ± standard deviations, $n = 5$. † Within a row, values not sharing a letter are significantly different ($p < 0.05$).

3.3. Gene Expression

The expression of the genes evaluated in leaf tissue is shown in Figure 1. The partial or complete overexpression and/or repression of a gene can be determined by comparing the expression of that gene against a critical threshold [37]. This threshold (value can be one) corresponds to the absolute value of a calibrator gene [38]. Based on this calibration, *LeHA1* was repressed to the highest degree in CA at 27 ($p < 0.01$) and 84 DAT (0.9- and 1.0-fold change, respectively), followed by WOA (1.0-fold change at 84 DAT, Figure 1a), whereas *LePEPC1* showed elevated expression in HLS at 27 DAT (0.6-fold change) compared to ID. However, *LePEPC1* was repressed under CA treatment at 27 and 49 DAT (up to 0.8-fold change; $p < 0.01$), whereas it showed elevated expression ($p < 0.01$) at 84 DAT (up to 1.1-fold change, Figure 1b). In the case of *LeFRO1*, high expression was observed at 27 and 49 DAT under the HLS treatment (4.10- and 4.19-fold change), whereas the CA treatment resulted in high expression at 49 DAT (2.33-fold change) and repression ($p < 0.01$) at 84 DAT. This gene exhibited elevated expression with WOA at 27 DAT (Figure 1c).

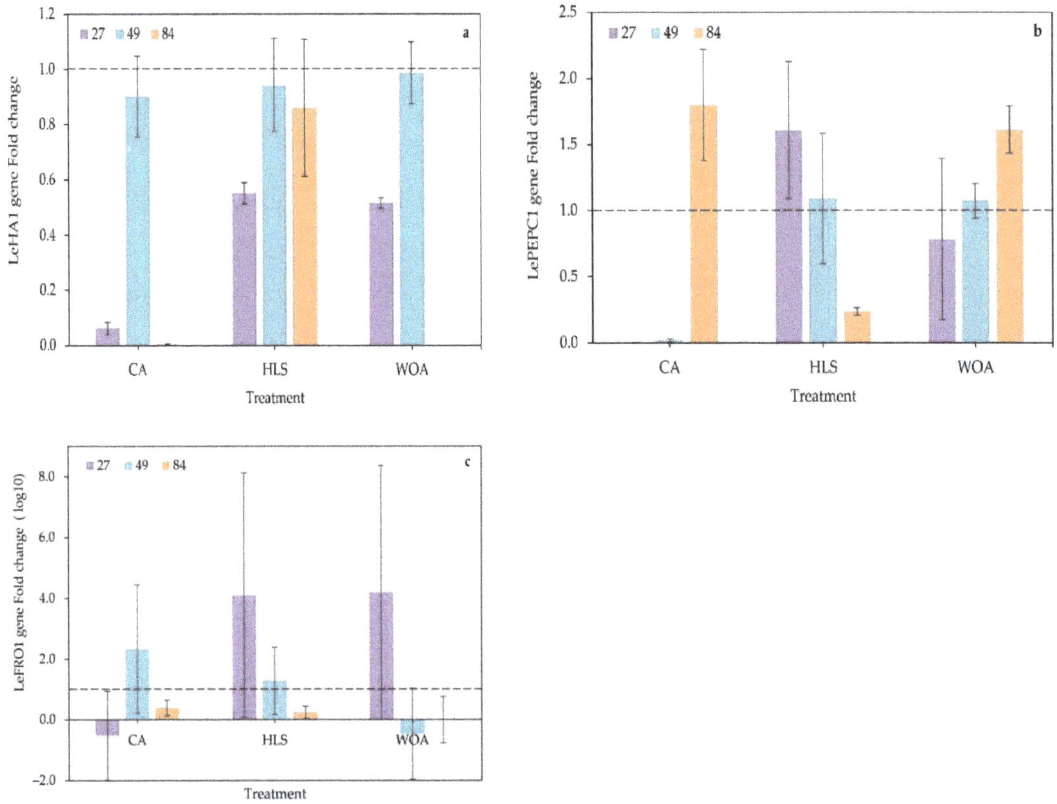

Figure 1. Expression of the *LeHA1* (**a**), *LePEPC1* (**b**), and *LeFRO1* (**c**) genes in tomato leaves of plants treated with CA and HLS. Values are means ± standard error ($n = 5$). The dashed line represents the constant value of the absolute control (ID).

Root gene expression was only evaluated at 49 and 84 DAT because increased expression of *FIT* (and probably *FRO1*, *HA1*, and *PEPC1*) has been associated with increased expression during the visible inflorescence stage (49 DAT) followed by a decrease during the fruit set-ripening stage (84 DAT) [18]. The behavior of gene expression in root tissue is presented in Figure 2. For *LeHA1*, elevated expression was documented at both sampling times; HLS at 49 DAT resulted in the greatest increase (3.8-fold), followed by WOA treatment (1.9-fold change), relative to ID. At 84 DAT, this gene showed 1.2- and 1.9-fold changes under HLS and WOA, respectively (Figure 2a). Regarding *LePEPC1* (Figure 2b), repression was documented to a greater extent at 84 DAT in CA (1.0-fold change) and HLS (0.8-fold change) relative to ID. *LeFRO1* exhibited repression; HLS application repressed ($p < 0.01$) *LeFRO1* up to 1.2- and 0.9-fold relative to ID at 49 and 84 DAT, respectively (Figure 2c).

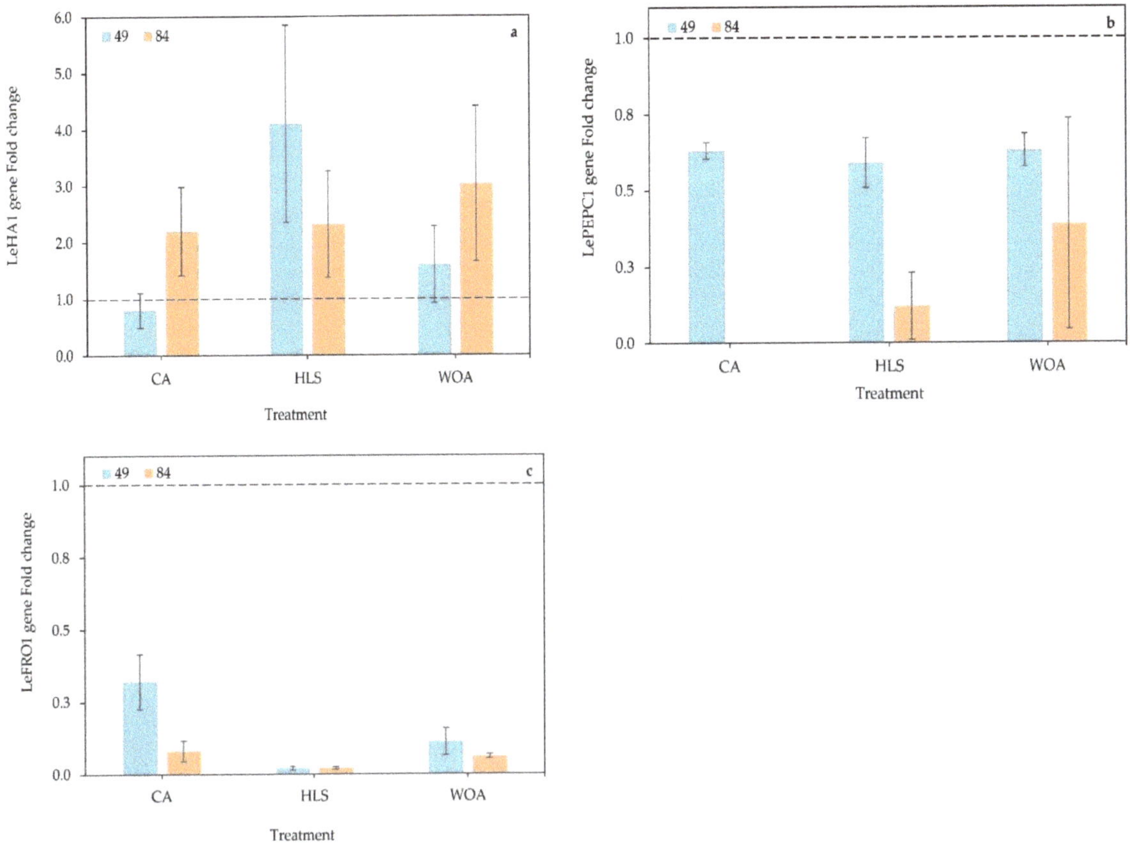

Figure 2. Expression of the *LeHA1* (**a**), *LePEPC1* (**b**), and *LeFRO1* (**c**) genes in tomato roots of plants treated with CA and HLS. Values are means ± standard error (n = 5). The dashed line represents the constant value of the absolute control (ID).

4. Discussion

Previous studies have found that the application of CA [39] and leonardite-derived compounds [21] improved the growth and yield of tomatoes grown in calcareous soil, probably via plant stimulation [40,41] and punctually by increasing photosynthetic parameters [42]. Similarly, the use of humic-Fe materials in soybean may be a suitable strategy for the management of Fe nutrition in calcareous soil [43], consequently leading to a reduction in Fe deficiency and an improvement in plant growth, development, and productivity. This is because the addition of Fe together with humic complexes act as "organic chelators" that boost plant growth [41]. CA and leonardite-derived complexes exhibit various levels of complexation with Fe, promoting plant growth by inducing a physiological response [44]. In addition, CA can reduce abiotic stress resulting from phosphorus deficiency in plants grown in calcareous soils [45–47]. Likewise, it has been described that the use of humic substances in calcareous soils promotes nutrient bioavailability by generating soluble complexes with minerals [23,48] and increasing nutrient uptake [40,41], and that their combination with chelate Fe (Fe-EDDHA) can increase the yield and nutrient content of plants [49].

Abiotic stress from Fe deprivation is signaled through the phloem from sinks in the root system, where there is an alteration in the absorption and transport of this nutrient [50]. In nongrass monocotyledons and in dicotyledons, such as tomato, that develop under conditions of Fe deprivation and/or in calcareous soils, strategy I presents an enzymatic mechanism for coping with Fe-deficient conditions; according to the results obtained, H^+-ATPase and FRO enzymes may be involved in this mechanism and act together with enzymes of the antioxidant system in addition to the regulation of iron transporters (IRT1, ZIP and NRAMP1) and probably coumarin excretion [13]. Although our study does not show a situation of iron deficiency recovery, we can clarify the individual impact of CA and HLS on the enzymatic mechanisms related to iron metabolism in tomato plants grown in calcareous soil.

The H^+-ATPase enzyme generates electrochemical potential gradients by energizing ion channels and transporter proteins in the plasma membrane via the extrusion of H^+, with the consequent consumption of ATP [51]. Partial or total nutrient (e.g., Fe) deprivation in the growth medium, high carbonate–bicarbonate concentration, and/or high soil pH can induce an increase in root H^+-ATPase activity, acidifying the rhizosphere, solubilizing Fe, and preventing Fe precipitation [52]. Our results show higher activity of this enzyme in leaf tissue, which is probably due to its physiological role in the regulation of intra- and extracellular pH, loading of assimilates in the phloem, and redistribution of nutrients [51].

H^+-ATPase activity can be stimulated by humic substances [53]. Applications of humic acid purified from leonardite increased plasma membrane H^+-ATPase activity in cucumber roots and shoots [54,55]. This contrasts with what was found in the present study, where a reduction in tomato root tissue H^+-ATPase activity via HLS was documented; the reduction was probably derived from the promotion of soluble complexes that allowed Fe to exist in a bioavailable form in the soil [23,48] or via the acidifying effect of HLS [56]. This may also have been due to feedback control mechanisms and the concentration of NO_3^- in the medium [55]. In contrast, the increase in the activity of this enzyme in leaf tissue (at 49 DAT) could be a consequence of the biostimulation (promoting a redistribution of nutrients) induced by HLS and CA [40,41]. In addition, CA is associated with the energy metabolism of the plant [44]; thus, H^+-ATPase activity would be necessary to generate gradients in the different cellular compartments of the leaf. The increase in root H^+-ATPase activity at 84 DAT under CA application implies an alteration in the gradients of the root system, as in radical cells, this enzyme plays an active role in xylem loading [51].

Tomato plants are thought to implement both strategies I and II; thus, in addition to extruding protons, reducing Fe, and promoting its transport, tomato plants also induce the synthesis of organic compounds (e.g., citrate, malate, and phenols, among others) [57]. In particular, CA tends to accumulate in the root under moderate conditions of Fe deprivation [58,59], where it acts as a reserve for chelating Fe and can present absorption windows over time [13]. The PEPC enzyme, which catalyzes the fixation of bicarbonate (in the presence of Mg^{2+}) to phosphoenolpyruvate, generating oxaloacetate and releasing Pi in C4 plants, could play a crucial role in this context. In nonphotosynthetic tissues and in C3 plants, PEPC enables anaplerotic reactions that provide intermediates in the Krebs cycle [30,58,60]. According to the pH-stat theory, PEPC would have to be forcibly activated to balance the pH change generated by the H^+-ATPase activity [30]; in this sense, the addition of acidifying compounds (such as CA and HLS) may impact PEPC activity via the contribution of carbonaceous skeletons.

According to previous studies, PEPC activity increases in roots under Fe-deficient conditions [30,58]. However, severe conditions of Fe deprivation (without Fe in nutrient solution) cause a reduction in the PEPC activity of the root system [59]. However, in this study, it was observed that the addition of HLS (+Fe chelated) promoted an increase in this enzyme (49 and 84 DAT), whereas CA and WOA reduced its activity (27 and 84 DAT). In the case of leaf tissue, the addition of CA and HLS increased PEPC activity. The modification of PEPC activity in the root and leaf tissue of tomato plants in response to CA and HLS application may be due to an alteration of anaplerotic reactions via the contribution of exogenous carbon skeletons [44], via a reprogramming mechanism in carbon metabolism under inadequate Fe conditions that facilitate root exudation [58], or via the acidic environment that can promote its exogenous addition [56].

The intracellular enzyme FRO is associated with the cell membrane and is responsible for reducing Fe^{3+} to soluble, bioavailable Fe^{2+} [61]. In leaf tissue, FRO activity is highly dependent on the Fe content in the cytoplasmic solution [62]. In this case (leaf tissue), a reduction in FRO activity was found at 27 DAT (in all treatments), which may imply the contribution and presence of Fe in its assimilable state (Fe^{2+}), whereas the increase in activity (49 and 84 DAT) may have been because Fe was present as Fe^{3+} in the apoplast because of the content of HCO_3^- and pH [60]; however, high bicarbonate content and pH apoplastic in plants may decrease FRO activity, causing Fe deficiency symptoms [63]. An increase in FRO enzyme activity has been documented in the roots of pea and tomato plants under Fe-deficient conditions [14,15,64], as was found in this work (27 and 84 DAT), which may suggest that the iron applied via nutrient solution is in its assimilable state (Fe^{2+}) via complexation with HLS or CA. The increase in the activity of this enzyme may or may not be accompanied by a greater number of secondary roots, which would imply a greater number of reduction zones [14].

Humic substances can reduce Fe^{3+} to soluble forms due to their photocatalytic properties or their redox activity [65–67], contributing Fe to the chemical chelate and to the plant (as Fe-complex), which results in an increased concentration of soluble Fe in the soil and an enhanced translocation of Fe from the roots to the leaves [26]. However, the high HCO_3^- content could precipitate Fe^{2+} to Fe^{3+} again, raising its concentration in the rhizosphere and stimulating root FRO activity, as seen in our results (49 DAT). A humic soil environment can condition FRO activity in the root plant, which could partially explain the behavior of this enzyme under our HLS treatment. A study carried out on cucumber reported that the application of humic acid purified to leonardite enhanced FRO activity at 72 h [54]. Repeated application of leonardite Fe humates can precipitate in the root, blocking entry pores in the cell wall and reducing Fe transport [68] as well as probably reducing FRO activity. However, a previous study demonstrated that the repeated application of 400 $\mu L\ L^{-1}$ of HLS on tomato plants potted in calcareous soils had a positive effect on Fe uptake [21]. Therefore, the reduction in the activity of this enzyme could not respond in the aforementioned sense. Despite this, caution should be exercised when applying humic substances. CA—a 6-carbon tricarboxylic compound [69]—is able to generate an acidic microenvironment in the rhizospheric zone, keeping Fe in its soluble and plant-available form and resulting in the low FRO activity (at least at 49 DAT) observed in our study.

The critical threshold of the calibrator gene values [38] allows us to determine the level of overexpression and/or total or partial repression of a gene [37]. In this context, *LeHA1*, *LeFRO1*, and probably *LePEPC1* may act as indicators of metabolically active Fe in the plant and in the growth medium [70].

Considering that the most suitable pH for the development of tomato plants is 6.0 [50], plants developed in calcareous soils with a high pH (8.5) and high HCO_3^- content present a nutrient deficit response [13], triggering changes in the expression of specific genes [71]. Previous studies have documented increased expression of the H^+-ATPase enzyme promoter gene in Fe-deprived tomato, pea, and cucumber plants [15,16,25,72]. The data obtained here suggest that the modification of this gene can be assumed, in addition to nutrient input, to be a response to the soil pH and rhizospheric HCO_3^- content because via these treatments, Fe (3 mg L^{-1}) was made available to the plant. In that sense, the repression of *LeHA1* (at 27 and 84 DAT) in leaf tissue under CA treatment and its high expression in root tissue under HLS treatment (at 49 DAT) suggest spatiotemporal expression patterns [14]. Such alterations would imply a modification of H^+ extrusion from the root tissue into the rhizospheric medium or cellular compartments [17,52]. In addition, humic complexes can induce an acid reaction in the soil by way of their diversity of functional groups [56], affecting *LeHA1* expression.

The application of purified humic acid from leonardite (2–250 mg L^{-1} C) on cucumber grown under Fe-deficient conditions led to higher expression of the *CsHA2* gene (up to 4-fold over the control) and low expression of *CsHA1* (downregulated up to 5-fold) in the apical root [54]. These results show similar trends to ours; however, the present work showed higher *LeHA1* expression in tomato roots without Fe deficiency and with continuous application of HLS. This discrepancy may be due to the differential response of the gene isoform [54,55], which leads to transient changes in H^+-ATPase activity.

Similarly, the gene encoding the PEPC protein showed increased expression in the root system under Fe-deficient environments [30,58]. Here, we found that *LePEPC1* was repressed in the roots of plants treated with CA and HLS with an Fe supply (in the form of EDTA) (Figure 2b). This suggests a reduced requirement for anaplerotic reactions due to the Fe supply in the growth medium. In the case of elevated leaf expression under HLS at 27 DAT, this may be indicative of active Fe in the plant [70].

In tomatoes, the *PEPC* gene family may be associated with other types of stresses, such as salinity and cold [73], but its expression could also be a means of coping with nutrient deficits (e.g., Fe) [59]. Applications of a mixture of organic acids (succinic, citric, malic, and oxalic acids, 100 μM) on alfalfa were reported to regulate the expression of the *PEPC* gene in roots under conditions of abiotic stress via Al [74].

The expression of the gene encoding FRO is strongly influenced, in calcareous soils, by the Fe content in the rhizospheric solution, the exposure time, and the HCO_3^- content [14,75]. In plants, there is a higher expression of this gene in roots, shoots, and reproductive organs, whereas it is constitutively expressed in leaves [76]. In both cases, variations in the amino acid residues of *FRO1* are vital to maintaining the stability and high activity of the FRO enzyme [61].

Tomato and pea plants deprived of Fe show increased expression of *FRO1* in the root [15,64]. In this regard, several studies have shown that the application of leonardite humic acid in cucumber plants generates a higher expression of *CsFRO1* (between two- and ninefold) after the first days of application. Likewise, leonardite-derived substances applied to tomato plants subjected to Fe deficiency increase *LeFRO1* gene expression [70,77]. The use of an Fe complex with water-extractable humic substances (Fe-WEHS) or with citrate (Fe-citrate) generates a higher expression of *FRO*, *IRT1*, and *NRAMP* in cucumber [24], whereas in tomatoes under Fe deficiency, these compounds cause higher expression of *LeFRO1*, *LIRT1*, *LIRT2*, and *Ferritin2* in leaves [70,77].

In our study, we found repression of *LeFRO1* at 49 and 84 DAT (Figure 2c) in tomato roots with HLS application. This suggests that HLS enhances Fe bioavailability in addition to procuring soluble complexes in calcareous soil [23,48]. In the case of CA, its endogenous variations in tomato plants could be involved in the regulation of nuclear target gene expression [72]. In addition, the alteration in CA concentration affects the regulation of the tricarboxylic cycle [78]. In this sense, when applied exogenously, CA can act as a potential substrate of several metabolic pathways protecting the plant against abiotic stresses [69].

The discrepancy in transcriptional and post-transcriptional stimulation found in the present work may be due to the presence of a series of specific regulatory mechanisms for each level [54] or to associated feedback mechanisms prior to enzymatic activation or gene expression that increase Fe root concentration and Fe translocation [54]. However, the use of CA and/or HLS as an Fe chelate partner could supply stable Fe to the root, stabilize the chelated Fe, or generate an environment conducive to Fe absorption [70], thereby improving the growth and development of plants grown under calcareous soils. These mechanisms could reduce the power of the experimental design without affecting its validity, being necessary to establish unrestricted conditions to determine the precise effect on the supply of stable Fe of CA and HLS.

The enzymatic and gene expression responses observed in tomato plants (grown under calcareous soil) treated with CA and HLS offer a promising outlook in the prevention of Fe deficiency; however, as both compounds show complexation with Fe and stimulate the plant [44], it is necessary to verify possible synergistic effects when applied in combination.

5. Conclusions

Iron deprivation in tomato plants caused a reduction in growth, development, and yield. The application of CA or HLS (continuously or weekly, respectively) throughout the crop cycle through a fertilizer solution improved the SPAD units, number of fruits harvested, and yield per plant. These substances also induced temporary alterations in enzyme activity ($p < 0.05$), reducing PEPC and FRO activity in roots. In leaf tissue, the CA treatment resulted in lower expression of *LeHA1* and *LePEPC1* ($p < 0.01$) as well as induced overexpression of *LeFRO1* ($p < 0.01$). In root tissue, HLS treatment resulted in *LeHA1* overexpression and *LePEPC1* and *LeFRO1* repression ($p < 0.01$), whereas CA repressed *LePEPC1* expression ($p < 0.01$). Thus, the use of CA and HLS may be a potential strategy in the management of ferric nutrition in tomato plants, as it can cause temporal alterations in enzyme activity (PEPC and FRO) and gene expression (*LeHA1, LePEPC1*, and *LeFRO1*) associated with iron uptake.

Author Contributions: Conceptualization, S.G.-M. and S.S.-G.; methodology, F.P.-L.; software, F.P.-L.; validation, A.B.-M. and A.J.-M.; formal analysis, F.P.-L.; writing—original draft preparation, F.P.-L.; writing—review and editing, F.P.-L. and S.G.-M.; funding acquisition, S.S.-G. All authors have read and agreed to the published version of the manuscript.

Funding: This research received no external funding.

Data Availability Statement: Not applicable.

Acknowledgments: The authors thank CONACyT and Arysta LifeScience for financial support.

Conflicts of Interest: The authors declare no conflict of interest.

References

1. Murgia, I.; Marzorati, F.; Vigani, G.; Morandini, P. Plant iron nutrition: The long road from soil to seeds. *J. Exp. Bot.* **2022**, *73*, 1809–1824. [CrossRef] [PubMed]
2. Mori, S. Iron acquisition by plants. *Curr. Opin. Plant Biol.* **1999**, *2*, 250–253. [CrossRef] [PubMed]
3. Kobayashi, T.; Nozoye, T.; Nishizawa, N.K. Iron transport and its regulation in plants. *Free Radic. Biol. Med.* **2019**, *133*, 11–20. [CrossRef]
4. Wahba, M.M.; Labib, F.; Zaghloul, A. Management of Calcareous Soils in Arid Region. *Int. J. Environ. Pollut. Environ. Model.* **2019**, *2*, 248–258.
5. Marschner, H.; Römheld, V. Strategies of plants for acquisition of iron. *Plant Soil* **1994**, *165*, 261–274. [CrossRef]
6. Martín-Barranco, A.; Thomine, S.; Vert, G.; Zelazny, E. A quick journey into the diversity of iron uptake strategies in photosynthetic organisms. *Plant Signal. Behav.* **2021**, *16*, 1975088. [CrossRef] [PubMed]
7. Chao, Z.F.; Chao, D.Y. Similarities and differences in iron homeostasis strategies between graminaceous and nongraminaceous plants. *New Phytol.* **2022**, *236*, 1655–1660. [CrossRef]
8. Li, S.; Song, Z.; Liu, X.; Zhou, X.; Yang, W.; Chen, J.; Chen, R. Mediation of Zinc and Iron Accumulation in Maize by ZmIRT2, a Novel Iron-Regulated Transporter. *Plant Cell Physiol.* **2022**, *63*, 521–534. [CrossRef]
9. Santi, S.; Schmidt, W. Dissecting iron deficiency-induced proton extrusion in Arabidopsis roots. *New Phytol.* **2009**, *183*, 1072–1084. [CrossRef]
10. Gupta, P.K.; Balyan, H.S.; Sharma, S.; Kumar, R. Biofortification and bioavailability of Zn, Fe and Se in wheat: Present status and future prospects. *Theor. Appl. Genet.* **2021**, *134*, 1–35. [CrossRef]
11. Schwarz, B.; Bauer, P. FIT, a regulatory hub for iron deficiency and stress signaling in roots, and FIT-dependent and -independent gene signatures. *J. Exp. Bot.* **2020**, *71*, 1694–1705. [CrossRef] [PubMed]
12. Rosenkranz, T.; Oburger, E.; Baune, M.; Weber, G.; Puschenreiter, M. Root exudation of coumarins from soil-grown Arabidopsis thaliana in response to iron deficiency. *Rhizosphere* **2021**, *17*, 100296. [CrossRef]
13. Vélez-Bermúdez, I.C.; Schmidt, W. Plant strategies to mine iron from alkaline substrates. *Plant Soil* **2023**, *483*, 1–25. [CrossRef]
14. Jiménez, M.R.; Casanova, L.; Saavedra, T.; Gama, F.; Suárez, M.P.; Correia, P.J.; Pestana, M. Responses of tomato (*Solanum lycopersicum* L.) plants to iron deficiency in the root zone. *Folia Hortic.* **2019**, *31*, 223–234. [CrossRef]
15. Kabir, A.H.; Paltridge, N.G.; Able, A.J.; Paull, J.G.; Stangoulis, J.C.R. Natural variation for Fe-efficiency is associated with upregulation of Strategy I mechanisms and enhanced citrate and ethylene synthesis in *Pisum sativum* L. *Planta* **2012**, *235*, 1409–1419. [CrossRef] [PubMed]
16. Santi, S.; Cesco, S.; Varanini, Z.; Pinton, R. Two plasma membrane H^+-ATPase genes are differentially expressed in iron-deficient cucumber plants. *Plant Physiol. Biochem.* **2005**, *43*, 287–292. [CrossRef]
17. Fan, Z.; Wu, Y.; Zhao, L.; Fu, L.; Deng, L.; Deng, J.; Ding, D.; Xiao, S.; Deng, X.; Peng, S.; et al. MYB308-mediated transcriptional activation of plasma membrane H^+-ATPase 6 promotes iron uptake in citrus. *Hortic. Res.* **2022**, *9*, uhac088. [CrossRef]
18. Filiz, E.; Kurt, F. FIT (Fer-like iron deficiency-induced transcription factor) in plant iron homeostasis: Genome-wide identification and bioinformatics analyses. *J. Plant Biochem. Biotechnol.* **2019**, *28*, 143–157. [CrossRef]

19. Ferreira, C.M.H.; López-Rayo, S.; Lucena, J.J.; Soares, E.V.; Soares, H. Evaluation of the Efficacy of Two New Biotechnological-Based Freeze-Dried Fertilizers for Sustainable Fe Deficiency Correction of Soybean Plants Grown in Calcareous Soils. *Front. Plant Sci.* **2019**, *10*, 1335. [CrossRef]
20. Zanin, L.; Tomasi, N.; Cesco, S.; Varanini, Z.; Pinton, R. Humic substances contribute to plant iron nutrition acting as chelators and biostimulants. *Front. Plant Sci.* **2019**, *10*, 675. [CrossRef]
21. Pérez-Labrada, F.; Benavides-Mendoza, A.; Juárez-Maldonado, A.; Solís-Gaona, S.; González-Morales, S. Organic acids combined with Fe-chelate improves ferric nutrition in tomato grown in calcisol soil. *J. Soil Sci. Plant Nutr.* **2020**, *20*, 673–683. [CrossRef]
22. Al-Balawna, Z.A.; Abu-Abdoun, I.I. Fate of Citric Acid Addition on Mineral Elements Availability in Calcareous Soils of Jordan Valley. *Int. Res. J. Pure Appl. Chem.* **2021**, *22*, 82–89. [CrossRef]
23. Olego, M.Á.; Cuesta Lasso, M.; Quiroga, M.J.; Visconti, F.; López, R.; Garzón-Jimeno, E. Effects of Leonardite Amendments on Vineyard Calcareous Soil Fertility, Vine Nutrition and Grape Quality. *Plants* **2022**, *11*, 356. [CrossRef]
24. Zanin, L.; Tomasi, N.; Rizzardo, C.; Gottardi, S.; Terzano, R.; Alfeld, M.; Janssens, K.; De Nobili, M.; Mimmo, T.; Cesco, S. Iron allocation in leaves of Fe-deficient cucumber plants fed with natural Fe complexes. *Physiol. Plant.* **2015**, *154*, 82–94. [CrossRef] [PubMed]
25. Zamboni, A.; Zanin, L.; Tomasi, N.; Avesani, L.; Pinton, R.; Varanini, Z.; Cesco, S. Early transcriptomic response to Fe supply in Fe-deficient tomato plants is strongly influenced by the nature of the chelating agent. *BMC Genom.* **2016**, *17*, 35. [CrossRef] [PubMed]
26. Cieschi, M.T.; Lucena, J.J. Leonardite iron humate and synthetic iron chelate mixtures in *Glycine max* nutrition. *J. Sci. Food Agric.* **2021**, *101*, 4207–4219. [CrossRef]
27. Steiner, A.A. A universal method for preparing nutrient solutions of a certain desired composition. *Plant Soil* **1961**, *15*, 134–154. [CrossRef]
28. Rabotti, G.; Zocchi, G. Plasma membrane-bound H^+-ATPase and reductase activities in Fe-deficient cucumber roots. *Physiol. Plant.* **1994**, *90*, 779–785. [CrossRef]
29. Palmgren, M.G.; Askerlund, P.; Fredrikson, K.; Widell, S.; Sommarin, M.; Larsson, C. Sealed Inside-Out and Right-Side-Out Plasma Membrane Vesicles. *Plant Physiol.* **1990**, *92*, 871–880. [CrossRef]
30. Nisi, P.D.; Zochi, G. Phosphoenolpyruvate carboxylase in cucumber (*Cucumis sativus* L.) roots under iron deficiency: Activity and kinetic characterization. *J. Exp. Bot.* **2000**, *51*, 1903–1909. [CrossRef]
31. Romera, F.; Welch, R.; Norvell, W.; Schaefer, S.; Kochian, L. Ethylene involvement in the over-expression of Fe(III)-chelate reductase by roots of E107 pea [*Pisum sativum* L. (brz, brz)] and *chloronerva* tomato (*Lycopersicon esculentum* L.) mutant genotypes. *Biometals* **1996**, *9*, 38–44. [CrossRef]
32. Bradford, M.M. A rapid and sensitive method for the quantitation of microgram quantities of protein utilizing the principle of protein-dye binding. *Anal. Biochem.* **1976**, *72*, 248–254. [CrossRef] [PubMed]
33. Rio, D.C.; Ares, M.; Hannon, G.J.; Nilsen, T.W. Purification of RNA Using TRIzol (TRI Reagent). *Cold Spring Harb. Protoc.* **2010**, *2010*, pdb-prot5439. [CrossRef] [PubMed]
34. Paolacci, A.R.; Celletti, S.; Catarcione, G.; Hawkesford, M.J.; Astolfi, S.; Ciaffi, M. Iron deprivation results in a rapid but not sustained increase of the expression of genes involved in iron metabolism and sulfate uptake in tomato (*Solanum lycopersicum* L.) seedlings. *J. Integr. Plant Biol.* **2014**, *56*, 88–100. [CrossRef]
35. Diamantopoulos, P.D.; Aivalakis, G.; Flemetakis, E.; Katinakis, P. Expression of three β-type carbonic anhydrases in tomato fruits. *Mol. Biol. Rep.* **2013**, *40*, 4189–4196. [CrossRef]
36. Larionov, A.; Krause, A.; Miller, W. A standard curve based method for relative real time PCR data processing. *BMC Bioinform.* **2005**, *6*, 62. [CrossRef]
37. Dembélé, D.; Kastner, P. Fold change rank ordering statistics: A new method for detecting differentially expressed genes. *BMC Bioinform.* **2014**, *15*, 14. [CrossRef]
38. Love, M.I.; Huber, W.; Anders, S. Moderated estimation of fold change and dispersion for RNA-seq data with DESeq2. *Genome Biol.* **2014**, *15*, 550. [CrossRef]
39. Pérez-Labrada, F.; Mendoza, A.B.; Valdez-Aguilar, L.A.; Robledo-Torres, V. Citric acid in the nutrient solution increases the mineral absorption in potted tomato grown in calcareous soil. *Pakistan J. Bot.* **2016**, *48*, 67–74.
40. Massimi, M.; Radócz, L.; Csótó, A. Impact of Organic Acids and Biological Treatments in Foliar Nutrition on Tomato and Pepper Plants. *Horticulturae* **2023**, *9*, 413. [CrossRef]
41. Sharma, S.; Anand, N.; Bindraban, P.S.; Pandey, R. Foliar Application of Humic Acid with Fe Supplement Improved Rice, Soybean, and Lettuce Iron Fortification. *Agriculture* **2023**, *13*, 132. [CrossRef]
42. Zia-ur-Rehman, M.; Bani Mfarrej, M.F.; Usman, M.; Azhar, M.; Rizwan, M.; Alharby, H.F.; Bamagoos, A.A.; Alshamrani, R.; Ahmad, Z. Exogenous application of low and high molecular weight organic acids differentially affected the uptake of cadmium in wheat-rice cropping system in alkaline calcareous soil. *Environ. Pollut.* **2023**, *329*, 121682. [CrossRef] [PubMed]
43. Cieschi, M.T.; Polyakov, A.Y.; Lebedev, V.A.; Volkov, D.S.; Pankratov, D.A.; Veligzhanin, A.A.; Perminova, I.V.; Lucena, J.J. Eco-friendly iron-humic nanofertilizers synthesis for the prevention of iron chlorosis in soybean (*Glycine max*) grown in calcareous soil. *Front. Plant Sci.* **2019**, *10*, 413. [CrossRef] [PubMed]
44. Justi, M.; Silva, C.A.; Rosa, S.D. Organic acids as complexing agents for iron and their effects on the nutrition and growth of maize and soybean. *Arch. Agron. Soil Sci.* **2022**, *68*, 1369–1384. [CrossRef]

45. Jalali, M.; Jalali, M. Effect of Low-Molecular-Weight Organic Acids on the Release of Phosphorus from Amended Calcareous Soils: Experimental and Modeling. *J. Soil Sci. Plant Nutr.* **2022**, *22*, 4179–4193. [CrossRef]
46. Karadihalli Thammaiah, M.; Pandey, R.N.; Purakayastha, T.J.; Chobhe, K.A.; Vashisth, A.; Chandra, S.; Pawar, A.B.; Trivedi, A. Impact of Low Molecular Weight Organic Acids on Soil Phosphorus Release and Availability to Wheat. *Commun. Soil Sci. Plant Anal.* **2022**, *53*, 2497–2508. [CrossRef]
47. Zhao, K.; Wang, C.; Xiao, X.; Li, M.; Zhao, W.; Wang, Y.; Yang, Y. The Hormetic Response of Soil P Extraction Induced by Low-Molecular-Weight Organic Acids. *Processes* **2023**, *11*, 216. [CrossRef]
48. Sun, Q.; Liu, J.; Huo, L.; Li, Y.C.; Li, X.; Xia, L.; Zhou, Z.; Zhang, M.; Li, B. Humic acids derived from Leonardite to improve enzymatic activities and bioavailability of nutrients in a calcareous soil. *Int. J. Agric. Biol. Eng.* **2020**, *13*, 200–205. [CrossRef]
49. Abdulla, A.A.; Esmai, A.O.; Yaseen, H.S. Combination Influence of Humic Acid and Chelated Iron on yield and quality of Broccoli (*Brassica oleracea* L.) in Erbil, Iraqi Kurdistan Region. *ZANCO J. Pure Appl. Sci.* **2023**, *35*, 126–135. [CrossRef]
50. Gayomba, S.R.; Zhai, Z.; Jung, H.; Vatamaniuk, O.K. Local and systemic signaling of iron status and its interactions with homeostasis of other essential elements. *Front. Plant Sci.* **2015**, *6*, 716. [CrossRef]
51. Palmgren, M.G. Plant plasma membrane H^+-ATPases: Powerhouses for Nutrient Uptake. *Annu. Rev. Plant Physiol. Plant Mol. Biol.* **2001**, *52*, 817–845. [CrossRef] [PubMed]
52. Chen, L.; Zhao, R.; Yu, J.; Gu, J.; Li, Y.; Chen, W.; Guo, W. Functional analysis of plasma membrane H^+-ATPases in response to alkaline stress in blueberry. *Sci. Hortic.* **2022**, *306*, 111453. [CrossRef]
53. Canellas, L.P.; Olivares, F.L.; Aguiar, N.O.; Jones, D.L.; Nebbioso, A.; Mazzei, P.; Piccolo, A. Humic and fulvic acids as biostimulants in horticulture. *Sci. Hortic.* **2015**, *196*, 15–27. [CrossRef]
54. Elena, A.; Diane, L.; Eva, B.; Marta, F.; Roberto, B.; Zamarreño, A.M.; García-Mina, J.M. The root application of a purified leonardite humic acid modifies the transcriptional regulation of the main physiological root responses to Fe deficiency in Fe-sufficient cucumber plants. *Plant Physiol. Biochem.* **2009**, *47*, 215–223. [CrossRef]
55. Mora, V.; Bacaicoa, E.; Zamarreño, A.-M.; Aguirre, E.; Garnica, M.; Fuentes, M.; García-Mina, J.-M. Action of humic acid on promotion of cucumber shoot growth involves nitrate-related changes associated with the root-to-shoot distribution of cytokinins, polyamines and mineral nutrients. *J. Plant Physiol.* **2010**, *167*, 633–642. [CrossRef]
56. Alghamdi, S.A.; Al-Ghamdi, F.A.M.; El-Zohri, M.; Al-Ghamdi, A.A.M. Modifying of calcareous soil with some acidifying materials and its effect on *Helianthus annuus* (L.) growth. *Saudi J. Biol. Sci.* **2023**, *30*, 103568. [CrossRef]
57. Astolfi, S.; Pii, Y.; Mimmo, T.; Lucini, L.; Miras-Moreno, M.B.; Coppa, E.; Violino, S.; Celletti, S.; Cesco, S. Single and Combined Fe and S Deficiency Differentially Modulate Root Exudate Composition in Tomato: A Double Strategy for Fe Acquisition? *Int. J. Mol. Sci.* **2020**, *21*, 4038. [CrossRef]
58. Martinez-Cuenca, M.-R.; Iglesias, D.J.; Talon, M.; Abadia, J.; Lopez-Millan, A.-F.; Primo-Millo, E.; Legaz, F. Metabolic responses to iron deficiency in roots of Carrizo citrange [*Citrus sinensis* (L.) Osbeck. × *Poncirus trifoliata* (L.) Raf.]. *Tree Physiol.* **2013**, *33*, 320–329. [CrossRef]
59. Covarrubias, J.I.; Rombolà, A.D. Organic acids metabolism in roots of grapevine rootstocks under severe iron deficiency. *Plant Soil* **2015**, *394*, 165–175. [CrossRef]
60. Alhendawi, R.A.M.; Mohamed, A.A.M. The influence of high pH on maize growth and utilization of micronutrients under various concentrations of bicarbonates. *Am. J. Agric. Environ. Sci.* **2015**, *15*, 259–264. [CrossRef]
61. Kong, D.; Chen, C.; Wu, H.; Li, Y.; Li, J.; Ling, H.-Q. Sequence Diversity and Enzyme Activity of Ferric-Chelate Reductase LeFRO1 in Tomato. *J. Genet. Genomics* **2013**, *40*, 565–573. [CrossRef]
62. Larbi, A.; Morales, F.; Abadía, A.; Abadía, J. Changes in iron and organic acid concentrations in xylem sap and apoplastic fluid of iron-deficient Beta vulgaris plants in response to iron resupply. *J. Plant Physiol.* **2010**, *167*, 255–260. [CrossRef] [PubMed]
63. Zhao, Y.; Liu, S.; Li, F.; Sun, M.; Liang, Z.; Sun, Z.; Yu, F.; Li, H. The low ferric chelate reductase activity and high apoplastic pH in leaves cause iron deficiency chlorosis in 'Huangguan' pears grafted onto quince A grown in calcareous soil. *Sci. Hortic.* **2023**, *310*, 111754. [CrossRef]
64. Zamboni, A.; Zanin, L.; Tomasi, N.; Pezzotti, M.; Pinton, R.; Varanini, Z.; Cesco, S. Genome-wide microarray analysis of tomato roots showed defined responses to iron deficiency. *BMC Genom.* **2012**, *13*, 101. [CrossRef] [PubMed]
65. Skogerboe, R.K.; Wilson, S.A. Reduction of ionic species by fulvic acid. *Anal. Chem.* **1981**, *53*, 228–232. [CrossRef]
66. Struyk, Z.; Sposito, G. Redox properties of standard humic acids. *Geoderma* **2001**, *102*, 329–346. [CrossRef]
67. Yang, F.; Tang, C.; Antonietti, M. Natural and artificial humic substances to manage minerals, ions, water, and soil microorganisms. *Chem. Soc. Rev.* **2021**, *50*, 6221–6239. [CrossRef]
68. Cieschi, M.T.; Lucena, J.J. Iron and Humic Acid Accumulation on Soybean Roots Fertilized with Leonardite Iron Humates under Calcareous Conditions. *J. Agric. Food Chem.* **2018**, *66*, 13386–13396. [CrossRef]
69. Tahjib-Ul-Arif, M.; Zahan, M.I.; Karim, M.M.; Imran, S.; Hunter, C.T.; Islam, M.S.; Mia, M.A.; Hannan, M.A.; Rhaman, M.S.; Hossain, M.A.; et al. Citric Acid-Mediated Abiotic Stress Tolerance in Plants. *Int. J. Mol. Sci.* **2021**, *22*, 7235. [CrossRef]
70. Tomasi, N.; De Nobili, M.; Gottardi, S.; Zanin, L.; Mimmo, T.; Varanini, Z.; Römheld, V.; Pinton, R.; Cesco, S. Physiological and molecular characterization of Fe acquisition by tomato plants from natural Fe complexes. *Biol. Fertil. Soils* **2013**, *49*, 187–200. [CrossRef]
71. Lee, S.; Rahman, M.M.; Nakanishi, H.; Nishizawa, N.K.; An, G.; Nam, H.G.; Jeon, J.-S. Concomitant Activation of OsNAS2 and OsNAS3 Contributes to the Enhanced Accumulation of Iron and Zinc in Rice. *Int. J. Mol. Sci.* **2023**, *24*, 6568. [CrossRef] [PubMed]

72. Vigani, G.; Pii, Y.; Celletti, S.; Maver, M.; Mimmo, T.; Cesco, S.; Astolfi, S. Mitochondria dysfunctions under Fe and S deficiency: Is citric acid involved in the regulation of adaptive responses? *Plant Physiol. Biochem.* **2018**, *126*, 86–96. [CrossRef] [PubMed]
73. Waseem, M.; Ahmad, F. The phosphoenolpyruvate carboxylase gene family identification and expression analysis under abiotic and phytohormone stresses in *Solanum lycopersicum* L. *Gene* **2019**, *690*, 11–20. [CrossRef]
74. An, Y.; Zhou, P.; Xiao, Q.; Shi, D. Effects of foliar application of organic acids on alleviation of aluminum toxicity in alfalfa. *J. Plant Nutr. Soil Sci.* **2014**, *177*, 421–430. [CrossRef]
75. Hsieh, E.-J.; Waters, B.M. Alkaline stress and iron deficiency regulate iron uptake and riboflavin synthesis gene expression differently in root and leaf tissue: Implications for iron deficiency chlorosis. *J. Exp. Bot.* **2016**, *67*, 5671–5685. [CrossRef]
76. Li, L.; Cheng, X.; Ling, H.-Q. Isolation and characterization of Fe(III)-chelate reductase gene LeFRO1 in tomato. *Plant Mol. Biol.* **2004**, *54*, 125–136. [CrossRef]
77. Tomasi, N.; Rizzardo, C.; Monte, R.; Gottardi, S.; Jelali, N.; Terzano, R.; Vekemans, B.; De Nobili, M.; Varanini, Z.; Pinton, R.; et al. Micro-analytical, physiological and molecular aspects of Fe acquisition in leaves of Fe-deficient tomato plants re-supplied with natural Fe-complexes in nutrient solution. *Plant Soil* **2009**, *325*, 25–38. [CrossRef]
78. Drincovich, M.F.; Voll, L.M.; Maurino, V.G. Editorial: On the Diversity of Roles of Organic Acids. *Front. Plant Sci.* **2016**, *7*, 1592. [CrossRef]

horticulturae

MDPI

Article

Combined Use of TiO_2 Nanoparticles and Biochar Produced from Moss (*Leucobryum glaucum* (Hedw.) Ångstr.) Biomass for Chinese Spinach (*Amaranthus dubius* L.) Cultivation under Saline Stress

Ivan Širić [1], Sadeq K. Alhag [2], Laila A. Al-Shuraym [3], Boro Mioč [1], Valentino Držaić [1], Sami Abou Fayssal [4,5], Vinod Kumar [6], Jogendra Singh [6], Piyush Kumar [7], Rattan Singh [8], Rakesh Kumar Bachheti [9,10], Madhumita Goala [11], Pankaj Kumar [6,12,*] and Ebrahem M. Eid [13,*]

1 University of Zagreb, Faculty of Agriculture, Svetosimunska 25, 10000 Zagreb, Croatia
2 Biology Department, College of Science and Arts, King Khalid University, Muhayl Asser 61913, Saudi Arabia
3 Biology Department, Faculty of Science, Princess Nourah bint Abdulrahman University, Riyadh 12834, Saudi Arabia
4 Department of Agronomy, Faculty of Agronomy, University of Forestry, 10 Kliment Ohridski Blvd, 1797 Sofia, Bulgaria
5 Department of Plant Production, Faculty of Agriculture, Lebanese University, Beirut 1302, Lebanon
6 Agro-Ecology and Pollution Research Laboratory, Department of Zoology and Environmental Science, Gurukula Kangri (Deemed to Be University), Haridwar 249404, India
7 Department of Science, Vivek College of Education, Moradabad Road, Bijnor 246701, India
8 Department of Food Technology, Guru Jambeshwar University of Science & Technology, Hisar 125001, India
9 Department of Industrial Chemistry, College of Applied Sciences, Addis Ababa Science and Technology University, Addis Ababa P.O. Box 16417, Ethiopia
10 Department of Allied Sciences, Graphic Era Hill University, Dehradun 248002, India
11 Department of Environment Science, Graphic Era (Deemed to be University), Dehradun 248002, India
12 Research and Development Division, Society for AgroEnvironmental Sustainability, Dehradun 248007, India
13 Botany Department, Faculty of Science, Kafrelsheikh University, Kafr El-Sheikh 33516, Egypt
* Correspondence: rs.pankajkumar@gkv.ac.in (P.K.); ebrahem.eid@sci.kfs.edu.eg (E.M.E.)

Citation: Širić, I.; Alhag, S.K.; Al-Shuraym, L.A.; Mioč, B.; Držaić, V.; Abou Fayssal, S.; Kumar, V.; Singh, J.; Kumar, P.; Singh, R.; et al. Combined Use of TiO_2 Nanoparticles and Biochar Produced from Moss (*Leucobryum glaucum* (Hedw.) Ångstr.) Biomass for Chinese Spinach (*Amaranthus dubius* L.) Cultivation under Saline Stress. *Horticulturae* **2023**, *9*, 1056. https://doi.org/10.3390/horticulturae9091056

Academic Editor: Alberto Pardossi

Received: 22 August 2023
Revised: 15 September 2023
Accepted: 18 September 2023
Published: 21 September 2023

Abstract: Salinity-induced soil degradation poses a significant challenge to agricultural productivity and requires innovative crop-management strategies. In this study, the synergistic effect of biochar and TiO_2 nanoparticles (NPs) obtained from moss (*Leucobryum glaucum* (Hedw.) Ångstr.) biomass on the growth, yield, biochemical, and enzymatic response of Chinese spinach (*Amaranthus dubius* L.) grown under salinity stress was investigated. Purposely, *A. dubius* was grown under different combinations of arable soil, biochar, TiO_2 NPs, and saline soils. The produced biochar and TiO_2 NPs were characterized using microscopy image analysis, X-ray diffraction patterns (XRD), energy-dispersive X-ray spectroscopy (EDX), zeta potential, particle size distribution, and Fourier-transform infrared spectroscopy (FTIR). The results showed that saline stress caused a significant ($p < 0.05$) decline in growth, yield, and biochemical constituents of *A. dubius* compared to control treatments. However, the combined application of biochar and TiO_2 NPs significantly ($p < 0.05$) alleviated the saline stress and resulted in optimum fresh weight (30.81 g/plant), dry weight (4.90 g/plant), shoot and root length (28.64 and 12.54 cm), lead number (17.50), leaf area (12.50 cm^2/plant), chlorophyll (2.36 mg/g), carotenoids (2.85 mg/g), and relative water content (82.10%). Biochar and TiO_2-NP application helped to reduce the levels of stress enzymes such as catalase (2.93 μmol/min/mg P), superoxide dismutase (SOD: 2.47 EU/g P), peroxidase (POD: 40.03 EU/min/g P), and ascorbate peroxidase (3.10 mM/mg P) in saline soil. The findings of this study suggest that the combination of nanotechnology and biochar derived from unconventional biomass can be a viable option to mitigate salinity-related challenges and enhance crop yield.

Keywords: circular economy; nanotechnology; pyrolysis; sustainable agriculture; saline stress

1. Introduction

Titanium dioxide (TiO_2) is a white, opaque, and naturally occurring inorganic compound; it is composed of barium cation and sulfate anion and has low toxicity and negligible biological effects [1]. This has led to the production of TiO_2 nanoparticles (TiO_2 NPs), which consist of anatase and rutile crystal forms [2]. These NPs are biocompatible, chemically stable, inexpensive, and reusable features that make them suitable for agricultural use, especially in developing countries [3]. Out of several physical and chemical methods previously employed for TiO_2-NP synthesis, chemical synthesis methods were found to have harmful environmental impacts [4]. Thus, the green synthesis method has been adopted as the most eco-friendly, sustainable, and non-expensive way to biosynthesize NPs [5]. Generally, the average TiO_2-NP size ranges between 25 and 100 nm [6], and their structure arrangement is crystalline [7]. Many studies investigated plant extracts for potential TiO_2-NP biosynthesis. For instance, the use of Aloe vera leaf extract resulted in TiO_2 NPs with large surface area and irregular size [8]. TiO_2 NPs biosynthesized from *Anona squamosa* fruit peel extract showed a spherical nature and 21–25 nm size [9]. An irregular shape and 25–110 nm particle size were obtained from the biosynthesis of TiO_2 NPs using *Catharanthus roseus* leaf extract [10]. Vimala et al. [11] biosynthesized silver NPs (AgNPs) from *Campylopus flexuosus* (Hedw.) bird moss and revealed a 58 nm particle size and –25 mv zeta potential (good stability). Iron (FeNPs) and AgNPs biosynthesized from *S. fallax* moss were spherical with 120 and 100 nm sizes, respectively [12]. However, there is very scarce information regarding the biosynthesis of TiO_2 NPs from moss biomass.

Biochar is a carbon-rich solid product resulting from the thermal conversion of unstable carbon-enriched biomass into a stable form via pyrolysis. Biochar is generally characterized by high pH, porosity, surface area, and the availability of both micro- and macrospores [13]. Agricultural, forest, and food wastes (mainly lignocellulosic biomass sources) constitute the main substrate sources used for biochar production. The latter can be a promising strategy to mitigate the increasing environmental pollution induced by disposed wastes. For instance, a recent study found that the production of biochar from spent mushroom substrate is feasible and yields a qualitative product with very few impurities [14]. Moreover, the type of substrate used directly affects the surface area of produced biochar [15]. It is worth noting that the lower the impurities in produced biochar, the higher its quality is. Moreover, pyrolysis temperature plays a key role in determining biochar physicochemical properties. Increased pyrolysis temperature results in increased pH, carbon content, structural pores, and decreased biochar yield and volatile matter [16,17].

Chinese spinach (*Amaranthus dubius* L.; Amaranthaceae) is a 30–150 cm annual plant with a branched stem, native to Central and South American regions and imported to India from Africa for around 500 years. The global amaranth market was valued at around USD 9.1 billion in 2021 and is forecasted to reach USD 21.7 billion by 2029 with a compound annual growth rate (CAGR) of 11.51% in the 2022–2029 period [18]. In India, this plant is mainly grown in the mid and high hills of the Himalayan valley; it is estimated that around 40–50 thousand hectares of *A. dubius* are grown [19]. Although it is one of the most popularly consumed leafy vegetables in India (also known as the food of the poor), accurate information related to its production and market is still scarce [20,21]. Indian cuisine has several uses of amaranth, e.g., "Raab", which is a broth made from amaranth flour and other basic ingredients, and curries and salads (grain and fresh forms, respectively), among others. The amaranth plant is generally considered a highly tolerant crop to saline abiotic stress [22]. This is attributed to the low number of stomata of the plant, resulting in lower basal stomatal conductance [23]. Such a hypothesis was confirmed by Bellache et al. [24], who mentioned that the growth of both *A. alba* and *A. hybridus* species was severely affected at 450 and 600 mM NaCl with unpronounced effects at lower concentrations. However, a recent study [25] mentioned that this plant can be affected by saline stress at even relatively low to moderate concentrations. For instance, NaCl concentrations higher than 25 mM (50 and 100 mM) resulted in decreased growth and yield parameters due to saline accumulation in the shoots and roots. On the other hand, antioxidant activity, total flavonoids, and total phenolics increased at 50 mM NaCl, as mentioned by Hoang et al. [25].

In the last decade, numerous studies confirmed the role of NPs in saline stress mitigation of various leafy crops, e.g., lettuce [26], cauliflower [27], alfalfa [28], and Chinese cabbage [29]. The main observed improvements were increased plant biomass, leaf chlorophyll pigment contents, and systemic acquired resistance plant defense, along with decreased leaf malondialdehyde (MDA) content. CuO and ZnO NPs application to *A. hybridus* resulted in improved plant germination rate and growth along with increased antioxidant activity within shoots and roots, as reported by Francis et al. [30]. Furthermore, the polycyclic aromatic hydrocarbon uptake by *A. tricolor* L. was reduced by around 20–55% as a result of SiO2 NPs and ZnO NPs amendment [31]. However, no previous studies investigated the application of TiO_2 NPs on *A. dubius*. On the other hand, only a few studies evaluated the effect of biochar on *Amaranthus* spp. For instance, the amendment of *A. tricolor* L. with bamboo biochar resulted in 34.4% higher yields; however, no significant improvements were noticed in terms of plant leaf number and height [32]. The amendment of *A. tricolor* L. with biochar produced from litchi wastes resulted in increased chlorophyll pigments (chlorophyll 'a' and chlorophyll 'b') and yield and reduced leaf MDA content as reported by Jiang et al. [33]. However, no studies evaluated the potential of biochar as an amendment for *A. dubius*. The co-application of TiO_2 NPs with biochar for plant response evaluation has been scarcely investigated. To our knowledge, the co-amendment of *Sorghum bicolor* with TiO_2 NPs and biochar was the very first experiment performed. These authors reported increased chlorophyll pigments (Chl 'a' and Chl 'b') in co-treated plants compared to TiO_2 NP treated [34]. Therefore, it was evidenced from the above studies that the physiological effects can be positive and increase crop productivity and tolerance to stressful conditions [28].

Considering the aforementioned, it was hypothesized that the growth and yield characteristics of *A. dubius* grown in saline soils would be differentially affected by applying *L. glaucum* biomass-derived biochar and TiO_2 NPs, both separately and in combination. In addition, it is hypothesized that these treatments will induce different biochemical and enzymatic responses in the plant. Therefore, the current study aimed to investigate the separate and combined effects of TiO_2 NPs and biochar produced from *L. glaucum* biomass on growth and yield attributes and biochemical and enzyme response of *A. dubius* cultivation under saline soils.

2. Materials and Methods

2.1. Collection of Experimental Materials

For this study, moss (*Leucobryum glaucum* (Hedw.) Ångstr.) biomass was obtained from natural rocks of the Chilla Forest Range in Rajaji National Park, Haridwar, India (29°57′55.1″ N and 78°12′02.9″ E). The moss attached to the rock was carefully cut using a sharp knife, and the adhering soil was carefully removed. Then, moss biomass was cut into small pieces and placed in zip-locking plastic bags for further transportation to the laboratory. For the synthesis of TiO_2 NPs, AR-grade titanium-isopropoxide was procured from Sigma–Aldrich, India, having a purity of >97%. Moreover, *A. dubius* seeds were obtained from the local market of Jwalapur, Haridwar, India (29°55′20.4″ N and 78°06′19.5″ E).

2.2. Biosynthesis of TiO_2 NPs and Biochar Production

The moss biomass was carefully washed in triple distilled water to remove adhering dirt and then oven-dried at 60 °C until a constant weight was obtained. The dried moss biomass was converted into a fine powder using a mechanical grinder. Further, 10 g of dried moss powder was mixed in 50 mL of triple distilled water and heated for 60 min at 80 °C. After cooling down, the solution was filtered with Whatman filter paper number 1. For the synthesis of TiO_2 NPs, 50 mL of 5 mM titanium-isopropoxide was added to 50 mL of moss extract (1:1 v/v) and continuously stirred for 8 h at 25 °C. The solution was centrifuged for 10 min at 10,000 rpm, and formed nanoparticles were carefully separated, dried at 100 °C for 12 h, and calcinated at 570 °C using a muffle furnace for 3 h. On the other hand, the moss biochar was prepared using a slow pyrolysis process. For this, biomass was thoroughly washed and dried under sunlight to remove moisture content. The biomass was placed in a crucible disc and pyrolyzed in a muffle furnace (NSW-101, Narang Scientific, New Delhi, India) for 60 min, and the temperature was raised to 600 °C (10 °C/min). Finally, the prepared TiO_2 NPs and biochar were characterized and used for *A. dubius* cultivation experiments.

2.3. Experimental Design for Chinese Spinach Cultivation

The *A. dubius* cultivation experiments were conducted in the multipurpose laboratory of the Department of Zoology and Environmental Science, Gurukula Kangri (Deemed to be University), Haridwar, India (29°55′10.60″ N and 78°07′07.90″ E) during the winter season (September to October 2021). The average ambient temperature and humidity of the experimental site were 26 °C and 81%, respectively. For this purpose, the arable soil from the nearby agricultural fields was collected and filled in 25 kg capacity pre-sterilized plastic pots. A total of 19.50 kg arable soil and 0.50 kg vermicompost were filled in each pot and mixed thoroughly. This study involved a series of seven different treatments, namely T0 (arable soil only), T1 (arable soil + biochar), T2 (arable soil + TiO_2 NPs), T3 (arable soil + biochar + TiO_2 NPs), T4 (saline soil only), T5 (saline soil + biochar), T6 (saline soil + TiO_2 NPs), and T7 (saline soil + biochar + TiO_2 NPs), respectively, as depicted in Figure 1.

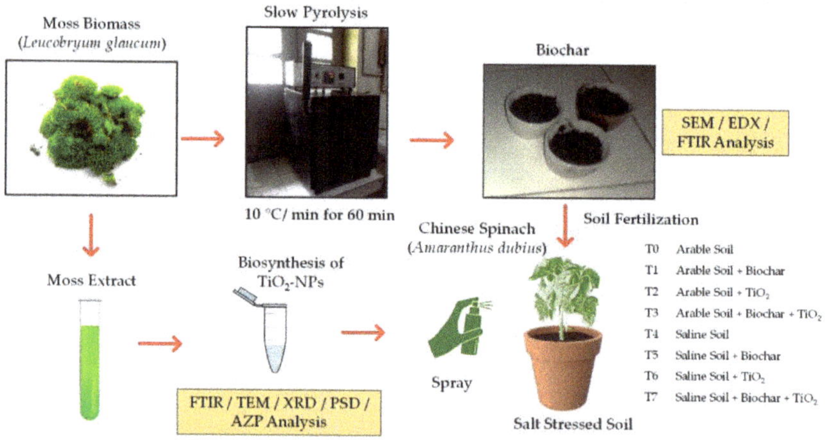

Figure 1. The layout of the experimental design used for *A. dubius* cultivation.

To achieve saline conditions in arable soil, a 150 mM NaCl (AR grade) solution was added to T4, T5, T6, and T7 treatments and mixed carefully to ensure uniform distribution of the saline solution. A 20 g dose of moss biochar per kg of soil was added to the designated pots and mixed carefully. In addition, 2.50 mg of TiO_2 NPs was added to 1000 mL of distilled water and placed in an ultrasonic water bath for 25 min to achieve uniform dissolution. The prepared solution was hand sprayed (foliar application) on *A. dubius* plants at an interval of every 10 days. For this, 25 mL of TiO_2 NPs solution was sprayed into each pot on days 10, 20, 30, and 40. Thus, a total of 100 mL of solution was sprayed into respective pots during the cultivation period. All treatments had five identical replicates to minimize experimental and analytical bias. Before their sowing, the seeds of *A. dubius* were soaked in water for 2 h and then gently dried using blotting paper to remove excess water content. Afterward, a total of three seeds were shown in each pot at a depth of 0.5 cm and covered gently with loose soil. The pots were equally watered using a borewell water supply with the help of a hand sprayer and placed under natural sunlight conditions (16/8 h dark/light period). After germination began, the pots were watered after each third day or as per requirements to maintain appropriate soil moisture content. The first harvesting was performed on the 40th day after showing when leaves were fully matured. A total of three harvestings were performed at an interval of the fifth day to represent average plant yield. During the experiments, regular weeding and pest monitoring were performed to minimize variations due to management.

2.4. Growth, Yield, and Biochemical Analyses

The effect of TiO_2 NPs and biochar on *A. dubius* grown under saline stress was studied by analyzing the selected biochemical and enzyme responses. In this regard, the chlorophyll content (mg/g) of *A. dubius* was determined using 80% acetone extract and the spectrophotometric (Cary 60, Agilent Technologies, Santa Clara, CA, USA) method at 645 and 663 nm. Similarly, carotenoid contents (mg/g) were measured at 480, 645, and 663 nm while using acetone, chloroform, and petroleum ether as extraction reagents. Relative water content (%) of *A. dubius* was estimated via floating leaf samples on deionized water in Petri dishes for 24 h to achieve full turgidity, followed by oven drying at 80 °C to achieve constant dry weight [35]. The relative water content (%) of the plant was then calculated based on the difference between its initial hydrated state and the final dry weight, providing valuable information about the plant's water content relative to its maximum capacity for water absorption. Similarly, catalase (CAT: micromoles per minute per milligram of protein or µmol/min/mg P) activity was estimated using a 0.50 g leaf sample homogenized in a 50 mM phosphate buffer (pH: 7.8) followed by spectrophotometric estimated at 240 nm as previously described by Uma et al. [36]. Similarly, peroxidase (POD: µmol/min/mg P) activities were also determined using a 50 mM phosphate buffer, 20 mM guaiacol, 40 mM H_2O_2, and 0.10 mL extract followed by estimation at 470 nm. Moreover, superoxide dismutase (SOD: units per milligram of protein or U/mg P) contents in *A. dubius* leaves were estimated using a modified method of nitroblue tetrazolium (NBT) assay at 560 nm. In comparison, ascorbate peroxidase (APX: µmol/min/mg P) activities were recorded at 290 nm while homogenizing the sample in a 0.10 M potassium phosphate buffer (pH 7.0) [37].

2.5. Analytical and Instrumental Methods

Before its use in NP synthesis and biochar production, the moss biomass was analyzed for selected proximate and ultimate elemental analyses as percent (%). In this, moisture content, dry weight, crude protein, volatile matter, crude lipid, and total ash were determined based on the methodologies adopted by Toor et al. [38]. The ultimate analysis of selected elements (carbon, nitrogen, and oxygen) was also determined using an elemental analyzer (CE 440, Exeter Analytical Inc., Chelmsford, MA, USA). On the other hand, the functional group characteristics of synthesized TiO_2 NPs and biochar produced from moss biomass were analyzed using Fourier's transform infrared (FTIR) spectroscopy (8400S, Shimadzu, Carlsbad, CA, USA). For this, KBr pellets were used to prepare a thin disc under high pressure and then subjected to FTIR spectroscopy to scan the sample in a range of 400–4000 1/cm. The final FTIR spectra were obtained by subtracting the KBr background and 3% smoothening in order to identify the bonding modes and functional groups present in TiO_2 NPs and biochar separately. The particle size distribution of TiO_2 NPs was determined using the scattered light intensity method, while transmission electron microscopy (TEM, FEI Tecnai G2 20 S-Twin, Hillsboro, OR, USA) was used to visualize the size and morphology of NPs. The X-ray diffraction patterns of powdered NP were analyzed using an XRD instrument (D8-Advance, Bruker, Billerica, MA, USA). Also, the apparent zeta potential (AZP) of TiO_2 NPs was analyzed using a zeta potential analyzer (Delsa™Nano C, Beckman Coulter, Brea, CA, USA). Moreover, biochar obtained from the pyrolysis of moss biomass was analyzed using scanning electron microscopy (SEM; Zeiss Gemini SEM, Carl Zeiss, Oberkochen, Germany). The Biochar sample was carefully coated and subjected to SEM imaging to visualize the surface morphology. Whereas energy-dispersive X-ray spectroscopy (EDX; Octane Eliter Plus, Mahwah, NJ, USA) analysis was simultaneously performed to understand the basic elemental composition of biochar.

2.6. Data Analysis and Software

The data generated in this study were analyzed using Microsoft Excel 365 (Microsoft Corp., Redmond, WA, USA) and OriginPro 2023b (OriginLab Corp., Northampton, MA, USA) software packages. The significant differences among treatment groups were studied based on the analysis of variance (ANOVA) while comparing the means using Tukey's post hoc test. The level of significant difference was adjusted to a 95% confidence interval or $p < 0.05$.

3. Results and Discussion

3.1. Characteristics of Moss Biomass, Synthesized TiO_2 NPs, and Biochar

3.1.1. Properties of Moss Biomass

The chemical composition of *L. glaucum* used in the present investigation is reported in Table 1. Moss moisture content was found to be 63.3 ± 2.04%; it is well known that most species cannot tolerate moisture levels lower than 35% as this leads to desiccation occurrence [39]. The dry weight of *L. glaucum* (36.7 ± 0.31%) was relatively higher than the range previously set by Hoekstra et al. [40] and Alpert [39] (10–30%); this can be attributed to the type of moss species studied. Ihl and Barboza [41] found a dry weight of 17.7–31.5% for Alaska moss species, which is close but lower than observed in *L. glaucum* (36.70 ± 0.31%). This may be related to the species type and ecosystem in which it thrives. A crude protein content of 7.25 ± 0.07% was found in *L. glaucum*, thus being in the range mentioned by Orlov and Sadovnikova [42,43] (5–10%) for Bryophytes. The volatile matter content of *L. glaucum* was 38.15 ± 1.78%, which was more than two-fold higher than attributed for other moss species (0.04–14.5%) [44]. A higher volatile matter production by mosses may be attributed to the attraction of flies for spore dispersal [45]. A crude lipid content of 2.41 ± 0.04% was found in *L. glaucum*; generally, such a component is scantly researched in moss species. Most attention goes to fatty acids that account for around 70% of the total lipid content of moss species [46]. Total ash content (2.97 ± 0.02%) was well below the range detected in Alaska moss species (8.1–19.6%) [41] and higher by 2–20-folds than found (0.1–1.2%) by Zaccone et al. [47]. Carbon (49.28 ± 2.60%) and nitrogen (3.10 ± 0.04%) contents detected in *L. glaucum* were higher than those reported by Klavina [48] on 16 moss species (40–43% and 0.4–2%, respectively) and in the ranges depicted by Zaccone et al. [47] (45–63% and 0.4–5.8%, respectively). On the other hand, the oxygen content of *L. glaucum* (28.05 ± 1.52%) was well below the range (48–53%) observed by Klavina [48]. CO_2 uptake (amount) directly affects the amount of oxygen produced by mosses [49]. Thus, the chemical composition of *L. glaucum* investigated in this study exhibited unique characteristics, including a relatively high dry weight, volatile matter content, and carbon and nitrogen levels, which may be influenced by the specific moss species and its ecosystem.

Table 1. Proximate and ultimate analysis results of moss biomass used in this study.

Parameter	Value
Moisture content (%)	63.30 ± 2.04
Dry weight (%)	36.70 ± 0.31
Crude protein (%)	7.25 ± 0.07
Volatile matter (%)	38.15 ± 1.78
Crude lipid (%)	2.41 ± 0.04
Total ash (%)	2.97 ± 0.02
Carbon (%)	49.28 ± 2.60
Nitrogen (%)	3.10 ± 0.04
Oxygen (%)	28.05 ± 1.52

Values are mean followed by the standard deviation of three analyses.

3.1.2. Properties of Synthesized TiO_2 NPs

The stability of synthesized NPs can be confirmed by evaluating particle size and apparent zeta potential [50]. TiO_2 NPs had a size ranging from 1 to 83 mm. At a 67 a.u. intensity (highest intensity reached), TiO_2 NPs were of 20 mm size. It is worth noting that between 60 and 67 a.u. intensity, TiO_2 NPs showed a certain size uniformity (18–22 mm) (Figure 2a,c). This indicates a specific relationship between the intensity level and the size of the nanoparticles. However, TiO_2 NPs were generally a non-uniform size. Larue et al. [51] mentioned that 4–100 nm NPs have the ability to cross plant cuticles by disrupting the wax layer. Other researchers mentioned that 3.5–20 mm NPs were likely to penetrate plants [52,53]. Zeta potential explains the potential stability of nanoparticles in a solution. The negative apparent zeta potential peak of TiO_2 NPs (around −17 mV) explains that particles were of similar charges and tended to strongly repel each other within the extract (Figure 2b). Thus, there would be a lesser tendency for TiO_2 NPs to settle down or come together in any solution (positive result). In summary, the stability of the synthesized TiO_2 NPs was confirmed through the evaluation of particle size and apparent zeta potential.

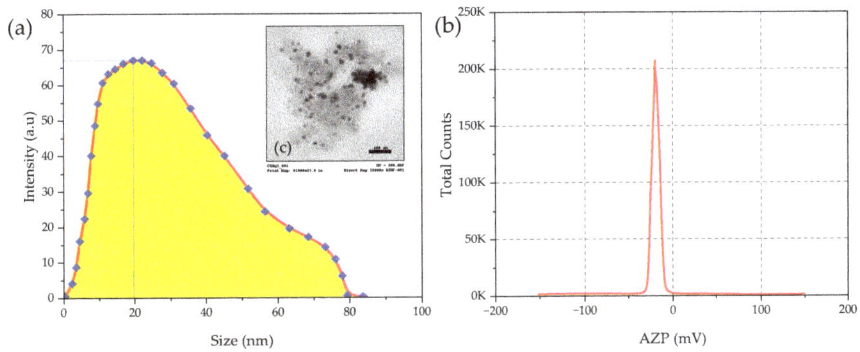

Figure 2. (**a**) Particle size distribution (nm), (**b**) apparent zeta potential (mV), and (**c**) TEM image of TiO_2 NPs synthesized using extract of moss biomass.

The FT-IR analysis was performed to determine the functional groups present in TiO_2 NPs synthesized using an extract of moss biomass. Positive wavelength peaks (1/cm) of absorption correspond to negative apparent peaks of IR transmittance [14]. Figure 3a shows sharp peaks at 520, 587, and 680 1/cm, referring to ring torsion of phenyl, strong C-I stretching between halo compounds, and C-H bending vibrations, respectively [54]. The peak observed at 828 1/cm can be attributed to strong C=C bending between mono- and di-substituted (alkane) compounds mostly found in carbohydrates and sugars. The peak detected at 1280 1/cm outlines the presence of collagen or amide III band components of protein. Aliphatic C-H stretching of cell wall polysaccharides can explain the peak observed at 1390 1/cm [48]. Figure 3b shows the patterns of TiO_2 NPs synthesized using an extract of moss biomass. XRD peaks observed at 2θ corresponding to 110°, 101°, 200°, 111°, 210°, 211°, 220°, 002°, 310°, and 301° plane indices indicated the crystalline structure of synthesized TiO_2 NPs. The determined results corroborate with previous reports of El-Desoky et al. [55] and Usgodaarachchi et al. [56]. Therefore, TiO_2 NPs synthesized using moss biomass extract had distinct peaks corresponding to specific chemical bonds and compounds. Additionally, XRD analysis confirmed the crystalline nature of the synthesized TiO_2 NPs.

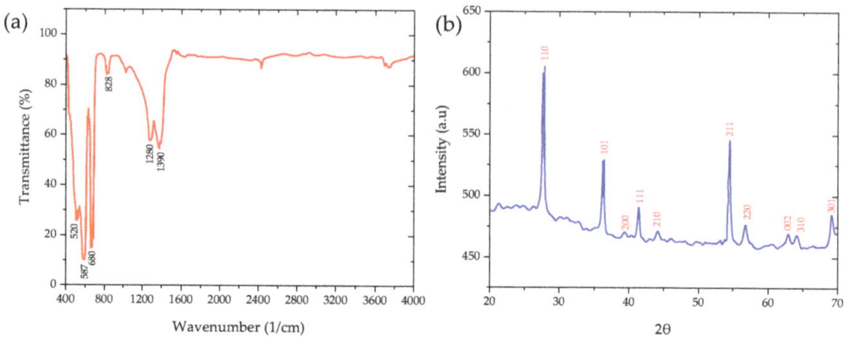

Figure 3. (**a**) FTIR spectra and (**b**) XRD patterns of TiO_2 NPs synthesized using extract of moss biomass.

3.1.3. Properties of Produced Biochar

Figure 4 shows (a) FTIR spectra, (b) EDX, and (c) SEM patterns of biochar produced from moss biomass. Four peaks were detected at 1020 (i), 1650 (ii), 1740 (iii), and 2940 (iv) 1/cm (Figure 4a) corresponding to (i) OH and C-O deformation in secondary alcohols and ethers induced by cell wall polysaccharides, (ii) amide I and C=O stretching in conjugated aryl ketones induced by proteins and phenolic compounds, (iii) C=O stretching in unconjugated ketones induced by carbohydrates, and (iv) C-H stretching in methyl and methylene groups mainly induced by lipids with contributions of proteins, carbohydrates, and phenolics, respectively [48]. SEM-EDX has been successfully used previously to evaluate the surface morphology and ultimate elemental composition of produced biochar [14]. Figure 4b depicted low amounts of calcium (Ca; 2.160%), magnesium (Mg; 2.09%), and potassium (K; 0.57%) detected at low (Mg) and moderate (Ca and K) counts. The highest ionization energy peaks were detected at low counts for carbon (C) and oxygen (O) (68.10% and 17.46%, respectively). Although C and O constituted around 85.5% of the total elements found in the produced biochar, around 9.6% of elements were not detected and may be potentially toxic elements. Therefore, more investigations should target the detection of these impurities. It is worth noting that biochar produced in this investigation had a higher C content than the one produced by Širić et al. [14] using spent mushroom substrate (68.10% and 62.06%, respectively), while the O content of the latter was higher than the former (25.10% and 17.46%, respectively). Although air oxidation is essential for the efficient increase of biochar's porosity [57], increased oxygen content can lead to biochar yield reduction [58]. Figure 2c shows visually good adsorption capacity and relative fineness of biochar produced from moss biomass. Hence, the biochar produced from moss biomass exhibited distinct chemical characteristics determined via FTIR, SEM-EDX, and SEM patterns, with specific functional groups and elemental composition identified.

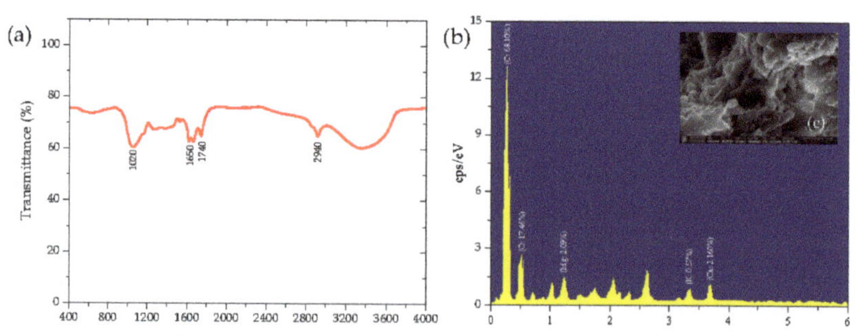

Figure 4. (**a**) FTIR spectra and (**b**,**c**) SEM-EDX patterns of biochar produced using moss biomass.

3.2. Effect of TiO_2 NPs and Biochar on Growth and Yield of A. dubius

A. dubius cultivation experiments showed that biochar mixing and foliar application of synthesized TiO_2 NPs showed a significant ($p < 0.05$) effect on growth and yield attributes (Table 2). The results showed that fresh weight in the control treatment (T0) was 31.10 g/plant, which was significantly reduced to 20.52 g/plant under saline stress (T4). However, the combined application of biochar and TiO_2 NPs application showed a significant incline in fresh weight of the T7 treatment, i.e., 30.81 (50.14%) as compared to the T4 treatment. However, the highest fresh weight (36.34 g/plant) was recorded in the T3 treatment, which had no saline stress with the combined application of moss biochar and TiO_2 NPs. A similar trend was also observed for the dry weight of *A. dubius*, i.e., maximum in T3 treatment (5.30 g/plant) while minimum in T4 treatment (3.42 g/plant). Likewise, shoot and root lengths (33.04 and 16.09 cm) were also found in the T3 treatment of non-saline soil, followed by the T7 treatment of saline soil (shoot: 28.64 and 12.54 cm). A notable observation was the average number of leaves (22.20) and leaf area (17.07 cm^2/plant), which were found to be the highest in the T3 treatment. Biochar mixing helps in increasing soil nutrient composition and water-holding capacity, which later results in higher crop yields. Moreover, biochar particles added to the soil help in the improved propagation of soil microbes, which is beneficial for efficient rhizosphere association development [59]. On the other hand, foliar application of different NPs has been found to alleviate osmotic stress and reactive oxygen species (ROS) scavenging under saline-stress soils [60]. Additionally, the immobilization of excess salts via biochar prevents their uptake by plant roots, reducing ion toxicity [61]. In particular, TiO_2 NPs have been found to offer biological mechanisms through which plants can ameliorate the harmful effects of salinity and promote better crop yield [62]. Both biochar and TiO_2 NPs contribute to the improvement of soil structure. Biochar improves soil porosity, water infiltration, and aeration, while TiO_2 NPs help reduce soil compaction. The synergistic effect leads to improved root penetration and nutrient uptake, which could have promoted better plant growth.

Table 2. Effect of TiO_2 NPs and biochar application on growth and yield attributes of *A. dubius*.

Parameters	Treatments							
	T0	T1	T2	T3	T4	T5	T6	T7
Fresh weight (g/plant)	31.10 ± 1.40 b	34.05 ± 1.07 a	32.51 ± 1.16 ab	36.34 ± 2.07 a	20.52 ± 2.25 d	22.29 ± 1.09 cd	24.09 ± 0.64 c	30.81 ± 0.72 b
Dry weight (g/plant)	5.09 ± 0.15 bc	5.23 ± 0.09 b	5.16 ± 0.08 bc	5.30 ± 0.06 a	3.42 ± 0.10 e	3.45 ± 0.03 de	3.53 ± 0.07 d	4.90 ± 0.14 c
Shoot length (cm)	30.57 ± 0.90 b	33.42 ± 1.08 a	30.90 ± 0.41 b	33.04 ± 0.57 ab	17.20 ± 1.01 f	19.16 ± 0.35 e	24.75 ± 0.81 d	28.64 ± 1.30 c
Root length (cm)	14.10 ± 0.41 c	15.36 ± 0.14 b	14.28 ± 0.20 c	16.09 ± 0.24 a	7.53 ± 0.12 g	8.10 ± 0.04 f	11.03 ± 0.09 e	12.54 ± 0.17 d
Leaf number	19.00 ± 0.56 c	20.50 ± 0.42 b	19.70 ± 0.28 c	22.20 ± 0.35 a	9.80 ± 0.49 g	10.20 ± 0.50 fg	15.40 ± 0.18 e	17.50 ± 0.44 d
Leaf area (cm^2/plant)	13.80 ± 0.35 c	14.02 ± 0.12 b	14.07 ± 0.15 b	17.08 ± 0.07 a	8.10 ± 0.14 g	8.95 ± 0.26 f	10.62 ± 0.31 e	12.50 ± 0.20 d

Values are mean followed by the standard deviation of five replicates; a–g: the same letters indicate no significant difference among treatment groups at $p < 0.05$.

In a study by Odjegba and Chukwunwike [63], *A. hybridus* was grown under salinity stress of 0.10 and 0.20 M NaCl. They observed that the salinity stress resulted in a significant ($p < 0.05$) decrease in plant weight. Similarly, Amukali et al. [64] also investigated the effect of 0–150 mM NaCl on germination and seedling growth of *A. hybridus*. They observed that germination was severely impacted by 150 mM NaCl dose while shoot length (7.97 cm), root length (4.18 cm), fresh weight (2.43 g), dry weight (0.49 g), and leaf area index (1.04) were significantly declined as compared to control treatment with no NaCl amendment. Similar outcomes were reported by Menezes et al. [65] when growing *A. cruentus* under saline stress (25 mM of NaCl). They reported that the major plant growth attributes and K^+/Na^+ ratio were significantly decreased in saline stress treatment compared to non-saline stress treatments. Therefore, it was observed that a combination of moss biochar mixing and foliar application of TiO_2 NPs showed a stimulatory impact on the growth and yield parameters of *A. dubius* in this study.

3.3. Effect of TiO_2 NPs and Biochar on Biochemical Response of A. dubius

Saline stress has deteriorating impacts on plant physiology and growth, which ultimately affects overall crop yield [66]. However, crop plants respond to saline stress by developing biochemical pathways and physiological adaptations that help them to survive high saline conditions. In this, several defense enzymes play a crucial role in plant survival. In this study, the saline stress caused by NaCl was significantly mitigated by the addition of moss biochar and TiO_2 NPs. Table 3 shows the effects of biochar and TiO_2-NP application on the biochemical and stress enzyme response of *A. dubius*. The findings showed that the chlorophyll content of *A. dubius* was significantly decreased in saline stress treatment (T4: 1.13 mg/g fwt.) compared to the control treatment (T0: 2.49 mg/g fwt.). A similar trend was observed for carotenoids (from 3.36 to 2.02 mg/g) and relative water content (from 87.01 to 72.50%). However, biochar and TiO_2-NP treatment resulted in a significant ($p < 0.05$) increase in chlorophyll, carotenoids, and relative water contents in both non-saline and saline treatments. Overall, the increasing order of these parameters was observed as: T4 < T5 < T6 < T7 < T0 < T1 < T2 < T3. On the other hand, the activities of selected plant enzymes varied significantly across the experimental treatments. SOD contents were highest in the T4 treatment, i.e., 4.05 U/mg P, which was reduced to 2.47 U/mg P in the T7 treatment. However, both CAT and SOD in non-saline stress treatments were found to be the lowest due to no specific stress. Specifically, CAT activities were highest (5.84 μmol/min/mg P) under T4 treatment, which was significantly ($p < 0.05$) reduced after the application of biochar and TiO_2 NPs. CAT acts as an important defense enzyme that helps in the breakdown of H_2O_2 molecules produced by external stresses [67]. Similarly, SOD helps in scavenging harmful superoxide radicles, which are produced by plants under saline stress [68]. Likewise, biochar and TiO_2-NP application also helped to reduce the contents of POD and APX in saline-stressed treatments, which indicates that the *A. dubius* plant could survive efficiently in these treatments. However, individual application of moss biochar and TiO_2 NPs did not show any significant ($p > 0.05$) improvement in biochemical parameters. Thus, in order to overcome the challenges of saline soils and increase *A. dubius* yields while ensuring environmental sustainability, the combined use of biochar and TiO_2 NPs offers a promising solution.

Several studies have confirmed that both biochar and NPs can ameliorate saline stress from agricultural soils [69–71]. Out of them, Farhangi-Abriz and Torabian [69] conducted laboratory experiments for French bean (*Phaseolus vulgaris* L. cv. Derakhshan) cultivation under saline-stressed soils. They found that the contents of chlorophyll were significantly reduced in saline treatments, which again increased in biochar-amended treatments. A report by Naz et al. [70] showed that foliar application of potassium mitigated saline stress conditions in *Spinacia oleracea* where activities of CAT and SOD enzymes were significantly ($p < 0.05$) affected. Moreover, González-García et al. [71] showed that foliar application of three types of nanoparticles (Se, Si, and Cu) showed a significant impact on SOD, CAT, POD, and APX activities of Bell pepper (*Capsicum* spp.) under saline stress. Therefore, as confirmed by the results of the current study, the combined use of biochar and TiO_2 NPs presents a promising and effective strategy for enhancing spinach cultivation in saline soils.

Table 3. Effect of TiO_2 NPs and biochar application on biochemical and enzyme response of *A. dubius*.

Parameters	Treatments							
	T0	T1	T2	T3	T4	T5	T6	T7
Chlorophyll content (mg/g)	2.49 ± 0.05 d	2.53 ± 0.03 c	2.94 ± 0.05 b	3.25 ± 0.07 a	1.13 ± 0.03 h	1.39 ± 0.04 g	1.87 ± 0.06 f	2.36 ± 0.05 e
Carotenoids (mg/g)	3.36 ± 0.07 a	3.42 ± 0.05 a	3.40 ± 0.06 a	3.49 ± 0.03 a	2.02 ± 0.03 e	2.23 ± 0.04 d	2.68 ± 0.03 c	2.85 ± 0.02 b
Relative water content (%)	87.01 ± 1.14 b	88.69 ± 0.64 b	89.05 ± 0.21 ab	91.40 ± 0.70 a	72.50 ± 1.63 e	73.14 ± 0.19 e	78.32 ± 0.24 d	82.10 ± 0.86 c
SOD (U/mg P)	2.20 ± 0.03 e	2.16 ± 0.02 ef	2.12 ± 0.03 f	2.04 ± 0.01 g	4.05 ± 0.04 a	3.50 ± 0.07 b	2.98 ± 0.03 c	2.47 ± 0.05 d
POD (µmol/min/mg P)	38.29 ± 0.58 e	37.65 ± 0.19 f	37.08 ± 0.35 fg	36.59 ± 0.52 g	58.72 ± 1.29 a	51.84 ± 0.91 b	48.01 ± 1.46 c	40.03 ± 2.49 d
CAT (µmol/min/mg P)	1.90 ± 0.03 e	1.82 ± 0.02 f	1.84 ± 0.04 ef	1.85 ± 0.02 ef	5.84 ± 0.07 a	5.10 ± 0.04 b	4.26 ± 0.10 c	2.93 ± 0.12 d
APX (µmol/min/mg P)	2.71 ± 0.05 e	2.66 ± 0.07 ef	2.64 ± 0.03 f	2.56 ± 0.04 g	7.45 ± 0.10 a	6.57 ± 0.09 b	4.74 ± 0.04 c	3.10 ± 0.05 d

Values are mean followed by the standard deviation of five replicates; a–g: the same letters indicate no significant difference among treatment groups at $p < 0.05$.

4. Conclusions

The findings of the present study indicated that moss biomass could act as a promising resource to produce biochar, while its extract can be used for the biosynthesis of TiO_2 NPs. The produced biochar and TiO_2 NPs were characterized for their suitability in agricultural use for *A. dubius* cultivation. The results showed a significant ($p < 0.05$) impact of the combined use of biochar and TiO_2 NPs on the growth, yield, biochemical, and stress enzyme response of *A. dubius* while alleviating saline stress. The combination of these amendments exhibited a synergistic effect, as evidenced by the significantly higher yields compared to the control group. This study suggests that moss could act as a biomass for the development of low-cost fertilizers and nanoparticles, which could benefit sustainable agriculture production. Further studies on optimization and economic feasibility of biochar and TiO_2 NP dose with unveiling the underlying mechanisms of plant, biochar, and TiO_2 NP interactions are highly recommended. Moreover, evaluating the field efficacy of the proposed method under different agro-climatic conditions and other cultivars is suggested for future works.

Author Contributions: Conceptualization, I.Š., S.A.F., P.K. (Pankaj Kumar) and E.M.E.; data curation, B.M., V.D., S.A.F., J.S. and R.S.; formal analysis, J.S., P.K. (Piyush Kumar), R.K.B. and M.G.; funding acquisition, S.K.A., L.A.A.-S. and E.M.E.; investigation, J.S., P.K. (Piyush Kumar), M.G. and P.K. (Pankaj Kumar); methodology, I.Š., S.K.A., L.A.A.-S., V.D., V.K., P.K. (Pankaj Kumar) and E.M.E.; project administration, S.K.A. and L.A.A.-S.; resources, B.M., V.K. and P.K. (Pankaj Kumar); software, S.K.A., L.A.A.-S., B.M., V.D., R.S. and R.K.B.; supervision, E.M.E.; validation, I.Š., B.M., V.K., P.K. (Piyush Kumar), R.S., R.K.B. and M.G.; visualization, I.Š., B.M., V.D., S.A.F., R.S. and R.K.B.; writing—original draft, S.A.F. and P.K. (Pankaj Kumar); writing—review and editing, I.Š., S.K.A., L.A.A.-S., B.M., V.D., V.K., J.S., P.K. (Piyush Kumar), R.S., R.K.B., M.G. and E.M.E. All authors have read and agreed to the published version of the manuscript.

Funding: This research was funded by Princess Nourah bint Abdulrahman University Researchers Supporting Project number (PNURSP2023R365), Princess Nourah bint Abdulrahman University, Riyadh, Saudi Arabia. This research was funded by the Deanship of Scientific Research at King Khalid University through the Large Groups Project under grant number (R.G.P.2/77/44).

Institutional Review Board Statement: Not applicable.

Informed Consent Statement: Not applicable.

Data Availability Statement: Not applicable.

Acknowledgments: The authors extend their appreciation to Princess Nourah bint Abdulrahman University Researchers Supporting Project number (PNURSP2023R365), Princess Nourah bint Abdulrahman University, Riyadh, Saudi Arabia. The authors extend their appreciation to the Deanship of Scientific Research at King Khalid University for funding this work through the Large Groups Project under grant number (R.G.P.2/77/44). All individuals included in this section have consented to the acknowledgement.

Conflicts of Interest: The authors declare no conflict of interest.

References

1. Grande, F.; Tucci, P. Titanium Dioxide Nanoparticles: A Risk for Human Health? *Mini-Rev. Med. Chem.* **2016**, *16*, 762–769. [CrossRef]
2. Shi, H.; Magaye, R.; Castranova, V.; Zhao, J. Titanium Dioxide Nanoparticles: A Review of Current Toxicological Data. *Part. Fibre Toxicol.* **2013**, *10*, 15. [CrossRef] [PubMed]
3. Gupta, N.; Rai, D.B.; Jangid, A.K.; Kulhari, H. Use of Nanotechnology in Antimicrobial Therapy. In *Methods in Microbiology*; Elsevier: Amsterdam, The Netherlands, 2019; Volume 46, pp. 143–172. ISBN 9780128149928.
4. Aravind, M.; Amalanathan, M.; Mary, M.S.M. Synthesis of TiO_2 Nanoparticles by Chemical and Green Synthesis Methods and Their Multifaceted Properties. *SN Appl. Sci.* **2021**, *3*, 409. [CrossRef]
5. Gonfa, Y.H.; Tessema, F.B.; Bachheti, A.; Tadesse, M.G.; Eid, E.M.; Abou Fayssal, S.; Adelodun, B.; Choi, K.S.; Širić, I.; Kumar, P.; et al. Essential Oil Composition of Aerial Part of *Pluchea ovalis* (Pers.) DC., Silver Nanoparticles Synthesis, and Larvicidal Activities against Fall Armyworm. *Sustainability* **2022**, *14*, 15785. [CrossRef]
6. Verma, V.; Al-Dossari, M.; Singh, J.; Rawat, M.; Kordy, M.G.M.; Shaban, M. A Review on Green Synthesis of TiO_2 NPs: Photocatalysis and Antimicrobial Applications. *Polymers* **2022**, *14*, 1444. [CrossRef] [PubMed]
7. Sett, A.; Gadewar, M.; Sharma, P.; Deka, M.; Bora, U. Green Synthesis of Gold Nanoparticles Using Aqueous Extract of *Dillenia indica*. *Adv. Nat. Sci. Nanosci. Nanotechnol.* **2016**, *7*, 025005. [CrossRef]
8. Srujana, S.; Anjamma, M.; Alimuddin; Singh, B.; Dhakar, R.C.; Natarajan, S.; Hechhu, R. A Comprehensive Study on the Synthesis and Characterization of TiO_2 Nanoparticles Using *Aloe vera* Plant Extract and Their Photocatalytic Activity against MB Dye. *Adsorpt. Sci. Technol.* **2022**, *2022*, 7244006. [CrossRef]
9. Roopan, S.M.; Bharathi, A.; Prabhakarn, A.; Abdul Rahuman, A.; Velayutham, K.; Rajakumar, G.; Padmaja, R.D.; Lekshmi, M.; Madhumitha, G. Efficient Phyto-Synthesis and Structural Characterization of Rutile TiO_2 Nanoparticles Using *Annona squamosa* Peel Extract. *Spectrochim. Acta A Mol. Biomol. Spectrosc.* **2012**, *98*, 86–90. [CrossRef]
10. Velayutham, K.; Rahuman, A.A.; Rajakumar, G.; Santhoshkumar, T.; Marimuthu, S.; Jayaseelan, C.; Bagavan, A.; Kirthi, A.V.; Kamaraj, C.; Zahir, A.A.; et al. Evaluation of Catharanthus Roseus Leaf Extract-Mediated Biosynthesis of Titanium Dioxide Nanoparticles against *Hippobosca maculata* and *Bovicola ovis*. *Parasitol. Res.* **2012**, *111*, 2329–2337. [CrossRef]
11. Vimala, A.; Sahaya Sathish, S.; Thamizharasi, T.; Palani, R.; Vijayakanth, P.; Kavitha, R. Moss (Bryophyte) Mediated Synthesis and Characterization of Silver Nanoparticles from *Campylopus flexuosus* (Hedw.) Bird. *J. Pharm. Sci. Res.* **2017**, *9*, 292–297.
12. Akhatova, F.; Konnova, S.; Kryuchkova, M.; Batasheva, S.; Mazurova, K.; Vikulina, A.; Volodkin, D.; Rozhina, E. Comparative Characterization of Iron and Silver Nanoparticles: Extract-Stabilized and Classical Synthesis Methods. *Int. J. Mol. Sci.* **2023**, *24*, 9274. [CrossRef]
13. Hernandez-Soriano, M.C.; Kerré, B.; Kopittke, P.M.; Horemans, B.; Smolders, E. Biochar Affects Carbon Composition and Stability in Soil: A Combined Spectroscopy-Microscopy Study. *Sci. Rep.* **2016**, *6*, 25127. [CrossRef] [PubMed]
14. Širić, I.; Eid, E.M.; Taher, M.A.; El-Morsy, M.H.E.; Osman, H.E.M.; Kumar, P.; Adelodun, B.; Abou Fayssal, S.; Mioč, B.; Andabaka, Ž.; et al. Combined Use of Spent Mushroom Substrate Biochar and PGPR Improves Growth, Yield, and Biochemical Response of Cauliflower (*Brassica oleracea* var. *botrytis*): *A Preliminary Study on Greenhouse Cultivation. Horticulturae* **2022**, *8*, 830. [CrossRef]
15. Tomczyk, A.; Sokołowska, Z.; Boguta, P. Biochar Physicochemical Properties: Pyrolysis Temperature and Feedstock Kind Effects. *Rev. Environ. Sci. Biotechnol.* **2020**, *19*, 191–215. [CrossRef]
16. Shaaban, A.; Se, S.-M.; Dimin, M.F.; Juoi, J.M.; Mohd Husin, M.H.; Mitan, N.M.M. Influence of Heating Temperature and Holding Time on Biochars Derived from Rubber Wood Sawdust via Slow Pyrolysis. *J. Anal. Appl. Pyrolysis* **2014**, *107*, 31–39. [CrossRef]
17. Zhang, J.; Liu, J.; Liu, R. Effects of Pyrolysis Temperature and Heating Time on Biochar Obtained from the Pyrolysis of Straw and Lignosulfonate. *Bioresour. Technol.* **2015**, *176*, 288–291. [CrossRef]
18. Maximize Market Research Amaranth Market: Global Industry Analysis and Forecast (2022–2029). Available online: https://www.maximizemarketresearch.com/market-report/global-amaranth-market/81299/ (accessed on 13 August 2023).
19. Raiger, H.L.; Phogat, B.S.; Dua, R.P.; Sharma, S.K. *Improved Varieties and Cultivation Practices of Grain Amaranth*; Indian Council of Agricultural Research, Krishi Bhavan: New Delhi, India, 2009.
20. Sarker, U.; Islam, M.T.; Oba, S. Salinity Stress Accelerates Nutrients, Dietary Fiber, Minerals, Phytochemicals and Antioxidant Activity in *Amaranthus tricolor* Leaves. *PLoS ONE* **2018**, *13*, e0206388. [CrossRef]
21. Sarker, U.; Oba, S. Protein, Dietary Fiber, Minerals, Antioxidant Pigments and Phytochemicals, and Antioxidant Activity in Selected Red Morph *Amaranthus* Leafy Vegetable. *PLoS ONE* **2019**, *14*, e0222517. [CrossRef]
22. Sarker, U.; Oba, S. Salinity Stress Enhances Color Parameters, Bioactive Leaf Pigments, Vitamins, Polyphenols, Flavonoids and Antioxidant Activity in Selected *Amaranthus* Leafy Vegetables. *J. Sci. Food Agric.* **2019**, *99*, 2275–2284. [CrossRef]
23. Estrada, Y.; Fernández-Ojeda, A.; Morales, B.; Egea-Fernández, J.M.; Flores, F.B.; Bolarín, M.C.; Egea, I. Unraveling the Strategies Used by the Underexploited Amaranth Species to Confront Salt Stress: Similarities and Differences with *Quinoa* Species. *Front. Plant Sci.* **2021**, *12*, 604481. [CrossRef]

24. Bellache, M.; Allal Benfekih, L.; Torres-Pagan, N.; Mir, R.; Verdeguer, M.; Vicente, O.; Boscaiu, M. Effects of Four-Week Exposure to Salt Treatments on Germination and Growth of Two *Amaranthus* Species. *Soil. Syst.* **2022**, *6*, 57. [CrossRef]
25. Hoang, L.H.; De Guzman, C.C.; Cadiz, N.M.; Tran, D.H. Physiological and Phytochemical Responses of Red Amaranth (*Amaranthus tricolor* L.) and Green Amaranth (*Amaranthus dubius* L.) to Different Salinity Levels. *Legume Res. An. Int. J.* **2019**, *43*, 206–211. [CrossRef]
26. Sardar, H.; Khalid, Z.; Ahsan, M.; Naz, S.; Nawaz, A.; Ahmad, R.; Razzaq, K.; Wabaidur, S.M.; Jacquard, C.; Širić, I.; et al. Enhancement of Salinity Stress Tolerance in Lettuce (*Lactuca sativa* L.) via Foliar Application of Nitric Oxide. *Plants* **2023**, *12*, 1115. [CrossRef] [PubMed]
27. Gour, T.; Sharma, A.; Lal, R.; Heikrujam, M.; Gupta, A.; Agarwal, L.K.; Chetri, S.P.K.; Kumar, R.; Sharma, K. Amelioration of the Physio-Biochemical Responses to Salinity Stress and Computing the Primary Germination Index Components in Cauliflower on Seed Priming. *SSRN Electron. J.* **2022**, *9*, e14403. [CrossRef]
28. El-Shal, R.M.; El-Naggar, A.H.; El-Beshbeshy, T.R.; Mahmoud, E.K.; El-Kader, N.I.A.; Missaui, A.M.; Du, D.; Ghoneim, A.M.; El-Sharkawy, M.S. Effect of Nano-Fertilizers on Alfalfa Plants Grown under Different Salt Stresses in Hydroponic System. *Agriculture* **2022**, *12*, 1113. [CrossRef]
29. Zhang, S. Mechanism of Migration and Transformation of Nano Selenium and Mitigates Cadmium Stress in Plants. Master's Thesis, Shandong University, Jinan, China, 2019.
30. Francis, D.V.; Sood, N.; Gokhale, T. Biogenic CuO and ZnO Nanoparticles as Nanofertilizers for Sustainable Growth of *Amaranthus hybridus*. *Plants* **2022**, *11*, 2776. [CrossRef]
31. Cai, Y.; Yuan, B.; Ma, X.; Fang, G.; Zhou, D.; Gao, J. Foliar Application of SiO_2 and ZnO Nanoparticles Affected Polycyclic Aromatic Hydrocarbons Uptake of Amaranth (*Amaranthus tricolor* L.): A Metabolomics and Typical Statistical Analysis. *SSRN Electron. J.* **2022**, *833*, 155258. [CrossRef]
32. Situmeang, Y.P.; Suarta, M.; Irianto, I.K.; Andriani, A.A.S.P.R. Biochar Bamboo Application on Growth and Yield of Red Amaranth (*Amaranthus tricolor* L). *IOP Conf. Ser. Mater. Sci. Eng.* **2018**, *434*, 012231. [CrossRef]
33. Jiang, S.; Dai, G.; Zhou, J.; Zhong, J.; Liu, J.; Shu, Y. An Assessment of Integrated Amendments of Biochar and Soil Replacement on the Phytotoxicity of Metal(Loid)s in Rotated Radish-Soya Bean-Amaranth in a Mining Acidy Soil. *Chemosphere* **2022**, *287*, 132082. [CrossRef]
34. Daryabeigi Zand, A.; Tabrizi, A.M.; Heir, A.V. Co-Application of Biochar and Titanium Dioxide Nanoparticles to Promote Remediation of Antimony from Soil by *Sorghum bicolor*: Metal Uptake and Plant Response. *Heliyon* **2020**, *6*, e04669. [CrossRef]
35. González, L.; González-Vilar, M. Determination of Relative Water Content. In *Handbook of Plant Ecophysiology Techniques*; Kluwer Academic Publishers: Dordrecht, The Netherlands, 2006; pp. 207–212.
36. Uma, S.; Karthic, R.; Kalpana, S.; Backiyarani, S. A Comparative Assessment of Photosynthetic Pigments and Defense Enzymes in Ex Vitro and In Vitro Propagated Plants of Banana (*Musa* spp.). *Biocatal. Agric. Biotechnol.* **2023**, *51*, 102799. [CrossRef]
37. Upadhyay, P.; Tewari, A.K.; Punetha, H. Biochemical Estimation of Polyphenol and Peroxidases in Resistant Sources against White Rust Disease of Rapeseed Mustard In-Cited by *Albugo candida*. *Pharma Innov. J.* **2023**, *12*, 4964–4974.
38. Toor, S.S.; Reddy, H.; Deng, S.; Hoffmann, J.; Spangsmark, D.; Madsen, L.B.; Holm-Nielsen, J.B.; Rosendahl, L.A. Hydrothermal Liquefaction of *Spirulina* and *Nannochloropsis salina* under Subcritical and Supercritical Water Conditions. *Bioresour. Technol.* **2013**, *131*, 413–419. [CrossRef] [PubMed]
39. Alpert, P. The Limits and Frontiers of Desiccation-Tolerant Life. *Integr. Comp. Biol.* **2005**, *45*, 685–695. [CrossRef]
40. Hoekstra, F.A.; Golovina, E.A.; Buitink, J. Mechanisms of Plant Desiccation. *Trends Plant Sci.* **2001**, *6*, 431–438. [CrossRef]
41. Ihl, C.; Barboza, P.S. Nutritional Value of Moss for Arctic Ruminants: A Test with Muskoxen. *J. Wildl. Manag.* **2007**, *71*, 752–758. [CrossRef]
42. Arnon, D.I. Copper Enzymes in Isolated Chloroplasts. Polyphenoloxidase in *Beta vulgaris*. *Plant Physiol.* **1949**, *24*, 1–15. [CrossRef]
43. Orlov, D.S.; Sadovnikova, L.K. Soil Organic Matter and Protective Functions of Humic Substances in the Bioshere. In *Use of Humic Substances to Remediate Polluted Environments: From Theory to Practice*; Springer: Dordrecht, The Netherlands, 2005; ISBN 978-1-4020-3252-3.
44. Ryde, I.; Davie-Martin, C.L.; Li, T.; Naursgaard, M.P.; Rinnan, R. Volatile Organic Compound Emissions from Subarctic Mosses and Lichens. *Atmos. Environ.* **2022**, *290*, 119357. [CrossRef]
45. McCuaig, B.; Dufour, S.C.; Raguso, R.A.; Bhatt, A.P.; Marino, P. Structural Changes in Plastids of Developing *Splachnum ampullaceum* Sporophytes and Relationship to Odour Production. *Plant Biol.* **2015**, *17*, 466–473. [CrossRef]
46. Resemann, H.C.; Lewandowska, M.; Gï¿½mann, J.; Feussner, I. Membrane Lipids, Waxes and Oxylipins in the Moss Model Organism *Physcomitrella patens*. *Plant Cell Physiol.* **2019**, *60*, 1166–1175. [CrossRef]
47. Zaccone, C.; Miano, T.M.; Shotyk, W. Qualitative Comparison between Raw Peat and Related Humic Acids in an Ombrotrophic Bog Profile. *Org. Geochem.* **2007**, *38*, 151–160. [CrossRef]
48. Klavina, L. Composition of Mosses, Their Metabolites and Environmental Stress Impacts. Ph.D. Thesis, University of Latvia, Riga, Latvia, 2018.
49. Serk, H.; Nilsson, M.B.; Bohlin, E.; Ehlers, I.; Wieloch, T.; Olid, C.; Grover, S.; Kalbitz, K.; Limpens, J.; Moore, T.; et al. Global CO_2 Fertilization of Sphagnum Peat Mosses via Suppression of Photorespiration during the Twentieth Century. *Sci. Rep.* **2021**, *11*, 24517. [CrossRef] [PubMed]

50. Ramdath, S.; Mellem, J.; Mbatha, L.S. Anticancer and Antimicrobial Activity Evaluation of Cowpea-Porous-Starch-Formulated Silver Nanoparticles. *J. Nanotechnol.* **2021**, *2021*, 5525690. [CrossRef]
51. Larue, C.; Castillo-Michel, H.; Sobanska, S.; Trcera, N.; Sorieul, S.; Cécillon, L.; Ouerdane, L.; Legros, S.; Sarret, G. Fate of Pristine TiO_2 Nanoparticles and Aged Paint-Containing TiO_2 Nanoparticles in Lettuce Crop after Foliar Exposure. *J. Hazard. Mater.* **2014**, *273*, 17–26. [CrossRef] [PubMed]
52. Palocci, C.; Valletta, A.; Chronopoulou, L.; Donati, L.; Bramosanti, M.; Brasili, E.; Baldan, B.; Pasqua, G. Endocytic Pathways Involved in PLGA Nanoparticle Uptake by Grapevine Cells and Role of Cell Wall and Membrane in Size Selection. *Plant Cell Rep.* **2017**, *36*, 1917–1928. [CrossRef]
53. Cai, L.; Cai, L.; Jia, H.; Liu, C.; Wang, D.; Sun, X. Foliar Exposure of Fe_3O_4 Nanoparticles on *Nicotiana benthamiana*: Evidence for Nanoparticles Uptake, Plant Growth Promoter and Defense Response Elicitor against Plant Virus. *J. Hazard. Mater.* **2020**, *393*, 122415. [CrossRef] [PubMed]
54. Movasaghi, Z.; Rehman, S.; Ur Rehman, I. Fourier Transform Infrared (FTIR) Spectroscopy of Biological Tissues. *Appl. Spectrosc. Rev.* **2008**, *43*, 134–179. [CrossRef]
55. El-Desoky, M.M.; Morad, I.; Wasfy, M.H.; Mansour, A.F. Synthesis, Structural and Electrical Properties of PVA/TiO_2 Nanocomposite Films with Different TiO_2 Phases Prepared by Sol–Gel Technique. *J. Mater. Sci. Mater. Electron.* **2020**, *31*, 17574–17584. [CrossRef]
56. Usgodaarachchi, L.; Thambiliyagodage, C.; Wijesekera, R.; Vigneswaran, S.; Kandanapitiye, M. Fabrication of TiO_2 Spheres and a Visible Light Active α-Fe_2O_3/TiO_2 -Rutile/TiO_2 -Anatase Heterogeneous Photocatalyst from Natural Ilmenite. *ACS Omega* **2022**, *7*, 27617–27637. [CrossRef]
57. Sun, Z.; Dai, L.; Lai, P.; Shen, F.; Shen, F.; Zhu, W. Air Oxidation in Surface Engineering of Biochar-Based Materials: A Critical Review. *Carbon Res.* **2022**, *1*, 32. [CrossRef]
58. Bakshi, S.; Banik, C.; Laird, D.A. Estimating the Organic Oxygen Content of Biochar. *Sci. Rep.* **2020**, *10*, 13082. [CrossRef] [PubMed]
59. Lehmann, J.; Rillig, M.C.; Thies, J.; Masiello, C.A.; Hockaday, W.C.; Crowley, D. Biochar Effects on Soil Biota—A Review. *Soil. Biol. Biochem.* **2011**, *43*, 1812–1836.
60. Tomar, R.S.; Kataria, S.; Jajoo, A. Behind the Scene: Critical Role of Reactive Oxygen Species and Reactive Nitrogen Species in Salt Stress Tolerance. *J. Agron. Crop Sci.* **2021**, *207*, 577–588. [CrossRef]
61. Shaheen, S.M.; Mosa, A.; Natasha; Arockiam Jeyasundar, P.G.S.; Hassan, N.E.E.; Yang, X.; Antoniadis, V.; Li, R.; Wang, J.; Zhang, T.; et al. Pros and Cons of Biochar to Soil Potentially Toxic Element Mobilization and Phytoavailability: Environmental Implications. *Earth Syst. Environ.* **2023**, *7*, 321–345. [CrossRef]
62. Gohari, G.; Mohammadi, A.; Akbari, A.; Panahirad, S.; Dadpour, M.R.; Fotopoulos, V.; Kimura, S. Titanium Dioxide Nanoparticles (TiO_2 NPs) Promote Growth and Ameliorate Salinity Stress Effects on Essential Oil Profile and Biochemical Attributes of *Dracocephalum moldavica*. *Sci. Rep.* **2020**, *10*, 912. [CrossRef]
63. Odjegba, V.J.; Chukwunwike, I.C. Physiological Responses of *Amaranthus hybridus* L. Under Salinity Stress. *Niger. J. Life Sci.* **2022**, *5*, 242–252. [CrossRef]
64. Amukali, O.; Obadoni, B.O.; Mensah, J.K. Effects of Different NaCl Concentrations on Germination and Seedling Growth of *Amaranthus hybridus* and *Celosia argentea*. *Afr. J. Environ. Sci. Tech.* **2015**, *9*, 301–306. [CrossRef]
65. Menezes, R.V.; De Azevedo Neto, A.D.; Ribeiro, M.d.O.; Cova, A.M.W. Growth and Contents of Organic and Inorganic Solutes in Amaranth under Salt Stress. *Pesqui. Agropecu. Trop.* **2017**, *47*, 22–30. [CrossRef]
66. Ferreira, J.; Sandhu, D.; Liu, X.; Halvorson, J. Spinach (*Spinacea oleracea* L.) Response to Salinity: Nutritional Value, Physiological Parameters, Antioxidant Capacity, and Gene Expression. *Agriculture* **2018**, *8*, 163. [CrossRef]
67. Seymen, M. Comparative Analysis of the Relationship between Morphological, Physiological, and Biochemical Properties in Spinach (*Spinacea oleracea* L.) under Deficit Irrigation Conditions. *Turk. J. Agric. For.* **2021**, *45*, 55–67. [CrossRef]
68. Das, K.; Roychoudhury, A. Reactive Oxygen Species (ROS) and Response of Antioxidants as ROS-Scavengers during Environmental Stress in Plants. *Front. Environ. Sci.* **2014**, *2*, 53. [CrossRef]
69. Farhangi-Abriz, S.; Torabian, S. Effect of Biochar on Growth and Ion Contents of Bean Plant under Saline Condition. *Environ. Sci. Pollut. Res.* **2018**, *25*, 11556–11564. [CrossRef] [PubMed]
70. Naz, T.; Mazhar Iqbal, M.; Tahir, M.; Hassan, M.M.; Rehmani, M.I.A.; Zafar, M.I.; Ghafoor, U.; Qazi, M.A.; EL Sabagh, A.; Sakran, M.I. Foliar Application of Potassium Mitigates Salinity Stress Conditions in Spinach (*Spinacia oleracea* L.) through Reducing NaCl Toxicity and Enhancing the Activity of Antioxidant Enzymes. *Horticulturae* **2021**, *7*, 566. [CrossRef]
71. González-García, Y.; Cárdenas-Álvarez, C.; Cadenas-Pliego, G.; Benavides-Mendoza, A.; Cabrera-de-la-Fuente, M.; Sandoval-Rangel, A.; Valdés-Reyna, J.; Juárez-Maldonado, A. Effect of Three Nanoparticles (Se, Si and Cu) on the Bioactive Compounds of Bell Pepper Fruits under Saline Stress. *Plants* **2021**, *10*, 217. [CrossRef] [PubMed]

horticulturae

MDPI

Article

Maternal Environment and Priming Agents Effect Germination and Seedling Quality in Pitaya under Salt Stress

Burcu Begüm Kenanoğlu [1], Kerem Mertoğlu [1,*], Melekber Sülüşoğlu Durul [2], Nazan Korkmaz [3] and Ayşen Melda Çolak [1]

1 Department of Horticulture, Usak University, Uşak 64200, Türkiye; burcu.kenanoglu@usak.edu.tr (B.B.K.); aysenmelda.colak@usak.edu.tr (A.M.Ç.)
2 Department of Horticulture, Kocaeli University, Kocaeli 54000, Türkiye; meleksl@kocaeli.edu.tr
3 Department of Plant and Animal Production, Muğla Sıtkı Koçman University, Muğla 48600, Türkiye; nkorkmaz@mu.edu.tr
* Correspondence: krmertoglum@gmail.com; Tel.: +90-533-2725691

Abstract: Lack of water and salinity are common problems in many parts of the world. Therefore, some types of cacti can present as promising crops. Therefore, the ability of cactus species to survive and adapt under natural stress conditions should be evaluated. The experiment was aimed at evaluating the effect of salt stress on germination and emergence of pitaya seeds obtained from different species (*Hyloceresu undatus* (Haw.) Britton and Rose and *Hylocereus polyrhizus* (Lem.) Britton and Rose), priming with plant growth regulators, namely salicylic acid (SA), oxalic acid (OA) and mepiquat chloride (MC). The experiment had a completely randomized design with a $2 \times 4 \times 3$ factorial scheme corresponding to two pitaya cultivars (white- and red-fleshed), four NaCl concentrations (0, 2500, 5000 and 10,000 ppm), and three PGRs (150 ppm/MC, SA, OC). According to the results, the maternal environment of the seed was important in salt stress resistance, while seeds matured in the environment with red fruit flesh were more tolerant to salt stress. Although Pitaya species are relatively salt-tolerant, growth (about 30%) was significantly reduced above 2500 ppm and germination (about 45%) above 5000 ppm. Germination percentage stood out as the most important trait determining seed quality and had positive effects on the germination stress tolerance index (r: 0.63), seedling length (r: 0.74) and fresh seedling weight (r: 0.56). This is the first study of how maternal environment affects germination and seedling quality under saline conditions in *Hylocereus*. The results obtained may contribute to pitaya cultivation and breeding.

Keywords: Cactaceae; seed viability; seedling quality; salinity; PGRs; *Hylocereus* sp.

Citation: Kenanoğlu, B.B.; Mertoğlu, K.; Sülüşoğlu Durul, M.; Korkmaz, N.; Çolak, A.M. Maternal Environment and Priming Agents Effect Germination and Seedling Quality in Pitaya under Salt Stress. *Horticulturae* **2023**, 9, 1170. https://doi.org/10.3390/horticulturae9111170

Academic Editors: Adalberto Benavides-Mendoza, Yolanda González-García, Fabián Pérez Labrada and Susana González-Morales

Received: 24 August 2023
Revised: 20 October 2023
Accepted: 23 October 2023
Published: 26 October 2023

1. Introduction

Pitaya, known as dragon fruit, is in the genus *Hylocereus*, which is represented by four species according to their peel and flesh color, namely: *H. undatus* (Haw.) Britton and Rose, *H. polyrhizus* (Syn, *H. monocanthus*), *H. costariscensis* (FAC Weber) Britton and Rose and *H. megalanthus* (K. Schum. ex Vaupel) Ralf Bauer (Syn. *Selenicereus megalanthus*) [1]. Hylocereus is a former genus of epiphytic cacti, commonly known as night-blooming cactus (though the word is also applied to many other cacti). Several species previously assigned to the genus produce big edible fruits known as pitayas, pitahayas, or dragon fruits. Fruits of these species are popularly consumed due to their aroma. In addition, in the current pandemic, the consumer's interest in pitaya fruit, which is rich in chemicals with high antioxidant effects such as phenols, organic acids, minerals and vitamins, has increased [1–3]. This antioxidative effect has been proven to prevent many diseases [4,5]. The fruit peel of pitaya, which contains more than 50% unsaturated fatty acids (linoleic and linolenic) in their seeds [6], is a source of pectin and is very important in food and textile dyeing with the natural colorants it contains [7]. The demands of pitaya, which has the potential for many different uses for nutraceutical and industrial purposes, has increased, which is reflected in its production. As a matter of fact, all countries where pitaya is cultivated emphasize that the amount of production has increased compared to the previous years [8].

Sustainable and economic continuation of this high demand is only possible if breeders develop genotypes that meet the demands of producers, consumers and industry. For this reason, recently, efforts have been accelerated to develop early, middle and late varieties with high yield and attractiveness, resistant to biotic and abiotic stress conditions in combination, self-fertile, rich in chemicals with high antioxidant effects, suitable for different cultivation systems (open/greenhouse) [9]. Due to the quantitative nature of the desired traits and the complexity of inheritance mechanisms, hybrid breeding is still the most widely used method [10]. However, information on the factors affecting the germination of cactus seeds from the genus *Hylocereus* is very limited [11].

Germination depends on the chemical composition and testa permeability of the seed and favorable environmental conditions [12]. Salinity is a major environmental problem worldwide, especially in arid and semi-arid regions, limiting or making cultivation economically unfeasible. With global climate change, saline areas are expected to increase due to irrigation irregularity and high evaporation [13]. It has been pointed out that cactus species such as pitaya, which are particularly adaptable to different soil conditions, may be an alternative cultivation option for utilizing these areas [14,15]. However, research on how pitaya species/cultivars/genotypes perform under these stress factors is scarce. Pitaya seeds germinated under saline conditions reduced germination rate and seedling length up to 33% and 38%, respectively, according to cultivars [16]. It was emphasized that not only the shoot but also the root development regressed in all investigated hybrid pitaya genotypes under salt concentrations [17]. Also, the germination speed index decreased as the salt stress increased regardless of salt source, such as NaCl, $MgCl_2$ and KCl [17,18]. But no reports were found in the literature on the effects of the maternal environment on germination and seedling quality.

By acting on the osmotic potential of the substrate, salts reduce the potential gradient between the substrate and the seed surface, limiting water uptake by the seed and adversely affecting the plants developing from these seeds. They disrupt the ion balance and negatively affect stomatal movement and photosynthesis [19]. Enzyme activity is impaired, and metabolic activity is restricted. Increased reactive oxygen species disrupt cell integrity and homeostasis [20–22]. For these reasons, it is not possible to maintain production sustainably. Although salt stress occurs in all plants, tolerance levels and reduction in growth and development rates differ among species and phenological stages [23]. Studies on the germination response of seeds exposed to artificial stress conditions are important ecophysiological tools that can be used to understand the ability of species to survive and adapt to natural stress conditions. These tools are also used to assess the sensitivity of these species and their ability to survive and adapt to suppression when exposed to unfavorable and novel environments [24]. Therefore, although salinity has been extensively analyzed in cultivated species, the effects of NaCl in cacti are poorly documented.

The positive effects of exogenously applied growth regulators on sustainability have been demonstrated in studies with different species to increase plant tolerance to various abiotic stress factors [25]. Salicylic acid (SA), oxalic acid (OA) or mepiquat chloride (MC) applied to plants or seeds regulate stomatal movements and contribute to ion uptake and transport as important signaling molecules [26–28]. Through morphological, physiological, and biochemical pathways, salicylic acid (SA), a plant hormone, plays a key role in inducing plant defense against a variety of biotic and abiotic stressors. Moreover, SA stimulated cell division in seedlings and roots, resulting in greater plant growth [29]. Mepiquat chloride is a synthetic inhibitor of endogenous hormone, whereas chlormequat chloride is a chemical inhibitor of gibberellin. It is irreversible, unlike gibberellin, which can be applied to plants for increased fertility after over use of MC, which is a key distinction between the two plant growth hormones [30]. OA is a natural organic acid with numerous physiological roles, the most important of which is the promotion of systemic resistance to fungal and viral infections via an increase in antioxidant enzymatic activities and phenolic compounds [31]. It was recently designated as a generally recognized safe (GRAS) substance. This molecule's antioxidant potential has been proven, and its role as a natural antioxidant compound in plant systems has been suggested [32]. It reduces the negative effects of reactive oxygen species (ROS) by protecting the structure and function of the cell membrane and increasing the synthesis of antioxidant-derived compounds. This allows the maintenance of osmolytes and homeostasis [33]. Plant growth regulators promote germination and development by maintaining photosynthetic activity [34–36], regulate the expression of genes involved in synthesizing vital amino acids under stress [37], restrict the development of different pathogens and have inducing effects on the acquisition of resistance [38–41].

This study evaluated the mitigating effect of some plant growth regulators on the performance of different pitaya species under different salt stress conditions during seed germination and seedling emergence periods under controlled laboratory conditions. Also, this is the first study conducted on how the maternal environment affects germination and seedling quality under saline conditions in *Hylocereus*.

2. Materials and Methods

2.1. Plant Materials

In the study, two different species, namely *H. undatus* (*Hu*) and *H. polyrhizus* (*Hp*) of the *Hylocereus* genus, were used as material. The plants were grown in Muğla province, which is located in the southwest of Turkey and has a subtropical climate. Ripe fruits of Siam Red (*Hp*, red peel color and red-fleshed) and Vietnam White (*Hu*, red peel color and white-fleshed) pitaya varieties. Seeds were obtained from fresh fruits of similar size that fully ripened on the plants. The fruits were peeled, and the seeds were separated from the pulp manually and washed with tap water several times to remove mucilage and the left-over pulp [16]. Seeds were treated with 2% sodium hypochlorite for 3 min for surface sterilization.

2.2. Preparing Solutions and Performing Seed Priming

The dried seeds were separated according to the treatment groups, weighed and placed in Petri dishes. Salicylic acid, mepiquat chloride or an oxalic acid solution with a concentration of 150 ppm (application concentration was determined by preliminary studies) was added to the Petri dishes at five times the seed weight [42]. The Petri dishes were wrapped with aluminum foil and kept in a growth chamber at 25 °C for 24 h.

2.3. Experimental Layout

Following the priming application, the seeds were counted and grouped. The study, carried out as a laboratory study, was designed with three replicates, and each replicate consisted of 50 seeds. The seeds of each replicate were homogeneously distributed in 3 layers of Whatman filter paper, 2 on the bottom and 1 on top of the seeds, and 7 mL of salt solution (2500 ppm, 5000 ppm, 10,000 ppm) or pure water (control) was applied to each filter paper (21 mL for each replicate at the start). The homogeneously soaked filter papers were folded into zip-lock bags to minimize moisture loss and placed in a growth chamber set at 25 °C, ~650 lux and 12/12 h photoperiod for germination [16,43]. Moisture was controlled, and the salt solutions were added to the filter papers according to their groups when necessary.

2.4. Germination-Related Traits

For 20 days, counts were made every day, and germination percentage (%), mean germination time and germination stress tolerance index were calculated at the end of the 20th day. On the 45th day following the establishment of the experiment, ten seedlings randomly selected from each replicate were measured for seedling shoot length (mm) using a caliper (Traceable—6, VWR International, Milan, Italy) sensitive to 0.01 mm and seedling fresh weight (mg) using a precision balance (Sartorius—CPA 16001S, Göttingen, Germany) sensitive to 0.001 g. Then, the samples were kept in an oven at 70 °C for 48 h, and their dry weights were determined with the help of the same precision balance. As a result of the ratio of dry weight to fresh weight multiplied by 100, the amount of dry matter per unit amount was found as a percentage (%) [16].

Seeds with radicles reaching 2 mm in length were considered for all parameters. The germination percentage was determined according to ISTA [44] rules.

Mean germination time (MGT) was calculated to evaluate the speed of germination as defined by ISTA [44] with the following Formula (1), where n is the seed number germinated on day D, and D is the number of days from the beginning of the germination test.

$$MGT = \Sigma(Dn)/\Sigma n \quad (1)$$

The germination stress tolerance index (GSTI) was calculated as a percentage (%) by using Formula (2), where n is the number of seeds germinated at day d [45].

$$GSTI = [nd2\ (1.00) + nd4\ (0.75) + nd6\ (0.5) + nd8\ (0.25) \text{ of stressed seeds}/nd2\ (1.00) + nd4\ (0.75) + nd6\ (0.5) + nd8\ (0.25) \text{ of control seeds}] \times 100 \quad (2)$$

2.5. Statistical Analysis

The experimental design was a 3-factor factorial, arranged in a completely randomized design with 3 replicates. Analysis of variance and comparison of means were performed by the MSTAT-C program (Michigan State University v. 2.10). Before running ANOVA, square root transformation was performed in order to ensure normal distribution and homogeneity of variances in germination percentage, germination stress tolerance index and seedling dry weight ratio. Determination of the relations among the characteristics was revealed with correlation analysis by using Minitab version 17 (Minitab Inc., State College, PA, USA), and results were expressed with correlation coefficients [46].

3. Results

The effects of flesh color and salt concentrations were found to be statistically significant in all of the characteristics investigated, while plant growth regulators had insignificant effects on germination ratio, shoot length and seedling dry matter ratio. In addition, in some of the traits examined, double or triple interactions of the factors also showed significant effects (Table 1).

Table 1. Distribution of germination and emergence performances according to factors.

Flesh Color (FC)	Germination (%)	MGT (Day)	Shoot Length (cm)	Fresh Seedling Weight (mg)	Seedling Dry Matter Ratio (%)	GSTI (%)
White	81.17 ± 24.49 B	11.24 ± 5.38 A	27.90 ± 9.87 B	12.29 ± 3.84 B	4.44 ± 1.54 B	36.08 ± 39.44 B
Red	85.33 ± 13.21 A	7.33 ± 3.02 B	31.49 ± 10.69 A	21.23 ± 7.54 A	4.65 ± 1.99 A	54.91 ± 34.08 A
			Salt Concentrations (SC)			
Control	98.44 ± 1.27 A	4.74 ± 0.58 D	41.56 ± 3.01 A	22.85 ± 6.16 A	2.87 ± 0.15 D	100.0 ± 0.00 A
2500 ppm	90.00 ± 6.45 B	7.06 ± 1.40 C	35.78 ± 3.71 B	20.76 ± 6.65 B	3.54 ± 0.31 C	51.32 ± 18.66 B
5000 ppm	89.50 ± 8.23 B	9.80 ± 2.40 B	26.18 ± 2.65 C	14.64 ± 4.06 C	4.45 ± 0.41 B	25.97 ± 17.67 C
10,000 ppm	55.06 ± 18.28 C	15.55 ± 4.22 A	15.26 ± 1.85 D	8.79 ± 1.99 D	7.33 ± 0.84 A	4.68 ± 4.34 D
			Plant Growth Regulators (PGRs)			
MC	83.54 ± 19.59	9.16 ± 4.78 B	30.09 ± 10.59	16.96 ± 7.73 A	4.46 ± 1.70	46.05 ± 37.90 B
SA	82.21 ± 22.04	9.62 ± 5.07 A	29.37 ± 10.69	16.28 ± 7.25 B	4.58 ± 1.77	43.22 ± 38.44 C
OA	84.00 ± 17.66	9.08 ± 4.52 B	29.63 ± 10.16	17.04 ± 7.51 A	4.60 ± 1.88	47.21 ± 38.12 A
			ANOVA Significance levels			
FC	**	***	***	***	*	***
SC	***	***	***	***	***	***
PGR	ns	*	ns	**	ns	***
FC*SC	***	***	ns	***	***	**
FC*PGR	ns	ns	ns	*	***	***
SC*PGR	*	ns	ns	**	***	***
FC*SC*PGR	ns	ns	ns	**	**	***

*, **, ***: Means statistical difference, respectively, at 0.05, 0.01 and 0.001. ns: non-significant. Means followed by different letters within the same columns are significantly different.

Increases in salt concentration slowed the seed germination and caused a decrease in all germination and seedling traits except dry matter content (Table 1). The germination ratio, which had been over 98% in the control, decreased to 55% in the highest saline environment. Simultaneously, mean germination time of seeds extended by 10 days. Additionally, losses in fresh seedling weight (around 6–14 mg) and shoot length (around 10–25 cm) were observed. The drop in GSTI value to 4.7% explains all of these unfavorable impacts, which are presented in Table 1.

Some species/varieties/genotypes can develop resistance or tolerance to certain stress factors through the mechanisms they possess. GSTI, which means seed vigor, was calculated to be much higher (18%) in seeds obtained from red-fleshed fruits. All seed (germination ratio and mean germination time) and seedling (shoot length, fresh seedling weight and seedling dry matter ratio) characteristics investigated obtained from red-fleshed fruits showed significantly better performance, as shown in Table 1.

OA, used as a priming agent, was effective in terms of germination stress tolerance, germination percentage, mean germination time and fresh seedling weight. At the same time, MC was effective regarding seedling length and dry matter accumulation (Table 1).

Findings on the effect of the flesh color of the fruit from which the seed was obtained, varying levels of salt stress and priming agents (SA, OA and MC) on germination percentage and mean germination time of pitaya seeds are given in Table 2. While the germination rate, determined as 98.44% in the control, was not affected much by 2500 ppm (90.00%) and 5000 ppm (89.50%) salt concentrations, it decreased sharply to 55.03% at 10,000 ppm (Table 2). When all the study factors were considered together, the germination percentage (85.33%) of the seeds of the Siam Red was higher than that of Vietnam White (81.16%) (Table 2). In contrast, the mean germination time decreased from 11.24 days in Vietnam White to 7.33 days in Siam Red (Table 2). No statistically significant differences were detected between SA, OA and MC applied for this purpose regarding germination rate and germination speed. At the same time, all three PGRs were observed to make a positive contribution. In both Siam Red (99.33%) and Vietnam White (98.33%) cultivars, SA treatment was prominent in triggering germination under non-stress conditions (Table 2). However, when germination occurred under saline conditions, OA (48.67%) was favored in Vietnam White and MC (72.00%) in Siam Red, and the same was observed for the mean germination time (Table 2). It has been reported that seeds of different species, including pitaya, showed positive effects on germination speed and rate after priming with SA, OA or MC.

Table 2. Variation of germination percentage and mean germination time according to factors.

		Germination Percentage (%)			
Flesh Color	Salt Concentrations	Hormones			
		MC	SA	OA	Mean
White	0	97.67 ± 1.36	98.33 ± 1.63	98.00 ± 1.26	98.00 ± 1.37 A
	2500	94.67 ± 3.27	89.33 ± 5.46	93.33 ± 5.47	92.44 ± 5.11 A
	5000	94.67 ± 3.26	92.33 ± 3.20	93.33 ± 4.84	93.44 ± 3.75 A
	10,000	39.33 ± 6.89	34.27 ± 14.96	48.67 ± 8.55	40.76 ± 11.79 B
	Mean	81.58 ± 25.25	78.57 ± 27.41	83.33 ± 21.20	81.16 ± 24.47 B
Red	0	98.67 ± 1.03	99.33 ± 1.03	98.67 ± 1.03	98.89 ± 1.02 A
	2500	83.33 ± 7.34	88.67 ± 6.89	90.67 ± 4.84	87.56 ± 6.84 B
	5000	88.00 ± 5.66	89.33 ± 8.26	79.33 ± 11.98	85.56 ± 9.61 B
	10,000	72.00 ± 10.43	66.00 ± 12.33	69.92 ± 10.79	69.31 ± 10.84 C
	Mean	85.50 ± 11.75	85.83 ± 14.62	84.65 ± 13.66	85.33 ± 13.22 A
Means of Flesh Colors	0	98.17 ± 1.27 A	98.83 ± 1.40 A	98.33 ± 1.15 A	98.44 ± 1.27 A
	2500	89.00 ± 8.02 A	89.00 ± 5.94 A	92.00 ± 5.12 A	90.00 ± 6.45 B
	5000	91.33 ± 5.61 A	90.83 ± 6.18 A	86.33 ± 11.37 A	89.50 ± 8.23 B
	10,000	55.67 ± 19.03 AB	50.14 ± 21.10 B	59.29 ± 14.47 A	55.03 ± 18.28 C
	Mean	83.54 ± 19.59	82.20 ± 22.04	83.99 ± 17.66	83.24 ± 19.71

Table 2. *Cont.*

Mean germination time (day)					
Flesh Color	Salt Concentrations	Hormones			
		MC	SA	OA	Mean
White	0	5.31 ± 0.19	5.25 ± 0.19	5.30 ± 0.21	5.29 ± 0.19 D
	2500	8.20 ± 0.55	8.79 ± 0.56	7.96 ± 0.20	8.35 ± 0.56 C
	5000	11.78 ± 0.72	12.32 ± 0.89	11.90 ± 1.49	12.00 ± 1.05 B
	10,000	19.51 ± 1.86	20.10 ± 1.67	18.39 ± 1.34	19.33 ± 1.70 A
	Mean	11.22 ± 5.50	11.61 ± 5.69	10.89 ± 5.12	11.24 ± 5.38 A
Red	0	4.20 ± 0.23	4.25 ± 0.21	4.15 ± 0.10	4.20 ± 0.18 D
	2500	5.73 ± 0.43	6.11 ± 0.27	5.46 ± 0.42	5.77 ± 0.46 C
	5000	7.18 ± 0.54	7.54 ± 0.65	8.06 ± 0.54	7.59 ± 0.66 B
	10,000	11.28 ± 0.57	12.56 ± 2.88	11.46 ± 1.43	11.77 ± 1.87 A
	Mean	7.10 ± 2.72	7.62 ± 3.44	7.28 ± 2.95	7.33 ± 3.02 B
Means of Flesh Colors	0	4.75 ± 0.61	4.75 ± 0.56	4.72 ± 0.62	4.74 ± 0.58 D
	2500	7.02 ± 1.42	7.45 ± 1.46	6.71 ± 1.34	7.06 ± 1.40 C
	5000	9.48 ± 2.48	9.93 ± 2.60	9.98 ± 2.27	9.78 ± 2.40 B
	10,000	15.40 ± 4.49	16.33 ± 4.53	14.92 ± 3.85	15.55 ± 4.22 A
	Mean	9.16 ± 4.77 B	9.62 ± 5.07 A	9.08 ± 4.51 B	9.29 ± 4.77

Means followed by different letters within the same columns and rows are significantly different.

The results of how the shoot length, fresh seedling weight and seedling dry weight ratio of the seedlings were affected by the combined effect of the factors are given in Table 3. According to the results, regardless of the treatments, an increase in seedling dry matter content and decreases in shoot length and fresh seedling weight were observed in parallel with the increase in salt concentration. At the same time, the values varied between 2.87–7.34%, 15.26–41.56 mm and 8.78–22.85 mg, respectively (Table 3).

Shoot length, fresh seedling weight and seedling dry matter content, which are defined as seedling development and quality criteria, were higher in Siam Red (31.48 mm, 21.23 mg and 4.65%, respectively) than in Vietnam White (27.90 mm, 12.23 mg and 4.44%, respectively) (Table 3), suggesting that *H. polyrhizus* is more tolerant to salt stress.

Table 3. Variation of shoot length and fresh seedling weight and seedling dry matter ratio according to factors.

Shoot Length (mm)					
Flesh Color	Salt Concentrations	Plant Growth Regulator			
		MC	SA	OA	Mean
White	0	39.83 ± 3.31	38.83 ± 2.56	41.00 ± 2.00	39.89 ± 2.68 A
	2500	33.00 ± 1.05	31.92 ± 1.53	33.92 ± 1.20	32.94 ± 1.46 B
	5000	25.00 ± 1.26	25.17 ± 2.06	24.00 ± 2.09	24.72 ± 1.82 C
	10,000	14.67 ± 1.17	13.33 ± 1.57	14.17 ± 1.13	14.06 ± 1.35 D
	Mean	28.12 ± 9.75	27.31 ± 9.78	28.27 ± 10.47	27.90 ± 9.87 B
Red	0	43.67 ± 2.50	43.83 ± 2.64	42.17 ± 1.94	43.22 ± 2.37 A
	2500	39.75 ± 1.44	39.75 ± 2.16	36.33 ± 3.92	38.61 ± 3.04 B
	5000	29.08 ± 2.31	25.50 ± 2.49	28.33 ± 1.54	27.64 ± 2.57 C
	10,000	15.75 ± 2.02	16.58 ± 0.74	17.08 ± 1.24	16.47 ± 1.46 D
	Mean	32.06 ± 11.23	31.42 ± 11.35	30.98 ± 9.87	31.48 ± 10.69 A
Means of Flesh Colors	0	41.75 ± 3.44	41.33 ± 3.60	41.58 ± 1.97	41.56 ± 3.01 A
	2500	36.37 ± 3.72	35.83 ± 4.46	35.12 ± 3.04	35.78 ± 3.71 B
	5000	27.04 ± 2.78	25.33 ± 2.19	26.17 ± 2.86	26.18 ± 2.65 C
	10,000	15.21 ± 1.67	14.96 ± 2.06	15.62 ± 1.89	15.26 ± 1.85 D
	Mean	30.09 ± 10.59	29.36 ± 10.68	29.62 ± 10.16	29.69 ± 10.41

Table 3. *Cont.*

Fresh seedling weight (mg)					
Flesh Color	Salt Concentrations	Plant Growth Regulator			
		MC	SA	OA	Mean
White	0	16.89 ± 1.01 BC	16.50 ± 0.95 BC	17.06 ± 0.74 BC	16.82 ± 0.89 A
	2500	14.71 ± 1.13 C	14.18 ± 0.08 C	14.48 ± 1.07 C	14.45 ± 0.87 B
	5000	11.39 ± 0.45 CD	10.97 ± 1.06 CD	10.51 ± 0.93 CD	10.96 ± 0.89 C
	10,000	7.07 ± 0.25 D	6.71 ± 0.31 D	6.97 ± 0.22 D	6.92 ± 0.29 D
	Mean	12.51 ± 3.86 A	12.09 ± 3.81 A	12.25 ± 3.99 A	12.29 ± 3.83 B
Red	0	28.94 ± 0.55 A	29.03 ± 0.54 A	28.69 ± 0.40 A	28.89 ± 0.49 A
	2500	28.28 ± 0.23 A	25.58 ± 3.14 AB	27.36 ± 2.41 A	27.07 ± 2.44 B
	5000	18.89 ± 1.41 B	16.14 ± 1.43 B	19.92 ± 1.38 B	18.32 ± 2.11 C
	10,000	9.53 ± 0.45 CD	11.11 ± 0.09 CD	11.31 ± 0.22 CD	10.65 ± 0.86 D
	Mean	21.41 ± 8.13 A	20.46 ± 7.50 B	21.82 ± 7.19 A	21.23 ± 7.54 A
Means of Flesh Colors	0	22.92 ± 6.34 A	22.76 ± 6.58 A	22.87 ± 6.10 A	22.85 ± 6.16 A
	2500	21.50 ± 7.13 A	19.88 ± 6.32 B	20.92 ± 6.96 A	20.76 ± 6.65 B
	5000	15.14 ± 4.04 C	13.56 ± 2.95 BC	15.21 ± 5.04 C	14.64 ± 4.06 C
	10,000	8.30 ± 1.33 D	8.91 ± 2.31 D	9.14 ± 2.27 D	8.78 ± 1.99 D
	Mean	16.96 ± 7.74 A	16.28 ± 7.25 B	17.04 ± 7.52 A	16.76 ± 7.46
Seedling dry matter ratio (%)					
Flesh Color	Salt Concentrations	Plant Growth Regulator			
		MC	SA	OA	Mean
White	0	2.99 ± 0.06 C	3.01 ± 0.05 C	2.97 ± 0.04 C	2.99 ± 0.05 D
	2500	3.46 ± 0.15 BC	3.66 ± 0.15 BC	3.78 ± 0.07 BC	3.64 ± 0.18 C
	5000	4.51 ± 0.07 B	4.39 ± 0.19 B	3.90 ± 0.48 BC	4.27 ± 0.39 B
	10,000	7.06 ± 0.17 A	7.37 ± 0.36 A	6.19 ± 0.87 A	6.88 ± 0.73 A
	Mean	4.51 ± 1.61 A	4.61 ± 1.71 A	4.21 ± 1.31 B	4.44 ± 1.54 B
Red	0	2.76 ± 0.12 C	2.77 ± 0.13 C	2.69 ± 0.09 C	2.74 ± 0.11 D
	2500	3.10 ± 0.11 BC	3.48 ± 0.35 BC	3.76 ± 0.28 BC	3.45 ± 0.38 C
	5000	4.90 ± 0.30 B	4.54 ± 0.43 B	4.46 ± 0.08 B	4.63 ± 0.35 B
	10,000	8.00 ± 0.70 A	7.43 ± 0.78 A	7.92 ± 0.54 A	7.78 ± 0.69 A
	Mean	4.69 ± 2.15 A	4.56 ± 1.87 A	4.71 ± 2.02 A	4.65 ± 1.99 A
Means of Flesh Colors	0	2.88 ± 0.15 B	2.89 ± 0.16 B	2.83 ± 0.16 B	2.87 ± 0.15 D
	2500	3.28 ± 0.23 B	3.57 ± 0.27 AB	3.77 ± 0.19 AB	3.54 ± 0.31 C
	5000	4.71 ± 0.29 AB	4.47 ± 0.33 AB	4.18 ± 0.44 AB	4.45 ± 0.41 B
	10,000	7.53 ± 0.69 A	7.40 ± 0.58 A	7.06 ± 1.14 A	7.34 ± 0.84 A
	Mean	4.46 ± 1.70	4.58 ± 1.77	4.60 ± 1.88	4.55 ± 1.78

Means followed by different letters within the same columns and rows are significantly different.

When factor effects were analyzed together, there was no difference between PGRs in terms of shoot length and seedling dry matter content, while OA (17.04 mg) and MC (16.96 mg) were more effective than SA (16.28 mg) in terms of fresh seedling weight (Table 3). At the highest salt stress dose of 10,000 ppm, OA was found to be the highest in shoot length (15.62 mm), fresh seedling weight (9.14 mg) and seedling dry matter content (4.60%) (Table 3).

The results of the stress tolerance index, which is the cumulative effect of the traits summarized above, detailed according to the factors, are given in Figure 1. When the effect of other factors is ignored, the increase in salt concentration causes the stress tolerance index to decrease, which was also reported by [45,47]. However, the fact that seeds obtained from fruits with red fruit flesh were more resistant to salinity-induced stress can be interpreted as the effect of the maternal environment. The maternal environment, defined as the maternal parent effects other than gene effects, has shown significant effects on resistance to different biotic and abiotic stress conditions [10,48]. Considering the fact that saline and arid areas will increase under the current global climate change, *H. polyrhizus* cultivars may be promising as parents in hybridizations in order to develop stress-resistant genotypes with superior fruit characteristics such as high antioxidant activity due to their diverse and higher phytochemicals [49], in addition to their resistance [16]. It was observed that all PGRs applied as priming agents were effective in reducing stress conditions, but they were ranked as OA > MC > SA. However, the interactions of the factors on the stress tolerance index were found to be significant. According to the results, MC was more effective as a priming agent in seeds obtained from fruits with white fruit flesh, while a similar effect was found in seeds with red fruit flesh due to OA treatment (Figure 1). Until new salt-tolerant genotypes are developed, identification of species-specific priming agents in the short term for saline environments is crucial in terms of contributing to research and cultivation of pitaya. SA, OA and MC are widely used to alleviate the negative effects caused by different stress conditions [34–36,40,41].

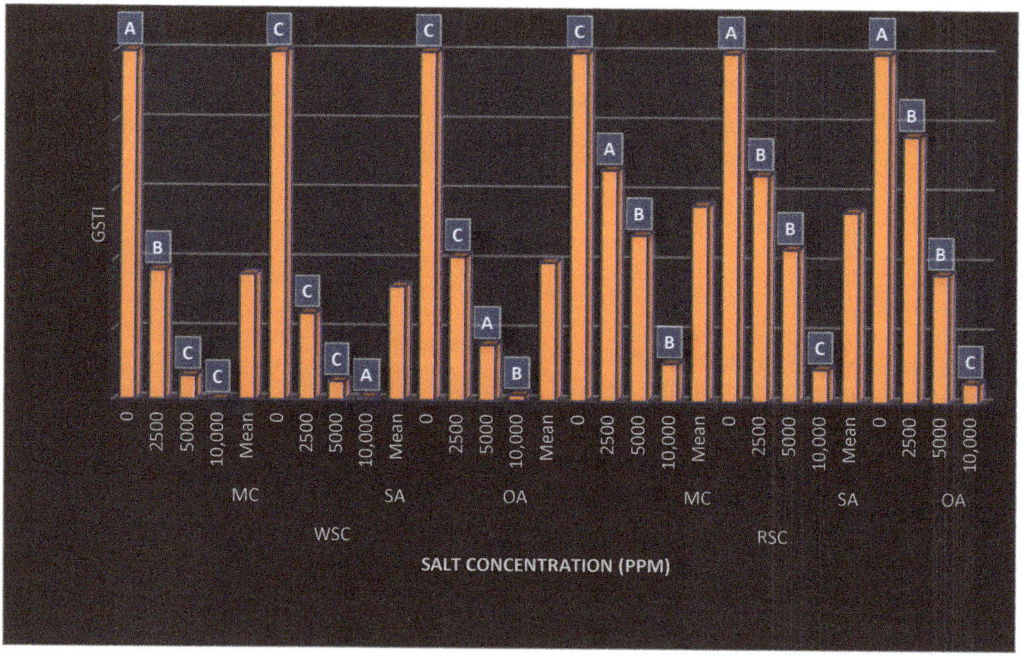

Figure 1. Variation of germination stress tolerance index according to factors (LSD (%): Seed C: 0.51, Salt C: 0.69, Horm: 0.64, Seed CxSalt C: 0.98, Seed CxHorm: 0.96, Salt CxHorm: 1.20, Seed CxSalt CxHorm: 1.70, F Values (%): Seed C: 3223.38, Salt C: 30,224.38, Horm: 521.50, Seed CxSalt C: 1458.50, Seed CxHorm: 193.87, Salt CxHorm: 48.00, Seed CxSalt CxHorm: 30.80, Means followed by different letters on different bars are significantly different.).

The results of the relations between the traits examined in the study are given in Figure 2. Since the development in the plumule leads to an increase in biomass, there was a high positive correlation between seedling length and fresh seedling weight (r: 0.84). However, since the development in tissues caused an increase in intercellular spaces, a high negative correlation was found between shoot length and seedling dry matter content (r: −0.89). It is known that an increase in intercellular spaces leads to a decrease in the amount of dry matter accumulated per unit area [50]. Mean germination time is one of the important criteria for determining seed strength. If the time is prolonged, a decrease in germination rate and developmental retardation in germinated seedlings are observed [45]. The results of the study were in this direction, and strong negative relationships were found between mean germination time and germination percentage (r: −0.83), seedling length (r: −0.88) and fresh seedling weight (r: −0.81). Conversely, the increase in germination stress tolerance positively affected shoot length (r: 0.90), fresh seedling weight (r: 0.80) and germination percentage (r: 0.63), which are expressed as quality seed parameters, while shortening the mean germination time (r: −0.84). However, a negative correlation (r: −0.77) was calculated between GSTI and seedling dry matter ratio due to the increase in intercellular space in seedlings with vigorous growth. Germination percentage, which is one of the most important parameters determining seed quality, contributed to the increase in the stress tolerance index (r: 0.63) and plant biomass (r: 0.56) while decreasing the mean germination time (r: −0.84). The results are consistent with previous studies [45,51].

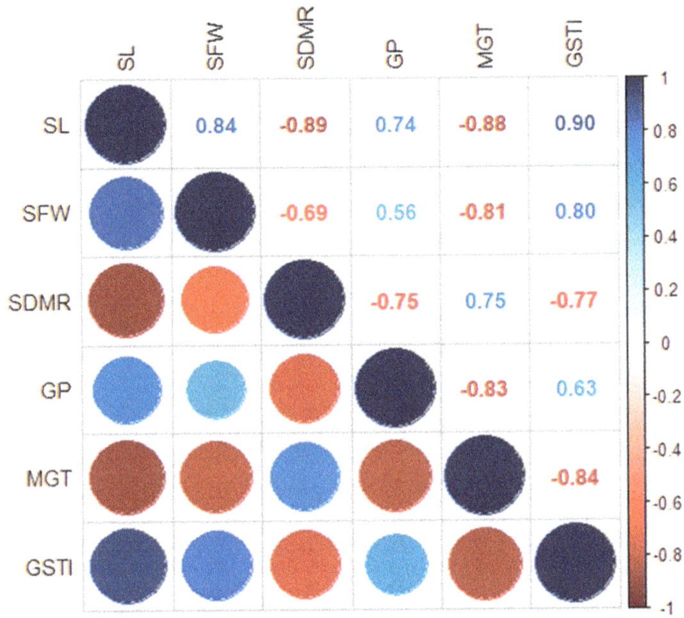

Figure 2. Correlations among investigated characteristics (SL: Seedling length, SFW: Seedling fresh weight, SDMR: Seedling dry matter ratio, GP: Germination percentage, MGT: Mean germination time, GSTI: Germination stress tolerance index).

4. Discussion

Agricultural production depends on the suitability of ecological factors, and sometimes it is necessary to cultivate under different stress conditions [52]. In these cases, it can be critical to know the tolerance of the material to the prevailing stress conditions or the practices to counteract this stress [53]. Due to water deficiency (osmotic stress), ion toxicity, and ion imbalance (ionic stress), or a combination of these factors [54,55], high salinity levels have an adverse effect on seed germination and plant growth, ultimately inhibiting germination and preventing crop production. Because of the low osmotic potential caused by high soil salinity, which hinders water uptake by the seed, seed germination is typically inhibited [56]. In addition, excessive soil salt and chloride ion concentrations may be hazardous to seeds [57].

An increase in salt concentration increases the osmotic pressure of the seed and inhibits water uptake. Due to the lower water flux caused by the osmotic effect, salt stress affects seed metabolism and can lead to the inhibition of reserve mobilization [58]. Disruption of ion balance and increased reactive oxygen species lead to cell integrity and homeostasis disruption; the latter acts on different chemical components of the seed, disrupting the metabolic cycle. All these conditions lead to loss of seed viability or reduced performance [19–22]. The study findings were similar to those of the previous studies. In parallel with the increase in salt concentration, there was a decrease in the germination rate and an increase in the mean germination time. There are reports of decreased seed germination at increasing salt concentrations in different species such as cotton [41] and cucumber [47], including pitaya [18]. Ascending salt concentrations not only prevent seed germination but also extend the germination time by delaying the start of germination [59].

Increased salt concentration in the seed interferes with the conversion of macromolecules, which are a source for respiration, limiting the formation of soluble sugars necessary for embryo development. This disruption of the energy flow causes germination not to occur at all, to stop at a certain stage or to retard development [19,20,25]. Based on these reasons, shoot length and fresh seedling weight, which are defined as seedling development and quality criteria, were also decreased under saline conditions. Our results are in agreement with other reports conducted on different species, such as sunflower [59], sugar beet [51] and pitaya [16].

Genotypes show high variation in salt tolerance through different adaptation pathways. In studies examining pitaya varieties with different flesh colors, the juices of varieties with red flesh color were found to have higher values in terms of pH and antioxidant activity [60,61]. Pitaya seeds with red fruit flesh have higher oil and sugar reserves [61]. Betalain derivatives, a natural osmolyte and antioxidant, K, which has a high ion exchange potential with Na, hydrolytic enzymes that catalyze metabolic events and antioxidants that eliminate the negative effects caused by excessive ROS are found at higher levels in *H. polyrhizus* with red fruit flesh [49,62,63].

The better performance of the Siam Red under saline conditions may be because its seeds are realized in a halophytic environment from the formation and development stage. Indeed, in *Anabasis setifera*, the maternal environment has been proven to affect resistance to salt stress significantly [48]. In addition, the high level of antioxidant activity may have contributed to lower negative effects caused by ROS. Seeds of different pitaya species were germinated under salt stress [16], and a 33% decrease in the germination rate of *H. undatus* with white fruit flesh was observed, while no decrease was reported in *H. polyrhizus* with red fruit flesh. Similarly, the reduction in the germination rate of seeds from white fruit flesh (62%) was higher than in that of red (46%). In conclusion, it was emphasized that *H. polyrhizus* is more tolerant to salt stress than *H. undatus*, which is similar to the results of the current study.

The cumulative effect of both seed composition and maternal environment factors that favor tolerance to salt stress allowed for better results in *H. polyrhizus* seedlings growing under salt stress.

Seeds that have been primed before being sown have improved imbibition capability, pre-germinative metabolic processes, seedling emergence, growth, vigor, productivity, and salinity adaptation. To maintain genome integrity, seed priming improves DNA repair, stabilizes RNA, and boosts new protein synthesis. Priming practices may allow sustainable cultivation in areas where soil or irrigation water has high salinity before sowing. In these areas where ecological factors are unsuitable, it is vital to determine species-specific priming agents [25,45]. Using PGRs that regulate different physiological events can successfully alleviate stress effects [64,65]. These positive effects have been proven to be because these chemicals act as signaling agents, maintain ion homeostasis, maintain membrane stability, suppress ROS and are involved in different physiological process [25–28,33].

The prominence of OA, an important osmoregulatory and antioxidant known to be positively correlated with K availability in stress conditions [66], may be interpreted as the seed trying to maintain ion homeostasis primarily to relieve salt stress. In the remediation of different heavy metals, exogenous application of OA to plants increased tolerance, enabling higher levels of heavy metal accumulation. In addition, among the grapevine rootstocks, it was emphasized that the genotype with the highest resistance to Na stress was the one containing the highest level of OA, and it was reported to show significant effects in eliminating Na stress [67].

5. Conclusions

Nowadays, priming techniques, which are used especially by commercial seed companies with coating technology, is widely used against stress factors such as salinity. This study investigated the effect of some plant growth regulators as priming agent and maternal environment as a first report on the performance of seeds obtained from pitaya fruits with different fruit flesh colors under different salt stress conditions during germination and seedling emergence periods. According to the results, it was determined that the environment in which the seed was grown was important in salt stress resistance, while seeds matured in the environment from red fruit flesh were more tolerant to salt stress than seeds from white fruit flesh. Although pitaya species are relatively salt-tolerant, growth of seedlings was significantly reduced above 2500 ppm and germination percentage started to decrease at 5000 ppm. In reducing the stress caused by NaCl, MC can be used as a priming agent in seeds obtained from fruits with white fruit flesh, and OA can be used as a priming agent in seeds obtained from fruits with red fruit flesh. It is planned to expand the scope of the study by increasing the number of species and cultivars. If the maternal effect is found to be significant still, reciprocal hybridizations will be planned in order to reveal the inheritance pattern of salt resistance and obtain new genotypes with resistance and superior fruit characteristics. It is thought that all the results obtained and that will be obtained could contribute to the cultivation and breeding of pitaya.

Author Contributions: Design of the research was done by K.M. and N.K. K.M. carried out the research. K.M. and M.S.D. made the statistical analyzes and interpretation of the results. A.M.Ç., B.B.K. and K.M. wrote the original draft. B.B.K. and K.M. edited the final version of the article. All authors have read and agreed to the published version of the manuscript.

Funding: This research received no external funding.

Data Availability Statement: The data that support the findings of the study are available from the corresponding author upon reasonable request.

Conflicts of Interest: The authors declare no conflict of interest.

References

1. Crane, J.H.; Balerdi, C.F. *Pitaya Growing in the Florida home Landscape*; IFAS Extension: Lake City, FL, USA, 2005.
2. Wu, L.C.; Hsu, H.W.; Chen, Y.C.; Chiu, C.C.; Lin, Y.I.; Ho, J.A.A. Antioxidant and antiproliferative activities of red pitaya. *Food Chem.* **2006**, *95*, 319–327. [CrossRef]
3. Esquivel, P.; Stintzing, F.C.; Carle, R. Comparison of morphological and chemical fruit traits from different pitaya genotypes (*Hylocereus* sp.) grown in Costa Rica. *J. Appl. Bot. Food Qual.* **2007**, *81*, 7–14.
4. Cheok, A.; Xu, Y.; Zhang, Z.; Caton, P.W.; Rodriguez-Mateos, A. Betalain-rich dragon fruit (pitaya) consumption improves vascular function in men and women: A double-blind, randomized controlled crossover trial. *Am. J. Clin. Nutr.* **2022**, *115*, 1418–1431. [CrossRef]
5. Nishikito, D.F.; Borges, A.C.A.; Laurindo, L.F.; Otoboni, A.M.B.; Direito, R.; Goulart, R.D.A.; Nicolau, C.C.T.; Fiorini, A.M.R.; Sinatora, S.M.; Barbalho, S.M. Anti-inflammatory, antioxidant, and other health effects of dragon fruit and potential delivery systems for ıts bioactive compounds. *Pharmaceutics* **2023**, *15*, 159. [CrossRef]
6. Ariffin, A.A.; Bakar, J.; Tan, C.P.; Rahman, R.A.; Karim, R.; Loi, C.C. Essential fatty acids of pitaya (dragon fruit) seed oil. *Food Chem.* **2009**, *114*, 561–564. [CrossRef]
7. Mahmud, M.H.; Raihan, M.T.; Shakhik, M.T.Z.; Khan, F.T.; Islam, M.T. Dragon fruit (*Hylocereus polyrhizus*): A green colorant for cotton fabric. *Colorants* **2023**, *2*, 230–244. [CrossRef]
8. Mercado-Silva, E.M. Pitaya—*Hylocereus undatus* (haw). In *Exotic Fruits*; Academic Press: Cambridge, MA, USA, 2018; pp. 339–349.
9. Tel-Zur, N. Pitahayas: Introduction, agrotechniques, and breeding. *VII Int. Congr. Cactus Pear Cochineal* **2010**, *995*, 109–115. [CrossRef]
10. Evrenosoğlu, Y.; Mertoğlu, K.; Bilgin, N.A.; Misirli, A.; Özsoy, A.N. Inheritance pattern of fire blight resistance in pear. *Sci. Hortic.* **2019**, *246*, 887–892. [CrossRef]
11. Suarez-Roman, R.S.; Caetano, C.M.; Ramírez, H.; Morales, J.G. Caracterización morfoanatómica y fisiológica de semilla sexual de pitahaya amarilla *Selenicereus megalanthus* (Haw.) Britt & Rose. *Rev. De La Asoc. Colomb. De Cienc. Biológicas* **2012**, *24*, 97–111.
12. Carvalho, N.M.; Nakagawa, J. Sementes: Ciência, tecnologia e produção. In *Seeds: Science, Technology and Production Funep*, 5th ed.; Atena: Jaboticabal, Brazil, 2012.

13. Shrivastava, P.; Kumar, R. Soil salinity: A serious environmental issue and plant growth promoting bacteria as one of the tools for its alleviation. *Saudi J. Biol. Sci.* **2015**, *22*, 123–131. [CrossRef]
14. Mizrahi, Y.; Nerd, A.; Sitrit, Y. New Fruits for arid climates. In *Trends in New Crops and New Uses*; Janick, J., Whipkey, A., Eds.; ASHS Press: Alexandria, VA, USA, 2002.
15. Tel-Zur, N.; Abbo, S.; Bar-Zvi, D.; Mizrahi, Y. Genetic relationships among *Hylocereus* and *Selenicereus* vine cacti (Cactaceae): Evidence from hybridization and cytological studies. *Ann. Bot.* **2004**, *94*, 527–534. [CrossRef] [PubMed]
16. Carvalho, S.M.C.; Paiva, E.P.D.; Torres, S.B.; Neta, M.L.D.S.; Leite, M.D.S.; Benedito, C.P.; Albuquerque, C.C.D.; Sá, F.V.D.S. Pre-germination treatments in pitaya (*Hylocereus* spp.) seeds to attenuate salt stress. *Rev. Ciência Agronômica* **2022**, *53*, e20218121. [CrossRef]
17. Orozco, A.; Gardea, A.A.; Rascón-Chu, A.; Sánchez, A. Effect of salinity on seed germination, growth and metabolic activity of pitaya seedlings [*Stenocereus thurberi* (Engelm.) Buxb.]. *J. Prof. Assoc. Cactus Dev.* **2017**, *19*, 67–78. [CrossRef]
18. Ortiz, T.A.; Gomes, G.R.; Takahashi, L.U.S.A.; Urbano, M.R.; Strapasson, E. Water and salt stress in germinating seeds of pitaya genotypes (*Hylocereus* spp.). *Afr. J. Agric. Res.* **2014**, *9*, 3610–3619.
19. Parihar, P.; Singh, S.; Singh, R.; Singh, V.P.; Prasad, S.M. Effect of salinity stress on plants and its tolerance strategies: A review. *Environ. Sci. Pollut. Res.* **2015**, *22*, 4056–4075. [CrossRef]
20. Oliveira, A.B.; Alencar, N.L.M.; Gallão, M.I.; Gomes Filho, E. Avaliação citoquímica durante a germinação de sementes de sorgo envelhecidas artificialmente e osmocondicionadas, sob salinidade [Cytochemical evaluation during the germination of artificial aged and primed sorghum seeds under salinity]. *Rev. Ciênc. Agron.* **2011**, *42*, 223–231. [CrossRef]
21. Singh, A.L.; Hsriprasanna, K.; Chaudhari, V. Differential nutrients absorption an important tool for screening and identification of soil salinity tolerant peanut genotypes. *Indian J. Plant Physiol.* **2016**, *21*, 83–92. [CrossRef]
22. Isayenkov, S.V.; Maathuis, F.J. Plant salinity stress: Many unanswered questions remain. *Front. Plant Sci.* **2019**, *10*, 80. [CrossRef]
23. Parida, A.K.; Das, A.B. Salt tolerance and salinity effects on plants: A review. *Ecotoxicol. Environ. Saf.* **2005**, *60*, 324–349. [CrossRef]
24. Pereira, M.R.R.; Martins, C.C.; Souza, G.S.F.; Martins, D. Influência do estresse hídrico e salino na germinação de Urochloa decumbens e Urochloa ruziziensis [Influence of saline and water stress on germination of Urochloa decumbens and Urochloa ruziziensis]. *Biosci. J.* **2012**, *28*, 537–545.
25. Hussain, M.; Farooq, S.; Hassan, W.; Ul-Allah, S.; Tanveer, M.; Farooq, M.; Nawaz, A. Drought stress in sunflower: Physiological effects and its management throughout breeding and agronomic alternatives. *Agric. Water Manag.* **2018**, *201*, 152–166. [CrossRef]
26. Shaar-Moshe, L.; Blumwald, E.; Peleg, Z. Unique physiological and transcriptional shifts under combinations of salinity, drought, and heat. *Plant Physiol.* **2017**, *174*, 421–434. [CrossRef] [PubMed]
27. Prasad, R.; Shivay, Y.S. Oxalic acid/oxalates in plants: From self-defence to phytoremediation. *Curr. Sci.* **2017**, *112*, 1665–1667. [CrossRef]
28. Janda, T.; Szalai, G.; Pál, M. Salicylic acid signalling in plants. *Int. J. Mol. Sci.* **2020**, *21*, 2655. [CrossRef]
29. War, A.R.; Paulraj, M.G.; War, M.Y.; Ignacimuthu, S. Jasmonic acid- mediated induced resistance in groundnut (*Arachis hypogaea* L.) against *Helicoverpa armigera* (Hubner) (Lepidoptera: Noctuidae). *J. Plant. Growth. Regul.* **2011**, *30*, 512–523. [CrossRef]
30. Chia, L. What Are the Differences between Mepiquat Chloride and Chlormequat Chloride? Plant Hormones. 2018. Available online: https://www.plantgrowthhormones.com/info/ (accessed on 14 October 2023).
31. Tian, S.; Wan, Y.; Qin, G.; Xu, Y. Induction of defense responses against Alternaria rot by different elicitors in harvested pear fruit. *Appl. Microbiol. Biotechnol.* **2006**, *70*, 729–734. [CrossRef]
32. Pareek, S. Novel postharvest treatments of fresh produce. *Innov. Postharvest. Technol.* **2017**, *68*, 35–51.
33. Sakouhi, L.; Kharbech, O.; Massoud, M.B.; Munemasa, S.; Murata, Y.; Chaoui, A. Oxalic acid mitigates cadmium toxicity in *Cicer arietinum* L. germinating seeds by maintaining the cellular redox homeostasis. *J. Plant. Growth. Regul.* **2022**, *41*, 697–709. [CrossRef]
34. Srivastava, L.M. Gibberellins. In *Plant Growth and Development*; Srivastava, L.M., Ed.; Academic Press: New York, NY, USA, 2022; pp. 172–181.
35. Siebert, J.D.; Stewart, A.M. Influence of plant density on cotton response to mepiquat chloride application. *Agron. J.* **2006**, *98*, 1634–1639. [CrossRef]
36. Mahdavian, K.; Ghorbanli, M.; Kalantari, K.M. Role of salicylic acid in regulating ultraviolet radiation ınduced oxidative stress in pepper leaves. *Russ. J. Plant Physiol.* **2008**, *55*, 560–563. [CrossRef]
37. Tong, R.; Zhou, B.; Cao, Y.; Ge, X.; Jiang, L. Metabolic profiles of moso bamboo in response to drought stress in a field investigation. *Sci. Total Environ.* **2020**, *720*, 137722. [CrossRef]
38. Lehner, A.; Meimoun, P.; Errakhi, R.; Madiona, K.; Barakate, M.; Bouteau, F. Toxic and signaling effects of oxalic acid: Natural born killer or natural born protector? *Plant Signal. Behav.* **2008**, *3*, 746–748. [CrossRef] [PubMed]
39. Wang, Q.; Lai, T.; Qin, G.; Tian, S. Response of jujube fruits to exogenous oxalic acid treatment based on proteomic analysis. *Plant Cell Physiol.* **2009**, *50*, 230–242. [CrossRef] [PubMed]
40. Kim, D.S.; Hwang, B.K. An important role of the pepper phenylalanine ammonia-lyase gene (PAL1) in salicylic acid-dependent signalling of the defence response to microbial pathogens. *J. Exp. Bot.* **2014**, *65*, 2295–2306. [CrossRef]
41. Zhang, Q.C.; Deng, X.X.; Wang, J.G. The effects of mepiquat chloride (DPC) on the soluble protein content and the activities of protective enzymes in cotton in response to aphid feeding and on the activities of detoxifying enzymes in aphids. *BMC Plant Biol.* **2022**, *22*, 213. [CrossRef] [PubMed]

42. Bahrabadi, E.; Tavakkol Afshari, R.; Mahallati, M.N.; Seyyedi, S.M. Abscisic, gibberellic, and salicylic acids effects on germination indices of corn under salinity and drought stresses. *J. Crop Improv.* **2022**, *36*, 73–89. [CrossRef]
43. Zerpa-Catanho, D.; Hernández-Pridybailo, A.; Madrigal-Ortiz, V.; Zúñiga-Centeno, A.; Porras-Martínez, C.; Jiménez, V.M.; Barboza-Barquero, L. Seed germination of pitaya (*Hylocereus* spp.) as affected by seed extraction method, storage, germination conditions, germination assessment approach and water potential. *J. Crop Improv.* **2019**, *33*, 372–394. [CrossRef]
44. ISTA. *International Rules for Seed Testing*; International Seed Testing Association: Wallisellen, Switzerland, 2003.
45. Ergin, N.; Kulan, E.; Gözükara, M.; Kaya, M.; Çetin, Ş.; Kaya, M.D. Response of germination and seedling development of cotton to salinity under optimal and suboptimal temperatures. *Kahramanmaraş Sütçü İmam Üniversitesi Tarım Doğa Derg.* **2021**, *24*, 108–115. [CrossRef]
46. Zar, J.H. *Biostatistical Analysis: Pearson New International Edition*; Pearson Higher, Ed.; Pearson Higher: Harlow, UK, 2013.
47. Demir, I.; Kuzucu, C.O.; Ermis, S.; Öktem, G. Radicle emergence as seed vigour test estimates seedling quality of hybrid cucumber (*Cucumis sativus* L.) cultivars in low temperature and salt stress conditions. *Horticulturae* **2022**, *9*, 3. [CrossRef]
48. El-Keblawy, A.; Gairola, S.; Bhatt, A. Maternal salinity environment affects salt tolerance during germination in Anabasis setifera: A facultative desert halophyte. *J. Arid. Land* **2016**, *8*, 254–263. [CrossRef]
49. Attar, Ş.H.; Gündeşli, M.A.; Urün, I.; Kafkas, S.; Kafkas, N.E.; Ercisli, S.; Ge, C.; Mlcek, J.; Adamkova, A. Nutritional analysis of red-purple and white-fleshed pitaya (*Hylocereus*) species. *Molecules* **2022**, *27*, 808. [CrossRef] [PubMed]
50. Çolak, A.M.; Mertoğlu, K.; Alan, F.; Esatbeyoglu, T.; Bulduk, İ.; Akbel, E.; Kahramanoğlu, I. Screening of naturally grown european cranberrybush (*Viburnum opulus* L.) genotypes based on physico-chemical characteristics. *Foods* **2022**, *11*, 1614. [CrossRef]
51. Kulan, E.G.; Arpacıoğlu, A.; Ergin, N.; Kaya, M.D. Evaluation of germination, emergence and physiological properties of sugar beet cultivars under salinity. *Trak. Univ. J. Nat. Sci.* **2021**, *22*, 263–274. [CrossRef]
52. Ibrahim, E.A. Seed priming to alleviate salinity stress in germinating seeds. *J. Plant Physiol.* **2016**, *192*, 38–46. [CrossRef] [PubMed]
53. Fernandes, A.C.; Coutinho, G. Nitrogênio no desenvolvimento inicial de mudas de pitaya vermelha. *Glob. Sci. Technol.* **2019**, *12*, 32–43.
54. Läuchli, A.; Grattan, S.R. Plant growth and development under salinity stress advances. In *Molecular Breeding toward Drought and Salt Tolerant Crops*; Jenks, M.A., Hasegawa, P.M., Jain, S.M., Eds.; Springer: Dordrecht, The Netherlands, 2007; pp. 285–315.
55. McNeil, S.D.; Nuccie, M.L.; Hanson, A.D. Betaines and related osmoprotectants. Targets for metabolic engineering of stress resistance. *Plant Physiol.* **1999**, *120*, 945–949. [CrossRef]
56. Welbaum, G.E.; Tissaoui, T.; Bradford, K.J. Water relations of seed development and germination in muskmelon (*Cucumis melo* L.) III. Sensitivity of germination to water potential and abscisic acid during development. *Plant. Physiol.* **1999**, *92*, 1029–1037. [CrossRef]
57. Khajeh-Hosseini, M.; Powell, A.; Bingham, I. The interaction between salinity stress and seed vigour during germination of soyabean seeds. *Seed Sci. Technol.* **2003**, *31*, 715–725. [CrossRef]
58. Freire, M.H.D.C.; Sousa, G.G.D.; de Souza, M.V.; de Ceita, E.D.; Fiusa, J.N.; Leite, K.N. Emergence and biomass accumulation in seedlings of rice cultivars irrigated with saline water. *Rev. Bras. De Eng. Agrícola E Ambient.* **2018**, *22*, 471–475. [CrossRef]
59. Kaya, M.D.; Akdoğan, G.; Kulan, E.G.; Dağhan, H.; Sari, A. Salinity tolerance classification of sunflower (*Helianthus annuus* L.) and safflower (*Carthamus tinctorius* L.) by cluster and principal component analysis. *Appl. Ecol. Environ. Res.* **2019**, *17*, 3849–3857. [CrossRef]
60. Escribano, J.; Pedreño, M.A.; García-Carmona, F.; Muñoz, R. Characterization of the antiradical activity of betalains from *Beta vulgaris* L. roots. *Phytochem. Anal. Int. J. Plant Chem. Biochem. Tech.* **1998**, *9*, 124–127. [CrossRef]
61. Tamby Chik, C.; Bachok, S.; Baba, N.; Abdullah, A.; Abdullah, N. Quality characteristics and acceptability of three types of pitaya fruits in a consumer acceptance test. *J. Tour. Hosp. Culin. Arts* **2011**, *3*, 89–98.
62. Nizamlıoğlu, N.M.; Ünver, A.; Kadakal, Ç. Mineral content of pitaya (*Hylocereus polyrhizus* and *Hylocereus undatus*) seeds grown in Turkey. *Erwerbs-Obstbau* **2021**, *63*, 209–213. [CrossRef]
63. Paśko, P.; Galanty, A.; Zagrodzki, P.; Ku, Y.G.; Luksirikul, P.; Weisz, M.; Gorinstein, S. Bioactivity and cytotoxicity of different species of pitaya fruits—A comparative study with advanced chemometric analysis. *Food Biosci.* **2021**, *40*, 100888. [CrossRef]
64. Guirra, K.S.; Torres, S.B.; Leite, M.D.S.; Guirra, B.S.; Nogueira Neto, F.A.; Rêgo, A.L. Phytohormones on the germination and initial growth of pumpkin seedlings under different types of water. *Rev. Bras. De Eng. Agrícola E Ambient.* **2020**, *24*, 827–833. [CrossRef]
65. Kerchev, P.; van der Meer, T.; Sujeeth, N.; Verlee, A.; Stevens, C.V.; Van Breusegem, F.; Gechev, T. Molecular priming as an approach to induce tolerance against abiotic and oxidative stresses in crop plants. *Biotechnol. Adv.* **2020**, *40*, 107503. [CrossRef] [PubMed]
66. Rubio, J.S.; García-Sánchez, F.; Rubio, F.; García, A.L.; Martínez, V. The importance of K+ in ameliorating the negative effects of salt stress on the growth of pepper plants. *Eur. J. Hortic. Sci.* **2010**, *75*, 33–41.
67. Guo, S.H.; Niu, Y.J.; Zhai, H.; Han, N.; Du, Y.P. Effects of alkaline stress on organic acid metabolism in roots of grape hybrid rootstocks. *Sci. Hortic.* **2018**, *227*, 255–260. [CrossRef]

horticulturae

MDPI

Article

Tomato Accumulates Cadmium to a Concentration Independent of Plant Growth

Xingyu Zhang [1], Cong Zhang [1] and Yuyang Zhang [1,2,3,4,*]

1 National Key Laboratory for Germplasm Innovation and Utilization of Horticultural Crops, Huazhong Agricultural University, Wuhan 430070, China; xyzhang_hzau@163.com (X.Z.); zhangcong_hzau@163.com (C.Z.)
2 Hubei Hongshan Laboratory, Wuhan 430070, China
3 Shenzhen Institute of Nutrition and Health, Huazhong Agricultural University, Wuhan 430070, China
4 Shenzhen Branch, Guangdong Laboratory for Lingnan Modern Agriculture, Genome Analysis Laboratory of the Ministry of Agriculture, Agricultural Genomics Institute at Shenzhen, Chinese Academy of Agricultural Sciences, Shenzhen 518000, China
* Correspondence: yyzhang@mail.hzau.edu.cn

Abstract: Cadmium (Cd) contamination is a growing concern, as exposure to the metal has been shown to inhibit plant growth and development. However, soil Cd pollution in China is typically mild, and thus its concentration often does not impede plant growth. On the other hand, it is unknown if increased plant growth impacts Cd uptake, movement, and accumulation. Here, we analyzed the relationship between Cd accumulation in 31 tomato cultivars and the impact on specific growth parameters in mild Cd contamination. The results showed that there are variations in the Cd distribution among the 31 tomato cultivars studied. There were higher Cd concentrations in shoots of the cultivar '*SV3557*', whereas root Cd concentrations were the lowest. The roots of the cultivar '*HF11*' recorded the lowest Cd content but had higher Cd content in the shoots. The Cd concentration in roots and shoots was not related to root length, plant height, and root weight. However, Cd accumulation in the shoots was markedly promoted by root length and plant height, and Cd accumulation in the roots was promoted by root weight. Subsequently, we imposed Cd on four selected tomato cultivars to ascertain their accumulation in the shoot tissues. The results revealed that, among the four tomato cultivars, Cd was highly accumulated in the leaves, followed by the stems, and the fruits (leaf > stem > fruit). When identifying significant loci associated with Cd accumulation in tomato plants, it is crucial to find a suitable indicator to assess the plant's ability to accumulate Cd. Thus, Cd concentration in shoots can be used as a reliable proxy for evaluating tomato plants' capacity for Cd accumulation. This study serves as a valuable reference in guiding the selection of such an index.

Keywords: cadmium; accumulation; distribution; correlation; *Solanum lycopersicum*

Citation: Zhang, X.; Zhang, C.; Zhang, Y. Tomato Accumulates Cadmium to a Concentration Independent of Plant Growth. *Horticulturae* **2023**, *9*, 1343. https://doi.org/10.3390/horticulturae9121343

Academic Editors: Simonetta Muccifora

Received: 18 November 2023
Revised: 11 December 2023
Accepted: 14 December 2023
Published: 15 December 2023

1. Introduction

The severity of heavy metal pollution has been on the rise as a consequence of various human activities, including mining, discharge of industrial wastewater, and excessive application of pesticides and fertilizers [1,2]. Cadmium (Cd) is a common heavy metal element that is highly toxic and not essential for plant growth [3]. Cadmium frequently engages in competition with Zn^{2+}, Fe^{2+}, Mn^{2+}, and other divalent metal ions for transport channels, resulting in detrimental effects such as inadequate nutrient uptake, reduced photosynthetic activity, and oxidative stress, thereby impeding plant development [4–6]. Furthermore, the expeditious migration and protracted half-life of Cd culminate in its bioaccumulation within organisms subsequent to its introduction into the ecological food web, thereby causing a gradual onset of chronic toxicity in humans [7]. Although the impact of Cd concentration on plant growth and development is minimal, it is important to note that the accumulation of Cd in edible plant parts may surpass the optimal safety thresholds [8]. Based on the findings of a survey [9], it has been determined that the prevalence of soil Cd pollution in China stands at approximately 7%. Within this percentage, soil that is categorized as slightly polluted and lightly polluted constitutes 6% (for dry land with pH ≤ 7.5, the concentration of cadmium was between 0.3~0.9 mg/kg), whereas moderately and severely polluted soil represents a mere 1% [9]. Hence, it is imperative to study plants cultivated in soils mildly contaminated with Cd in order to ensure the safety of agricultural produce.

The inhibitory effect of a high concentration of Cd on the growth of tomato and other plants has been widely investigated [10–12]. However, it has been observed that mild Cd pollution does not impact plant growth. The root serves as the primary means by which plants uptake mineral elements, making it a crucial component of nutrient uptake. Additionally, the root functions as the initial line of defense against Cd accumulation. Cadmium stress resulted in the inhibition of root length and lateral root formation of melon and pea, as observed in studies by Chen et al. [13] and Fusconi et al. [14]. According to Kubo et al. [15], wheat cultivars exhibiting high Cd accumulation demonstrate substantial Cd distribution in roots compared to cultivars with low Cd accumulation. Hence, this study employed a total of 31 distinct tomato cultivars characterized by varying root lengths and root weights. The objective was to investigate the potential influence of root growth, specifically under low Cd concentration, processes of Cd absorption, transport, and accumulation. Similarly, overabundance of Cd has been found to impede both shoot biomass and plant height in ornamental plants, as demonstrated by Liu et al. [16] and Wu et al. [17]. It is noteworthy that studies in wheat and *Leptoplax emarginata* have found that the transpiration rate affects plant height and shoot biomass, subsequently influencing the absorption of Cd [18,19]. This study aimed to examine the potential correlation between plant height and Cd accumulation in a sample of 31 tomato cultivars, and also to determine whether plants could influence Cd accumulation through physiological processes such as transpiration rate or photosynthesis.

The tomato (*Solanum lycopersicum*) is an economically valuable crop that is cultivated globally [20,21]. The tomato plant has a shorter growth cycle, a readily observable growth phenotype, and a comparatively compact genome [22,23]. The tomato plant, belonging to the Solanaceae family, holds significance not only as a food security vegetable crop but also as a prominent model plant for biological studies [22,24]. The distribution of Cd in tissues varies among different species or cultivars, as reported by An et al. [25], Hu et al. [26], and Wang et al. [27]. To the best of our knowledge, there is limited research examining potential variations in the distribution of Cd within different tomato cultivars. Hence, the present study is undertaken to assess the distribution of Cd within the tissues of four distinct tomato cultivars subjected to varying concentrations of Cd. This study provides valuable insights into the key characteristics associated with the accumulation of Cd in tomato plants. Moreover, there is a lack of literature regarding the potential impact of plant growth on the processes of absorption, transport, and accumulation of Cd in soils with low Cd concentrations. This will enable us to more accurately assess the capacity of tomato plants to accumulate Cd. As an illustration, in cloning major loci associated with Cd accumulation in tomato, the utilization of a genome-wide association study (GWAS) typically involves the examination of numerous tomato accessions, each exhibiting distinct growth potential. It is imperative to evaluate the potential impact of tomato growth on its ability to accumulate Cd, as well as determine the most suitable index that can effectively mitigate the interference caused by growth.

2. Materials and Methods

2.1. Experimental Materials and Design

Two independent experiments were conducted in a greenhouse at the Huazhong Agricultural University in Wuhan, China. In the first experiment, 31 tomato cultivars were sown in a substrate with 1 mg/kg Cd to evaluate the accumulation pattern. Regarding the risk control standard for soil contamination of agricultural land [28] (China) (GB 15618-

2018) and the standard level of pollution of soils in China, the concentration of Cd in the substrate was set to a light pollution level, 1 mg/kg. Cd was added as $CdCl_2 \cdot 2.5H_2O$, mixed well with the cultivation substrate, and loaded into the planting groove. Thirty-one commercially available tomato cultivars were germinated at 25 °C and sown in the planting groove, and five individual plants were retained for each cultivar (Supplementary Table S1).

In the second experiment, the tissue distribution of Cd in *'HX'*, *'HF12'*, *'HF15'*, and *'JNBL'* was investigated. Seeds of the four cultivars were planted in a greenhouse without Cd treatment at 25 °C for one month, and the seedlings with similar growth were selected and transplanted to the substrate with Cd concentration of 0, 0.5, 1.0, 1.5, 2.0, and 4.0 mg/kg. We watered the soil regularly to keep it moist and to ensure the growth of the plants. After the plant entered the vigorous growth period, we removed all the lateral branches, leaving only the main stems. The scheme of two independent experiments was presented (Supplementary Figure S1). Cadmium-contaminated substrates and tomato were treated professionally by the department of Hazardous Waste Recovery at Huazhong Agricultural University.

2.2. *Determination of Growth Parameters*

Thirty-one tomato cultivars were used in this study. Seedlings were harvested and divided into shoots and roots from the end of the stem. The plant height and root length were measured. The plants were then rinsed with deionized water, quickly dried with gauze, and their fresh weight was measured. Subsequently, the samples were dried at 110 °C for 15 min and then dried to a constant weight at 75 °C. The dry weights (DW) were recorded.

2.3. *Determination of Cd Concentration*

Tissue samples of 31 tomato cultivars were dried, ground, and passed through a 0.149 mm mesh nylon sieve for chemical analyses. Samples weighing 0.2 g were digested with 10 mL nitric acid at a gradient of temperatures spanning 120–180 °C for 1 h using a MARS6 microwave. After digestion, the samples were diluted to 25 mL with deionized water, and the content of Cd was determined by an Agilent graphite furnace (Agilent AA 240Z), with a detection limit of 0.019 μg/L. The Cd accumulation and the translocation factor (TF) were calculated as follows:

Cd accumulation (μg/plant) = Cd concentration × dry weight per plant [29]. The translocation factor (TF) is defined as the ration of Cd concentration in shoots and Cd concentration in roots [30].

2.4. *Statistical Analyses*

Statistical analyses were performed using IMB-SPSS statistical software (version 26.0). Data were analyzed using a one-way analysis of variance with the least significant difference test at a 5% significance level. The polar heatmaps with dendrogram were created using Origin software (version 2022; Figure 1). The scatter plots (Figures 2–4) and curve (Figure 5) were created using the GraphPad Prism software (version 8).

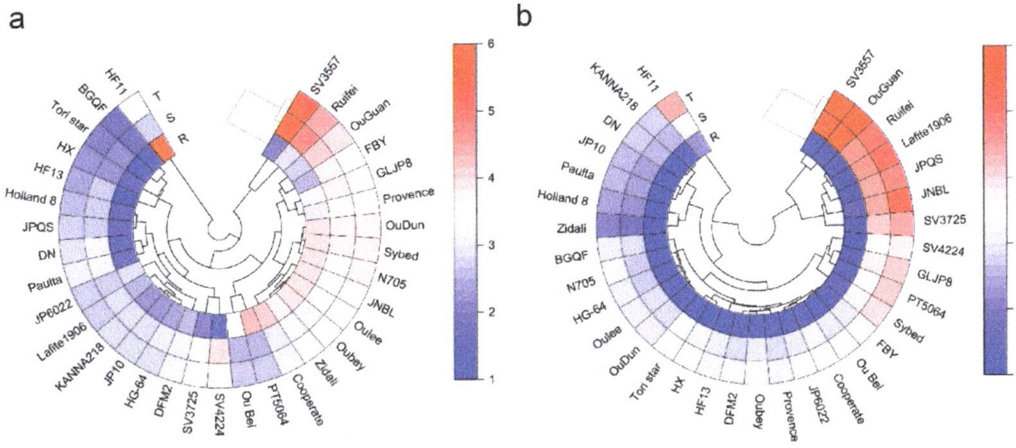

Figure 1. Cd concentration and accumulation in 31 tomato cultivars. Cd concentration (mg/kg DW) in root, shoot, and total plant of 31 tomato cultivars (**a**). Cd accumulation (µg/plant) in root, shoot, and total plant of 31 tomato cultivars (**b**). The "R" in the figure represents root, "S" represents shoot, and "T" represents the total plant.

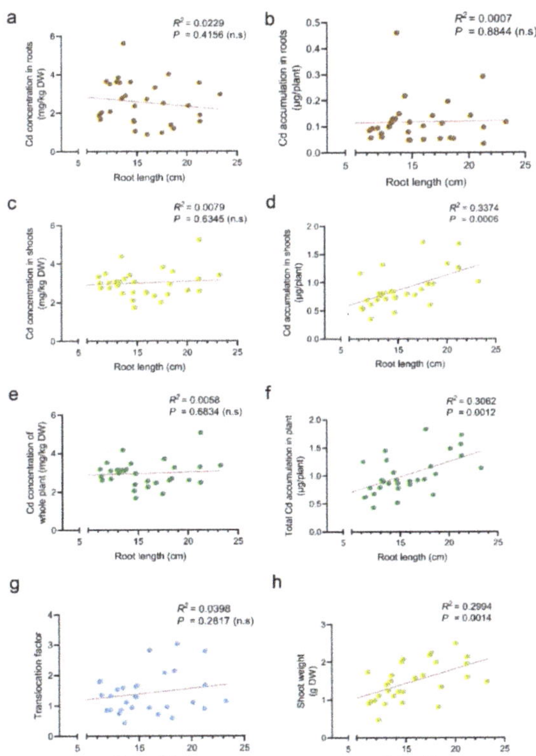

Figure 2. Effect of root length on Cd accumulation in tomato plants. Correlation between root length and Cd concentration in roots (**a**), Cd accumulation in roots (**b**), Cd concentration in shoots (**c**), Cd accumulation in shoots (**d**), Cd concentration of whole plant (**e**), total Cd accumulation in plant (**f**), translocation factor (**g**), and shoot weight (**h**). R^2, coefficient of determination. n.s, no significance.

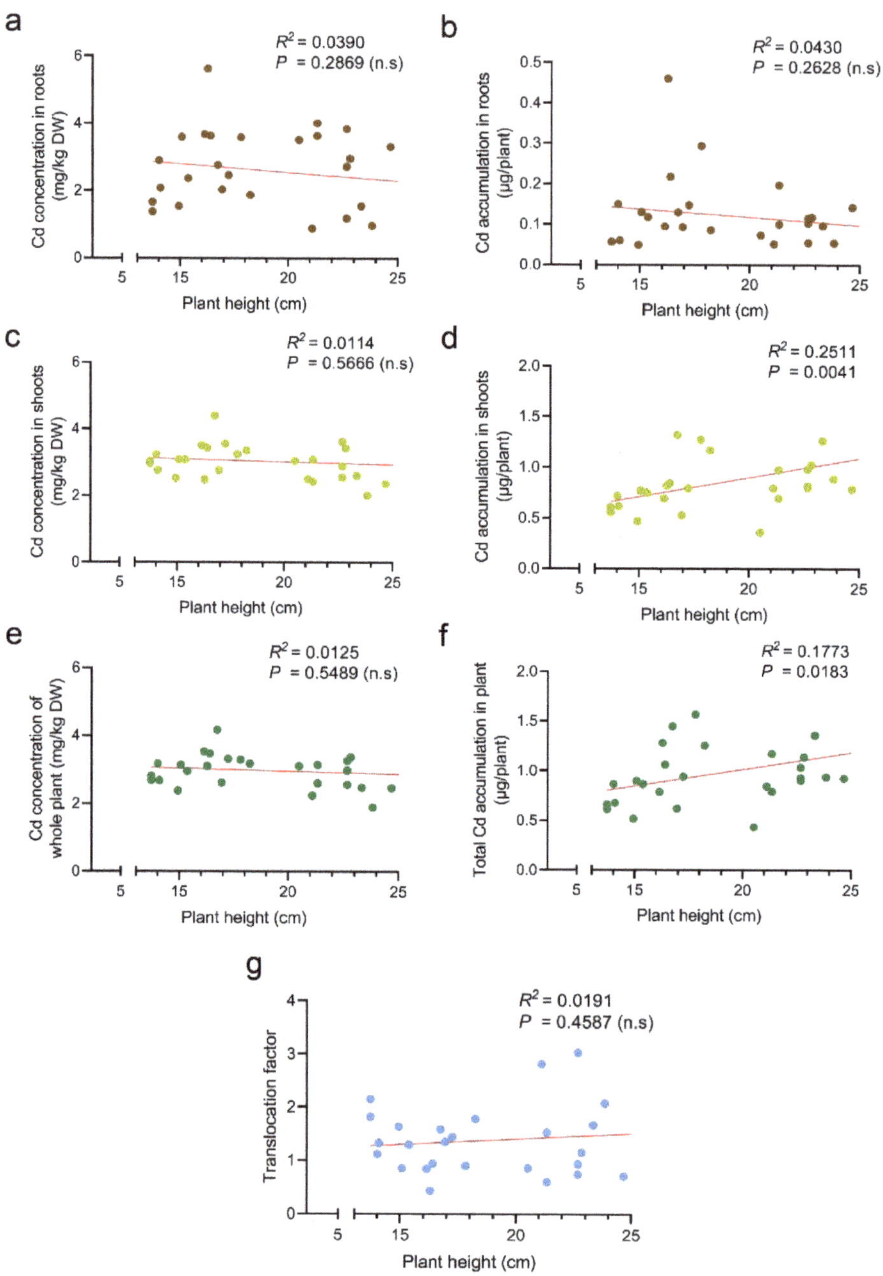

Figure 3. Effect of plant height on Cd accumulation in tomato plants. Correlation between plant height and Cd concentration in roots (**a**), Cd accumulation in roots (**b**), Cd concentration in shoots (**c**), Cd accumulation in shoots (**d**), Cd concentration of whole plant (**e**), total Cd accumulation in plant (**f**), and translocation factor (**g**). R^2, coefficient of determination. n.s, no significance.

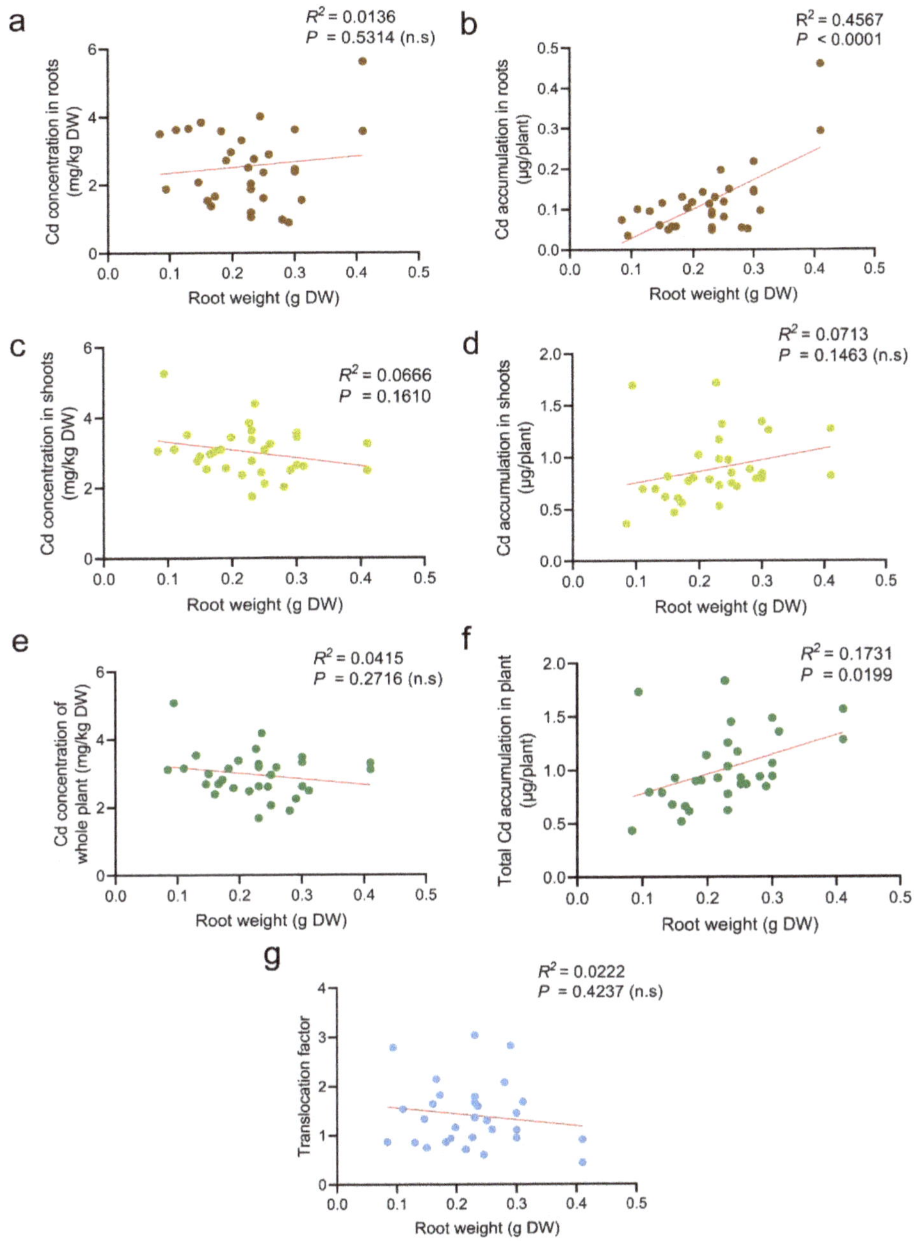

Figure 4. Effect of root weight on Cd accumulation in tomato plants. Correlation between root weight and Cd concentration in roots (**a**), Cd accumulation in roots (**b**), Cd concentration in shoots (**c**), Cd accumulation in shoots (**d**), Cd concentration of whole plant (**e**), total Cd accumulation in plant (**f**), and translocation factor (**g**). R^2, coefficient of determination. n.s, no significance.

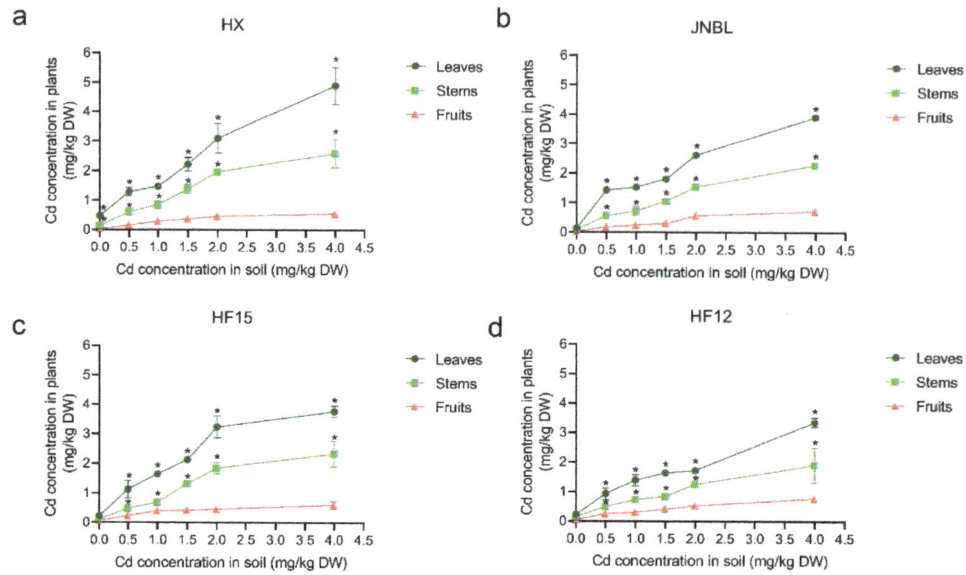

Figure 5. Distribution of Cd in tomato tissues. Cd content in stems, leaves, and fruits of *'HX'* (**a**), *'JNBL'* (**b**), *'HF15'* (**c**), and *'HF12'* (**d**) under 0, 0.5, 1.0, 1.5, 2.0, and 4.0 mg/kg Cd treatments. '*' means a significant difference with Cd concentration in fruits at the 5% significance level by one-way analysis of variance.

3. Results

3.1. Accumulation of Cd in Tomato

The distribution of Cd in different tomato cultivars was found to be different. For instance, the cultivar *'SV3557'* recorded the highest Cd concentration in the shoot but the lowest Cd concentration in the root. The cultivar *'HF11'* had a low Cd concentration in the shoot but higher Cd concentration in the root. Additionally, the Cd concentration in the shoots and roots of the cultivars *'OuDun'*, *'Sybed'*, *'JP6022'* and most of the cultivars was similar (Figure 1a). To facilitate a more comprehensive assessment of Cd accumulation capacity of the 31 tomato cultivars, we computed the Cd accumulation per plant in the tomato cultivars. It is noteworthy that the roots of tomato cultivars exhibited limited capacity for Cd accumulation, a characteristic that can be attributed to their relatively low root biomass (Figure 1b).

3.2. Effect of Root Length on Cd Accumulation in Tomato

To examine the potential influence of growth phenotype of the tomato cultivars on Cd accumulation, we analyzed the relationship between these two variables (Supplementary Table S2). Initially, we observed no significant correlation between the length of roots and the concentration of Cd in the roots, shoots, and overall tomato. This finding suggests that the length of the root does not impact on Cd concentration in tomato plants (Figure 2a,c,e). Additionally, further analysis was performed on the relationship between the length of the roots and Cd accumulation in each individual tomato cultivar. It is noteworthy that no significant correlation was observed between the length of tomato roots and the Cd accumulation in the roots. However, a significant positive correlation was established between the length of tomato roots and Cd accumulation in the shoots (Figure 2b,d,f). This finding indicates that while the length of tomato roots had no significant impact on Cd concentration in tomato plants, it may contribute to an increased accumulation of Cd per plant by enhancing the biomass of tomato plants.

3.3. Effect of Plant Height on Cd Accumulation in Tomato

We investigated the correlation between tomato plant height and Cd accumulation. The results showed that there was no significant correlation between plant height and Cd concentration in the root, shoot, and overall tomato, indicating that plant height may not impact Cd concentration in tomato plants (Figure 3a,c,e). We also analyzed the correlation between plant height and Cd accumulation per tomato plant. There was no significant correlation between tomato plant height and Cd accumulation in the roots, but a significant positive correlation was found between plant height and Cd accumulation in shoots (Figure 3b,d,f). This is because plant height significantly affects the biomass of tomato and directly affects the calculation of Cd accumulation per plant.

3.4. Effect of Root Weight on Cd Accumulation in Tomato

Finally, we analyzed the correlation between tomato root weight and Cd accumulation. The results showed that there was no significant correlation between root weight and Cd concentration in root, shoot, and the whole plant of the tomato, indicating that root weight had no significant effect on Cd concentration in tomato plants (Figure 4a,c,e). We also analyzed the correlation between root weight and Cd accumulation per plant. There was no significant correlation between tomato root weight and Cd accumulation in shoots, but there was a significant positive correlation between tomato root weight and Cd accumulation in roots and the whole plant (Figure 4b,d,f). This finding demonstrates that the accumulation of Cd in tomato roots is a significant component of the overall Cd accumulation process.

3.5. Differences of Cd Accumulation in Tomato Tissues

To investigate the distribution of Cd in tomato tissues, four tomato cultivars *'HX'*, *'HF12'*, *'HF15'*, and *'JNBL'*, were planted in soils with Cd concentrations of 0, 0.5, 1.0, 1.5, 2.0, and 4.0 mg/kg. The findings of this study demonstrate a positive correlation between Cd concentration in soils and the corresponding Cd concentration in tissues of tomato plants. Furthermore, it was observed that the distribution of Cd concentration in the four tomato cultivars followed the pattern of leaf > stem > fruit (Figure 5). The present findings illustrate the spatial movement of Cd from the soil, through the root, and ultimately reaching the shoot. This information contributes to a deeper understanding of the intricate dynamics between Cd and plants.

4. Discussion

Differential accumulation of Cd was observed among different crop cultivars [26,31]. In this study, we investigated the root length, root weight, plant height, and other growth traits of 31 tomato cultivars, as well as their capacity for Cd accumulation in lightly polluted soil (Figure 1 and Table S1). We found that there were differences in the growth phenotype and Cd accumulation ability of the 31 tomato cultivars. The Cd concentration in the *'HF11'* cultivar was high in roots, but low in shoots, whereas in the *'SV3557'* cultivar, it was low in roots and high in shoots. This therefore suggests that Cd concentration in the roots may not accurately represent the Cd absorption capacity of plants. Once more, the findings suggest the presence of certain genes that may be selectively activated in vascular bundles to govern the distribution of Cd across different tissues. This disparity in Cd accumulation among the 31 cultivars examined in this study can likely be attributed to this underlying genetic regulation. In a recent study by Szwalec et al. [32], the accumulation of heavy metals in plants, specifically European Aspen and silver birch, was predominantly observed in the roots. However, the findings of this study indicate that the majority of the Cd accumulation was in the shoot, while the roots exhibited a minimal absolute accumulation of Cd (Figure 1b). Our study suggests that the primary function of the roots is to absorb and transport Cd, rather than serving as a storage site for this element. The observed disparity between our study and previous studies may potentially be attributed to genetic variations in the study materials (tomato).

As previously established [33–35], in environments with high Cd concentrations, reactive oxygen species (ROS) impede plant root growth. Conversely, in environments with low Cd concentrations, the root length of plants is typically unaffected [33–35]. Nevertheless, the relationship between plant root length and Cd accumulation in environments with low Cd concentration remains uncertain. The current study found no significant correlation between the length of tomato roots and Cd concentration in both the root and shoot (Figure 2a,c,e). It is striking that a notable positive correlation was observed between root length and Cd accumulation in the shoots of tomato plants. However, no significant correlation was found between root length and Cd accumulation in the roots of tomato plants (Figure 2b,d,f). Previous studies indicate a correlation between longer root length and enhanced Cd absorption, resulting in higher total Cd accumulation in plants [36,37]. It is noteworthy that a direct correlation exists between the accumulation of Cd and biomass. Therefore, we analyzed the correlation between root length and shoot biomass. The findings indicated a significant positive association between root length and shoot biomass (Figure 2h). Hence, it is possible to argue that the assertion that increased root length directly facilitates the uptake of Cd may not be entirely accurate. An increased root length in plants may lead to a corresponding increase in biomass, thereby resulting in higher overall levels of Cd accumulation. Ultimately, root length did not exhibit any significant alteration in response to varying concentrations of Cd in tomato plants.

Cadmium stress can inhibit plant height and biomass. However, the concentration of cadmium stress varies for different species, and even the appropriate concentration of cadmium treatment can promote plant height and biomass [38–41]. The inhibition of cadmium stress on plant growth is serious. However, there are few studies on whether plant height and biomass affect cadmium uptake in low-cadmium environments that do not affect plant growth. In this study of 31 tomato cultivars treated with mild cadmium pollution, we found that a different plant height and root weight did not cause the difference in cadmium concentration in tomato plants (Figures 3a,c,e and 4,a,c,e). Plant height and root weight were significantly correlated with cadmium accumulation per tomato plant, which was caused by the calculation formula (Figures 3d,f and 4b,f).

Cadmium is absorbed, transported, and distributed from soil to plants. Species vary in their absorptive capacities and the distribution of Cd in different tissues. Wang et al. [27] found that the concentration of Cd in apple organs following Cd treatment ranked in the order of root > leaf > stem. An et al. [25] demonstrated that Cd distribution in maize plants following Cd treatment exhibited a pattern of leaf > stem > fruit. The concentration of Cd in the feeder roots of sweet potato plants was higher compared to other tissues, while the concentration of Cd in the leaf was lower than in the stem [42].Given that the plant growth and development is genetically and integrally controlled [43], this study aimed to investigate potential variations in the distribution of Cd among different tomato cultivars. To achieve this, four cultivars were selected for analysis to determine their respective Cd accumulation capacities. The study found that the concentration of Cd in the stems, leaves, and fruits of the four tomato cultivars increased as the treatment concentration increased. Additionally, the distribution of Cd in the four tomato cultivars followed the pattern of leaf > stem > fruit (Figure 5). In contrast to a study on pepper by Hu et al. [26], their findings indicate notable variations in the distribution of Cd among different pepper varieties. Specifically, the Cd distribution pattern in the pepper variety *'Luojiao 318'* was observed to be fruit > leaf > stem. The genetic background of the four tested tomato cultivars may not differ significantly, resulting in similar distribution of Cd. It is noteworthy that the Cd concentration in tomato fruits by fresh weight is much lower than the maximum permissible concentration of 0.05 mg/kg permitted in China.

5. Conclusions

This study revealed variations in the Cd accumulation capacity among different tomato cultivars. The study found that tomato roots primarily functioned in the absorption and transportation of Cd, rather than storage. The growth phenotypes, including root length, plant height, and root weight, did not have an impact on the concentration of Cd in tomato tissues. This implies that the Cd accumulation ability of tomato can be objectively assessed based on Cd concentration. The concentration of Cd in tomato was highest in the leaves, followed by the stems and then the fruit. This implies that leaves may serve as storage tissues for Cd, exhibiting higher Cd concentrations, while stems may act as transport tissues for Cd, displaying lower Cd concentrations. The low concentration of Cd in fruits may be attributed to the dilution caused by their high-water content.

In future studies, we will assess the Cd accumulation capacity of 506 tomato accessions by measuring their Cd concentrations. Subsequently, we will employ GWAS to identify and isolate the primary loci responsible for Cd accumulation in tomato.

Supplementary Materials: The following supporting information can be downloaded at: https://www.mdpi.com/article/10.3390/horticulturae9121343/s1, Figure S1: The experimental schemes; Table S1: Names and sources of 31 tomato varieties; Table S2: Growth phenotype of 31 tomato varieties.

Author Contributions: Conceptualization, Y.Z. and X.Z.; methodology, X.Z.; formal analysis, X.Z.; investigation, C.Z.; resources, Y.Z.; data curation, C.Z.; writing—original draft preparation, X.Z.; writing—review and editing, Y.Z.; supervision, Y.Z.; project administration, Y.Z.; funding acquisition, Y.Z. All authors have read and agreed to the published version of the manuscript.

Funding: This work was supported by grants from the National Key Research & Development Plan (2022YFD1200502; 2021YFD1200201); National Natural Science Foundation of China (32372696; 31991182); Wuhan Biological Breeding Major Project (2022021302024852); Funds for High Quality Development of Hubei Seed Industry (HBZY2023B004); HZAU-AGIS Cooperation Fund (SZYJY2023022); Hubei Key Research & Development Plan (2022BBA0066; 2022BBA0062); Fundamental Research Funds for the Central Universities (2662022YLPY001).

Data Availability Statement: Data are contained within the article and Supplementary Materials.

Acknowledgments: The authors thank Limei Zhang and Tingyan Zhang from the College of Resources & Environment of Huazhong Agricultural University for technical assistance.

Conflicts of Interest: The authors declare no conflict of interest.

References

1. Briffa, J.; Sinagra, E.; Blundell, R. Heavy metal pollution in the environment and their toxicological effects on humans. *Heliyon* **2020**, *6*, 26. [CrossRef] [PubMed]
2. Xu, M.M.; Wang, X.Y.; Liu, X.P. Detection of heavy metal ions by ratiometric photoelectric sensor. *J. Agric. Food Chem.* **2022**, *70*, 11468–11480. [CrossRef] [PubMed]
3. Khan, M.A.; Castro-Guerrero, N.; Mendoza-Cozatl, D.G. Moving toward a precise nutrition: Preferential loading of seeds with essential nutrients over non-essential toxic elements. *Front. Plant Sci.* **2014**, *5*, 7. [CrossRef]
4. Koleli, N.; Eker, S.; Cakmak, I. Effect of zinc fertilization on cadmium toxicity in durum and bread wheat grown in zinc-deficient soil. *Environ. Pollut.* **2004**, *131*, 453–459. [CrossRef] [PubMed]
5. Li, J.; Liu, J.C.; Yan, C.L.; Du, D.L.; Lu, H.L. The alleviation effect of iron on cadmium phytotoxicity in mangrove *A. marina*. Alleviation effect of iron on cadmium phytotoxicity in for mangrove *Avicennia marina* (Forsk.) Vierh. *Chemosphere* **2019**, *226*, 413–420.
6. Topperwien, S.; Behra, R.; Sigg, L. Competition among zinc, manganese, and cadmium uptake in the freshwater alga *Scenedesmus vacuolatus*. *Environ. Toxicol. Chem.* **2007**, *26*, 483–490. [CrossRef]
7. Pozgajova, M.; Navratilova, A.; Kovar, M. Curative potential of substances with bioactive properties to alleviate cd toxicity: A review. *Int. J. Environ. Res. Public Health.* **2022**, *19*, 42. [CrossRef]
8. Liu, J.; Su, J.Y.; Wang, J.; Song, X.; Wang, H.W. A case study: Arsenic, cadmium and copper distribution in the soil-rice system in two main rice-producing provinces in China. *Sustainability* **2022**, *14*, 14355. [CrossRef]
9. The Ministry of Environmental Protection. *The Ministry of Land and Resources Report on the National Soil Contamination Survey*; The Ministry of Environmental Protection: Beijing, China, 2014.
10. Liu, W.T.; Zhou, Q.X.; Sun, Y.B.; Liu, R. Identification of Chinese cabbage genotypes with low cadmium accumulation for food safety. *Environ. Pollut.* **2009**, *157*, 1961–1967. [CrossRef]
11. Selvam, A.; Wong, J.W.C. Cadmium uptake potential of *Brassica napus* cocropped with *Brassica parachinensis* and *Zea mays*. *J. Hazard. Mater.* **2009**, *167*, 170–178. [CrossRef]
12. Delperee, C.; Lutts, S. Growth inhibition occurs independently of cell mortality in tomato (*Solanum lycopersicum*) exposed to high cadmium concentrations. *J. Integr. Plant Biol.* **2008**, *50*, 300–310. [CrossRef]

13. Chen, X.M.; Shi, X.Y.; Ai, Q.; Han, J.Y.; Wang, H.S.; Fu, Q.S. Transcriptomic and metabolomic analyses reveal that exogenous strigolactones alleviate the response of melon root to cadmium stress. *Hortic. Plant J.* **2022**, *8*, 637–649. [CrossRef]
14. Fusconi, A.; Repetto, O.; Bona, E.; Massa, N.; Gallo, C.; Dumas-Gaudot, E.; Berta, G. Effects of cadmium on meristem activity and nucleus ploidy in roots of *Pisum sativum* L. cv. Frisson seedlings. *Environ. Exp. Bot.* **2006**, *58*, 253–260. [CrossRef]
15. Kubo, K.; Watanabe, Y.; Matsunaka, H.; Seki, M.; Fujita, M.; Kawada, N.; Hatta, K.; Nakajima, T. Differences in cadmium accumulation and root morphology in seedlings of japanese wheat varieties with distinctive grain cadmium concentration. *Plant Prod. Sci.* **2011**, *14*, 148–155. [CrossRef]
16. Liu, Z.L.; Chen, M.D.; Lin, M.S.; Chen, Q.L.; Lu, Q.X.; Yao, J.; He, X.Y. Cadmium uptake and growth responses of seven urban flowering plants: Hyperaccumulator or bioindicator? *Sustainability* **2022**, *14*, 12. [CrossRef]
17. Wu, M.X.; Luo, Q.; Liu, S.L.; Zhao, Y.; Long, Y.; Pan, Y.Z. Screening ornamental plants to identify potential Cd hyperaccumulators for bioremediation. *Ecotoxicol. Environ. Saf.* **2018**, *162*, 35–41. [CrossRef] [PubMed]
18. Bartoli, F.; Coinchelin, D.; Robin, C.; Echevarria, G. Impact of active transport and transpiration on nickel and cadmium accumulation in the leaves of the Ni-hyperaccumulator *Leptoplax emarginata*: A biophysical approach. *Plant Soil* **2012**, *350*, 99–115. [CrossRef]
19. Van der Vliet, L.; Peterson, C.; Hale, B. Cd accumulation in roots and shoots of durum wheat: The roles of transpiration rate and apoplastic bypass. *J. Exp. Bot.* **2007**, *58*, 2939–2947. [CrossRef]
20. Quinet, M.; Angosto, T.; Yuste-Lisbona, F.J.; Blanchard-Gros, R.; Bigot, S.; Martinez, J.P.; Lutts, S. Tomato fruit development and metabolism. *Front. Plant Sci.* **2019**, *10*, 23. [CrossRef] [PubMed]
21. Li, Y.; Zhang, X.C.; Jiang, J.B.; Zhao, T.T.; Xu, X.Y.; Yang, H.H.; Li, J.F. Virus-induced gene silencing of *SlPYL4* decreases the drought tolerance of tomato. *Hortic. Plant J.* **2022**, *8*, 361–368. [CrossRef]
22. Tomato Genome Consortium. The tomato genome sequence provides insights into fleshy fruit evolution. *Nature* **2012**, *485*, 635–641. [CrossRef]
23. Zhou, Z.; Yuan, Y.Q.; Wang, K.T.; Wang, H.J.; Huang, J.Q.; Yu, H.; Cui, X. Rootstock-scion interactions affect fruit flavor in grafted tomato. *Hortic. Plant J.* **2022**, *8*, 499–510. [CrossRef]
24. Zhao, T.T.; Pei, T.; Jiang, J.B.; Yang, H.H.; Zhang, H.; Li, J.F.; Xu, X.Y. Understanding the mechanisms of resistance to tomato leaf mold: A review. *Hortic. Plant J.* **2022**, *8*, 667–675. [CrossRef]
25. An, L.Y.; Pan, Y.H.; Wang, Z.B.; Zhu, C. Heavy metal absorption status of five plant species in monoculture and intercropping. *Plant Soil* **2011**, *345*, 237–245. [CrossRef]
26. Hu, X.T.; Li, T.; Xu, W.H.; Chai, Y.R. Distribution of cadmium in subcellular fraction and expression difference of its transport genes among three cultivars of pepper. *Ecotoxicol. Environ. Saf.* **2021**, *216*, 10. [CrossRef]
27. Wang, Q.; Huang, D.; Niu, D.S.; Deng, J.; Ma, F.W.; Liu, C.H. Overexpression of auxin response gene *MdIAA24* enhanced cadmium tolerance in apple (*Malus domestica*). *Ecotoxicol. Environ. Saf.* **2021**, *225*, 8. [CrossRef]
28. Ministry of Ecology and Environment of the People's Republic of China. *Soil Environmental Quality—Risk Control Standard for Soil Contamination of Agricultural Land*; Ministry of Ecology and Environment of the People's Republic of China: Beijing, China, 2018.
29. Zhang, X.F.; Xia, H.P.; Li, Z.A.; Zhuang, P.; Gao, B. Potential of four forage grasses in remediation of Cd and Zn contaminated soils. *Bioresour. Technol.* **2010**, *101*, 2063–2066. [CrossRef]
30. Rastmanesh, F.; Moore, F.; Keshavarzi, B. Speciation and phytoavailability of heavy metals in contaminated soils in sarcheshmeh area, Kerman Province, Iran. *Bull. Environ. Contam. Toxicol.* **2010**, *85*, 515–519. [CrossRef]
31. Kubo, K.; Kobayashi, H.; Fujita, M.; Ota, T.; Minamiyama, Y.; Watanabe, Y.; Nakajima, T.; Shinano, T. Varietal differences in the absorption and partitioning of cadmium in common wheat (*Triticum aestivum* L.). *Environ. Exp. Bot.* **2016**, *124*, 79–88. [CrossRef]
32. Szwalec, A.; Mundala, P.; Kedzior, R. Suitability of selected plant species for phytoremediation: A case study of a coal combustion ash landfill. *Sustainability* **2022**, *14*, 15. [CrossRef]
33. Waheed, S.; Ahmad, R.; Irshad, M.; Khan, S.A.; Mahmood, Q.; Shahzad, M. Ca_2SiO_4 chemigation reduces cadmium localization in the subcellular leaf fractions of spinach (*Spinacia oleracea* L.) under cadmium stress. *Ecotoxicol. Environ. Saf.* **2021**, *207*, 10. [CrossRef] [PubMed]
34. Yi, L.T.; Wu, M.Y.; Yu, F.; Song, Q.; Zhao, Z.H.; Liao, L.; Tong, J.L. Enhanced cadmium phytoremediation capacity of poplar is associated with increased biomass and Cd accumulation under nitrogen deposition conditions. *Ecotoxicol. Environ. Saf.* **2022**, *246*, 11. [CrossRef] [PubMed]
35. Zhang, Y.N.; Sa, G.; Zhang, Y.; Hou, S.Y.; Wu, X.; Zhao, N.; Zhang, Y.H.; Deng, S.R.; Deng, C.; Deng, J.Y.; et al. *Populus euphratica* annexin1 facilitates cadmium enrichment in transgenic *Arabidopsis*. *J. Hazard. Mater.* **2021**, *405*, 12. [CrossRef] [PubMed]
36. Yu, R.G.; Xia, S.L.; Liu, C.F.; Zhang, Z.; Shi, G.R. Variations in root morphology among 18 herbaceous species and their relationship with cadmium accumulation. *Environ. Sci. Pollut. Res.* **2017**, *24*, 4731–4740. [CrossRef] [PubMed]
37. Zhang, D.Z.; Zhou, H.; Shao, L.L.; Wang, H.R.; Zhang, Y.B.; Zhu, T.; Ma, L.T.; Ding, Q.; Ma, L.J. Root characteristics critical for cadmium tolerance and reduced accumulation in wheat (*Triticum aestivum* L.). *J. Environ. Manag.* **2022**, *305*, 11. [CrossRef] [PubMed]
38. Gu, J.Y.; Hu, C.M.; Jia, X.W.; Ren, Y.F.; Su, D.M.; He, J.Y. Physiological and biochemical bases of spermidine-induced alleviation of cadmium and lead combined stress in rice. *Plant Physiol. Biochem.* **2022**, *189*, 104–114. [CrossRef] [PubMed]
39. Guo, H.D.; Yang, H.L.; Guo, W.L.; Li, X.Z.; Chen, B.H. Defense response of pumpkin rootstock to cadmium. *Sci. Hortic.* **2023**, *308*, 11. [CrossRef]

40. Wang, F.J.; Tan, H.F.; Huang, L.H.; Cai, C.; Ding, Y.F.; Bao, H.; Chen, Z.X.; Zhu, C. Application of exogenous salicylic acid reduces Cd toxicity and Cd accumulation in rice. *Ecotoxicol. Environ. Saf.* **2021**, *207*, 9. [CrossRef]
41. Wu, F.B.; Dong, J.; Cai, Y.; Chen, F.; Zhang, G.P. Differences in Mn uptake and subcellular distribution in different barley genotypes as a response to Cd toxicity. *Sci. Total Environ.* **2007**, *385*, 228–234. [CrossRef]
42. Zhang, D.W.; Dong, F.; Zhang, Y.; Huang, Y.L.; Zhang, C.F. Mechanisms of low cadmium accumulation in storage root of sweetpotato (*Ipomoea batatas* L.). *J. Plant Physiol.* **2020**, *254*, 8. [CrossRef]
43. Zhang, T.Y.; Wang, Y.; Munir, S.; Wang, T.T.; Ye, Z.B.; Zhang, J.H.; Zhang, Y.Y. Cyclin gene *SlCycB1;2* alters plant architecture in association with histone H3.2 in tomato. *Hortic. Plant J.* **2022**, *8*, 341–350. [CrossRef]

horticulturae

MDPI

Article

Preliminary Analysis, Combined with Omics of Chilling Injury Mechanism of Peach Fruits with Different Cold Sensitivities during Postharvest Cold Storage

Wenduo Zhan [1,2,†], Yan Wang [1,2,†], Wenyi Duan [1,2], Ang Li [1,2], Yule Miao [1,2], Hongmei Wang [1,2], Junren Meng [1,2], Hui Liu [1,2], Liang Niu [1], Lei Pan [1,2], Shihang Sun [1,2], Guochao Cui [1], Zhiqiang Wang [1,*] and Wenfang Zeng [1,2,*]

1 Zhengzhou Fruit Research Institute, Chinese Academy of Agricultural Sciences, Zhengzhou 450009, China; zhanwd0628@163.com (W.Z.); 15222763167@163.com (Y.W.); duanwenyi@caas.cn (W.D.); 1125351664la@gmail.com (A.L.); 15670838308@163.com (Y.M.); whm8998@163.com (H.W.); 82101219216@caas.cn (J.M.); liuhui@caas.cn (H.L.); niuliang@caas.cn (L.N.); panley@126.com (L.P.); sunshihang@caas.cn (S.S.); spcgc@163.com (G.C.)
2 Zhongyuan Research Center, Chinese Academy of Agricultural Sciences, Zhengzhou 450009, China
* Correspondence: wangzhiqiang@caas.cn (Z.W.); zengwenfang@caas.cn (W.Z.)
† These authors contributed equally to this work.

Abstract: The storage of peach fruits at 4–5 °C can easily lead to chilling injury and greatly reduce the quality and commercial value of peach fruits. In this study, two kinds of peach fruits (CX and CM) were selected to analyze the mechanisms of chilling injury in fruits with different chilling sensitivity by means of their lipidomic, transcriptome, and dynamic changes in plant hormones. We found that the ethylene, abscisic acid (ABA), and lipid contents changed differently between CX and CM. The ABA and dilactosyl diacylglycerol (DGDG) contents significantly increased after refrigeration in CM fruit, leading to strong cold resistance. However, low temperatures induced a greater accumulation of ethylene, phospholipids, and ABA-GE in CX fruit than in CM fruit, eventually leading to more severe CI symptoms in CX fruit. Additionally, a transcriptional regulatory network for CM and CX fruits during cold storage was constructed, providing a new theoretical reference for the cultivation of cold-resistant peach cultivars and the development of postharvest preservation technology.

Keywords: peach fruits; chilling injury; plant hormones; plant lipids; transcriptional factors

Citation: Zhan, W.; Wang, Y.; Duan, W.; Li, A.; Miao, Y.; Wang, H.; Meng, J.; Liu, H.; Niu, L.; Pan, L.; et al. Preliminary Analysis, Combined with Omics of Chilling Injury Mechanism of Peach Fruits with Different Cold Sensitivities during Postharvest Cold Storage. *Horticulturae* **2024**, *10*, 46. https://doi.org/10.3390/horticulturae10010046

Academic Editors: Adalberto Benavides-Mendoza, Yolanda González-García, Fabián Pérez Labrada and Susana González-Morales

Received: 23 October 2023
Revised: 27 December 2023
Accepted: 28 December 2023
Published: 2 January 2024

1. Introduction

The postharvest climacteric fruit will produce a lot of ethylene if it is stored at room temperature [1]. As a commonly used postharvest preservative technology, low-temperature storage can effectively delay the release rate of ethylene so as to delay tissue senescence and prolong the storage life of fruits to a certain extent. However, some fruits exposed to low temperatures below a certain threshold for long enough will suffer irreversible damage, which is called "chilling injury" (abbreviated as CI) [2]. Fruits with chilling injury usually have sunken epidermises [3], browning peel or pulp [4], etc. The internal browning of peach fruit caused by CI was most serious when stored at between 4 °C and 5 °C, but CI was not obvious at 0 °C [5].

While the fruits exhibited symptoms of chilling injury, the levels of endogenous hormones and metabolites were also affected to a certain extent. Chilling injury can increase the activity of Acs, the rate-limiting enzyme of ethylene biosynthesis in fruits and vegetables, accelerate the transformation of SAM into ACC, and lead to the accumulation of ACC. For example, the accumulation of ACC in cucumber fruits at 2.5 °C is much higher than 12.5 °C. [6]. While it is an important hormone in plants, ABA participates in the response to low-temperature stress, plays an important role in regulating plants' cold resistance, and is considered to be the initiating factor of cold resistance gene expression [7]. ABA can eliminate excessive H_2O_2 by promoting antioxidant enzymes and the AsA-GSH pathway, thus reducing chilling injury in peach fruits [8]. Low temperatures or exogenous ABA treatment can reduce the degree of chilling injury to cold-sensitive fruits by changing the balance of ABA in plants [9]. In kiwifruit, appropriate concentrations of exogenous ABA promoted the synthesis of lignin in the pericarp during low-temperature storage, protected the fruit from the external environment, and reduced the accumulation of lignin in the pulp and the appearance of black spots on the pericarp [10].

Low-temperature stress leads to a decrease in cell membrane fluidity, a change in the cell membrane from a liquid crystal phase to a gel state, and, finally, to an increase in membrane permeability. Long-term low-temperature stress leads to symptoms of chilling injury [11,12]. The damage to cell membranes under low-temperature stress is accompanied by changes in fatty acid unsaturation and membrane lipid composition [13,14]. Under low-temperature stress, plants will synthesize more unsaturated fatty acids and improve the fluidity of their cell membranes [15]. As the most important component of the cell membrane, phospholipids mainly include PA (phosphatidic acid), PC (phosphatidylcholine), PE (phosphatidylethanol), PG (phosphatidylglycerol), PI (phosphatidylinositol), and PS (phosphoinositide synthase) [16]. Previous studies have shown that the increase in membrane lipid content is consistent with the increase in chilling tolerance, and the unsaturation of membrane fatty acids is positively correlated with the fluidity of membrane lipids [17]. The structure of MGDG in glycolipids changes from conical to cylindrical under low-temperature stress, thus maintaining membrane stability [18]. Peach fruits treated with CS can maintain high membrane fatty acid unsaturation and low phosphatidic acid content, which effectively alleviates CI and maintains the integrity of the cell membrane [19].

Metabolic group, transcriptome, and other combination techniques have been widely used in many horticultural crops to reveal their response mechanism to environmental stress during growth and development or maturation and softening For example, the physiological and biochemical changes in peach fruits in the dimensions of the transcriptome, proteome, and metabolic groups during postharvest chilling injury; the molecular mechanism of peach fruit responding to low-temperature stress; and a simple and non-destructive method for the detection and evaluation of fruit chilling injury during transportation and storage can be used [20]. In addition, the integration of metabolic groups and transcriptomics laid the foundation for elucidating the molecular mechanisms of plants' responses to low-temperature stress. Arabidopsis E3 ligase PUB25/26 participates in plants' responses to low-temperature signals by negatively regulating the protein stability of transcription factor *MYB15* in the CBF signaling pathway [21]. Some fruits are prone to chilling injury when they are stored for a long time at unsuitably low temperatures. However, there have been few studies of the mechanism of fruit chilling injury combined with multi-group analysis.

Peach is a kind of fruit that is sensitive to cold, and its CI sensitivity depends on its genetic background, ripening stage, and orchard factors, such as its agronomic management and environmental conditions. The sensitivity levels of different peach varieties to 0 °C and 5 °C storage are very different [20]. In this study, we conducted a comprehensive analysis of lipid groups and transcriptional groups regarding the occurrence of chilling injury in fruits induced by low temperatures. According to the degree of chilling injury of fruits after low-temperature storage, two peach varieties with different chilling sensitivities, namely Chunxue (CX) and Chunmei (CM), were selected. Combined with lipid group and transcriptome analysis, the metabolic network related to fruit chilling injury was constructed, and the key transcription factors (TFs) that directly activate structural genes to affect the occurrence of chilling injury were screened, in order to further clarify the occurrence mechanism of fruit chilling injury and provide a theoretical basis for improving the commercial value of peach fruits.

2. Materials and Methods

2.1. Plant Materials and Treatments

Two peach varieties, namely CX and CM, were selected from the Zhengzhou Fruit Research Institute of the Chinese Academy of Agricultural Sciences. In the S4 I stage, fruits with the same size, uniform color, no diseases or insect pests, and no mechanical injury were picked, and at least 90 fruits were collected from each variety. The harvested fruits were immediately taken back to the laboratory and stored at 4 °C for 40 days. Samples were taken and physiological indices were measured at 0, 8, 16, 24, 32, and 40 days after storage. The pulp was then cut into small pieces, quickly frozen in liquid nitrogen, and stored at −80 °C for further use. Each sample had three biological replicas, and each repeat contained five fruits.

2.2. *Determination of Ethylene and ABA Content and Firmness*

Firstly, the fruit was transferred from 4 °C to room temperature (25 ± 2 °C) for 2 h, and the fruit was then placed in an airtight 2000 mL buckle box with a plug sealed at room temperature for 2 h; 1mL of gas was absorbed from the sample box using the injection needle, and the gas was then injected into a gas chromatograph (GC2010 Plus, Shimadzu, Kyoto, Japan) for the determination of ethylene production. Each box of samples was tested 3 times.

Two peach cultivars (CX 0 d, 8 d (ethylene release initiation L1), and 32 d (ethylene release peak L2) and CM 0 d, 24 d (ethylene release initiation L1), and 32 d (ethylene release peak L2)) were selected to determine the ABA contents. The 0.2 g liquid nitrogen frozen fruit was ground into powder using a high flux tissue grinder; then, a 1mL extract (methanol:isopropanol:acetic acid = 20:79:1 $v/v/v$) was added to extract overnight at 4 °C, 14,000× g centrifugal 5 min at 4 °C, precipitation and 1mL extract were extracted repeatedly, and the supernatants were combined twice and filtered through 0.45 μm disposable needle filter. Determination of ABA content by High performance liquid Chromatography (HPLC, LC-2030CD, Shimadzu, Japan).

The firmness of fruit repeatedly analyzed 5 times was detected using a TA-XTPLUS texture analyzer. A probe with a diameter of 5 mm (Godaming stable Microsystems, Godalming, UK) was also used. After removing the 1 mm peel, the firmness of each fruit was measured twice in the equatorial region.

2.3. *Lipidomics Analysis*

Three biological repetition lipid groups were sequenced in the 0 d, L1, and L2 periods for CX and CM. Lipid extraction was completed by Shanghai Applied Protein Technology Co., Ltd. (Shanghai, China). We ground the sample with liquid nitrogen, weighed the sample to 40 mg, added 200 μL water for mixing, added 240 μL pre-cooled methanol, performed vortex mixing, added 800 μL MTBE, repeated vortex mixing, placed the sample at room temperature for 20 min with 8000× g at 10 °C and centrifuged it for 15 min, took the upper organic phase, dried it with nitrogen, added 100 μL isopropanol solution for mass spectrometry analysis, vortexed it, centrifuged it for 15 min with 8000× g at 10 °C, and performed sample injection analysis. Lipids were separated via UHPLCNexeraLC-30A ultra-high-performance liquid chromatography. LipidSearch software version4.1 was used for peak identification, lipid identification (secondary identification), peak extraction, peak alignment, and quantification.

2.4. *Transcriptome Sequencing*

Transcript sequencing was used to analyze the differentially expressed genes of different cold-sensitive peaches during storage. According to the ethylene release rate and chilling injury symptoms of the two peach varieties during different storage periods, three repeated transcriptome sequences were carried out at the 0 d, L1, and L2 stages of CX and CM. The total RNA extraction, library construction, and RNA-Seq steps were completed by Nuoke Bioinformatics Co., Ltd. (Beijing, China). The library was sequenced via the Illumina NovaSeq6000 platform.

2.5. Analysis of Differential Genes by WGCNA

Based on the original data of ethylene release, firmness, ABA content, lipid content, and all gene expression determined via transcriptomics, the optimal beta value was determined via pretreatment using R language and Rstudio, and the visual cluster map and heatmap were constructed via a one-step method. According to the correlation degree between physiological data and gene expression, the color depth of different clustering modules in the heatmap was generated, i.e., the correlation degree. We used Gephi0.9.2 software to visualize the model network.

2.6. Prediction of Transcription Factors and Enrichment Analysis of Cis Motifs

The promoter sequence (2000 bp upstream of the transcriptional initiation site (TSS)) was retrieved from the peach genome (GDR; https://www.rosaceae.org/, accessed on 2 December 2023). Then, the TF-binding site (TFBS) of *Arabidopsis thaliana* was predicted using PlantTFDB v5.0 software (http://www.planttfdb.gao-lab.org/, accessed on 2 December 2023).

2.7. Data Analysis

The statistical significance of the differences was analyzed using Microsoft Office Excel 2019. We used GraphPad Prism 8 to draw the graphics and used TBtools to draw the heatmaps.

3. Results

3.1. Changes in Chilling Injury Symptoms and the Ethylene Contents of Peaches with Different Cold Sensitivities during Low-Temperature Storage

CM and CX, two different peach varieties, showed different cold sensitivities to low temperatures. CX began to release ethylene on day 8 after low-temperature storage, while CM began to release ethylene on day 24 after low-temperature storage. The fruits of the two peaches reached peak ethylene release at the same time, i.e., at 32 d (Figure 1B). At this time, the chilling injury symptoms of CX were more obvious than those of CM, and there was a watery change in the pulp (Figure 1A).

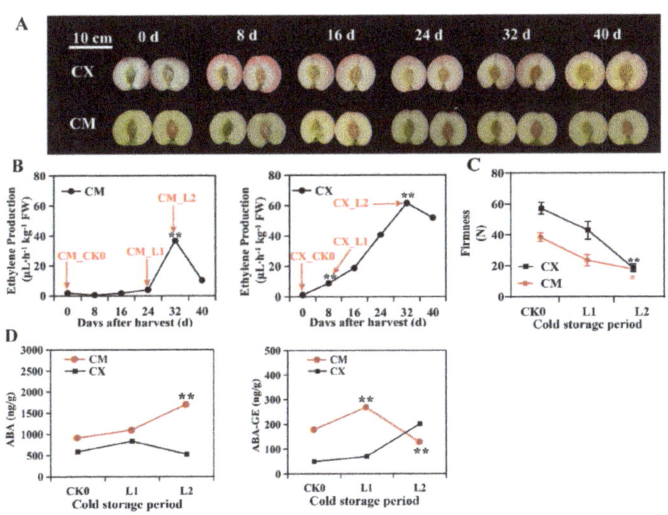

Figure 1. Phenotypic and physiological indices of different cold-sensitive peaches during low-temperature storage: (**A**) fruit profiles and chilling injury phenotypes of CX and CM during low-temperature storage; (**B**) ethylene release rate; (**C**) firmness change; (**D**) changes in ABA contents (* $p < 0.05$, ** $p < 0.01$).

3.2. Changes in ABA Contents and Fruit Firmness in Peaches with Different Cold Sensitivities during Low-Temperature Storage

During low-temperature storage, the ABA contents of CM peaches were always higher than those of CX peaches, and the ABA contents increased greatly in the L2 stage, but the ABA contents of CX fruit remained at low levels and exhibited no significant changes. ABA-GE first increased and then decreased in CM peaches, but it continued to rise in CX peaches, and there was a significant difference between the two varieties in the L2 period (Figure 1D). By measuring the changes in fruit firmness in the two kinds of cold-sensitive peaches during low-temperature storage, it was found that the decrease in the firmness of the CX peach was significantly higher than that of CM from L1 to L2, which was related to the degree of chilling injury, and the chilling injury degree of CX peach was significantly higher than that of CM peach fruit (Figure 1C).

3.3. Changes in Lipid Compositions and Contents of Peaches with Different Cold Sensitivities during Low-Temperature Storage

The lipids in cell membranes are very important for plants to maintain cell-membrane fluidity and adapt to low-temperature stress. In order to analyze the lipid changes in different cold-sensitive peaches during low-temperature storage, lipidomics analysis was carried out on two peach varieties at the 0 d, L1, and L2 stages. Phospholipids were the main compounds in the CX and CM 0 d samples, and their components changed to some extent during low-temperature storage. The proportions of PA and diglyceride (DG) in CM peaches during L1 and L2 increased, while those of PC and PS decreased; the proportion of PA slightly decreased in L1 and L2 for CX peaches, while PC increased significantly in L2, accounting for about 40% of the total lipid content (Figure 2A).

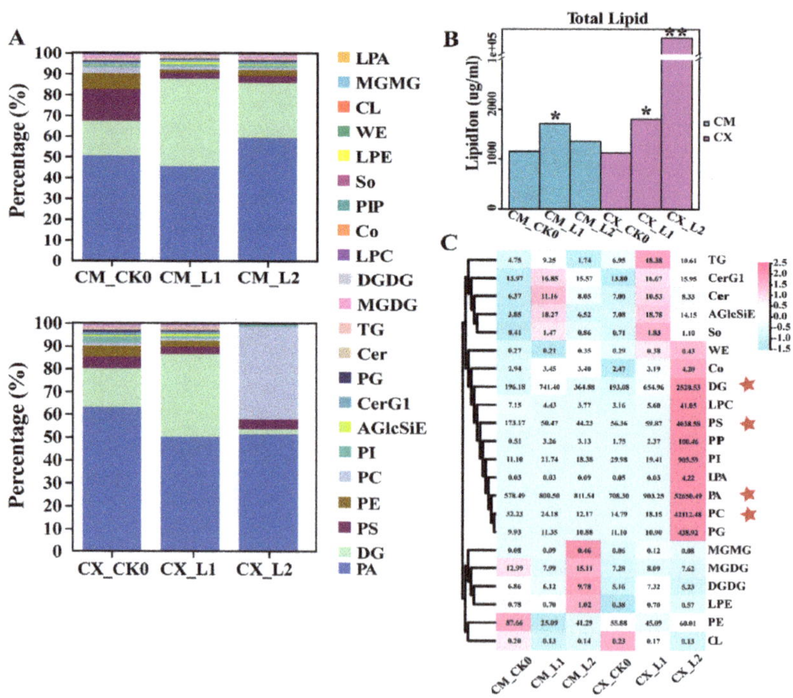

Figure 2. Changes in the lipid compositions and contents of different cold-sensitive peaches during low-temperature storage: (**A**) changes in the lipid composition of CX and CM fruits during low-temperature storage; (**B**) changes in the total lipid contents of CX and CM fruits; (**C**) thermographic analysis of the changes in lipid contents in CX and CM fruits (* $p < 0.05$, ** $p < 0.01$).

The total lipid contents of CX peach fruits changed greatly during low-temperature storage, especially during the L2 stage, where the contents of glycerol phospholipids (PA, PS, PC, PI, PG, PIP, PE) and DG significantly increased (Figure 2A), but the total lipid contents of CM peaches were always at low levels during low-temperature storage (Figure 2B). Specifically, there were great differences in some lipid contents between the two peach varieties. For example, the contents of PC and PS in CX fruits greatly increased during low-temperature storage, but they decreased in CM fruits. The contents of monolactosyl diacylglycerol (MGDG) and digalactosyl diacylglycerol (DGDG) increased in CM fruits. However, no significant changes in these compounds were observed in CX fruits (Figure 2C). There were no significant changes in unsaturated lipids in CM fruits during low-temperature storage. The unsaturated contents of CX increased significantly during L2 storage at low temperatures (Figure S1).

3.4. Transcriptional Analysis of Differentially Expressed Genes

In order to understand the expression of genes related to chilling injury in different peach varieties during low-temperature storage, transcriptome sequencing was carried out in CX and CM at the 0 d, L1, and L2 periods. Correlation analysis showed that the Q30 score was more than 93%. More than 94% of the sequence readings could be aligned with the peach reference genome, and the accuracy and quality of the sequencing data were sufficient for further analysis. We compared CX to CM during low-temperature storage, and a total of 7860 differentially expressed genes were identified at 0 d, including 3938 upregulated genes and 3922 downregulated genes (Figure 3A). In the L1 period, a total of 6758 differentially expressed genes were identified, including 3286 upregulated genes and 3472 downregulated genes (Figure 3B). In the L2 period, the number of differentially expressed genes reached 11,037, including 5595 upregulated genes and 5442 downregulated genes (Figure 3C). We explain the differentially expressed genes using the Venn diagram below. The results showed that there were 4565 common genes between CX and CM, of which 219 genes were differentially expressed between CX and CM during low-temperature storage (Figure 3D).

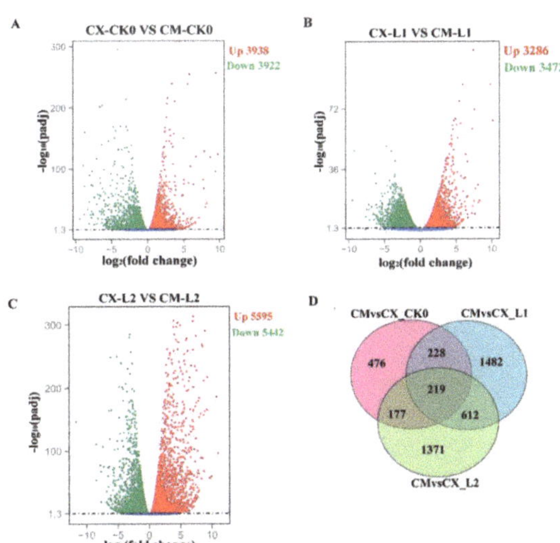

Figure 3. Transcriptome analysis of CX and CM: (**A**) analysis of differentially expressed genes between CX and CM during low-temperature storage for 0 d; (**B**) analysis of differentially expressed genes between CX and CM during the L1 stage of low-temperature storage; (**C**) analysis of differentially expressed genes between CX and CM during the L2 stage of low-temperature storage; (**D**) Venn diagram analysis of common differentially expressed genes of the different cold-sensitive peaches noted during low-temperature storage.

3.5. Analysis of Differential Expression of Genes Related to Ethylene Biosynthesis and Signal Transduction in Peach Fruits during Low-Temperature Storage

According to the DEGs obtained via RNA-Seq, 11 differentially expressed genes related to ethylene biosynthesis and signal transduction were screened. The transcriptional abundance of these genes was estimated via FPKM (thousands of base fragments per million pieces of localized transcripts) from the RNA-Seq data. The FPKM value was used to construct a heatmap to estimate the expression levels of these selected genes. Among the 19 differentially expressed genes related to the ethylene pathway, the mRNA transcription of the ethylene biosynthesis genes *SAMS*, *ACS1*, *ACO1*, and *ACO2* rapidly accumulated during L2 storage of CX, while the mRNA levels of *SAMS*, *ACS2*, and *ACO1* in CM increased during L1 storage, and the gene expression levels of *ACO2* and *ACS2* were opposite to those found in CX. The expression of the ethylene signal transduction genes *CTR1-2* and *EBF-1* significantly increased during L1 storage in CX, and *ETR* and *EBF-2* expression increased in the L2 stage, while *CTR-1* expression decreased during CX cold storage. In CM, the expression of *ETR* significantly increased in the L1 stage, and the expression levels of *CTR1*, *EIN2*, and *EBF* increased in the L2 stage (Figure 4A).

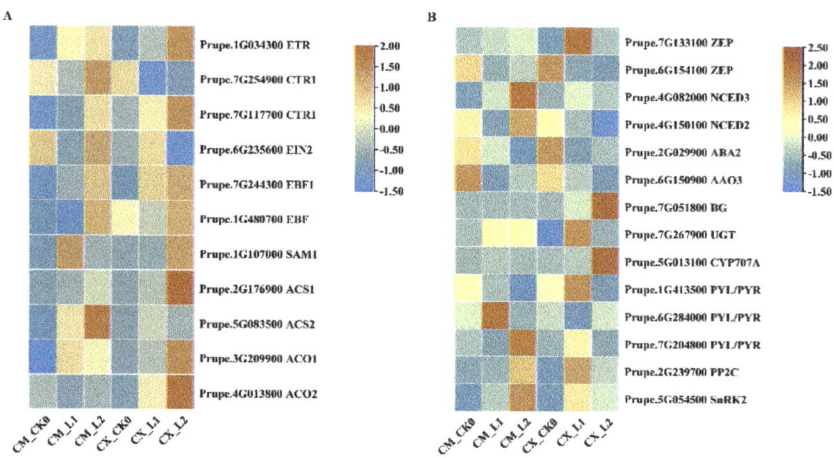

Figure 4. (**A**) Gene expression analysis of the ethylene biosynthesis and signal transduction pathway. (**B**) Analysis of gene expression in the ABA biosynthesis and signal transduction pathway.

3.6. Differential Expression Analysis of Genes Related to ABA Metabolism and Signal Transduction during Low-Temperature Storage

Through stress signal transduction, ABA initiates the stress response in plants and improves the ability of plants to resist various stress factors. After screening and analysis, the expression levels of the ABA-synthesis-related genes *NCED2* and *NCED3* in CM fruits significantly increased during low-temperature storage, but there were no significant differences in CX fruits. The ABA-metabolism-related genes *CYP707A* and *BG1* were significantly upregulated during the L2 stage in CX fruit, and the expression level of the *UGT* gene significantly increased during the L1 stage in CM fruit. The expression levels of genes related to the ABA signal transduction pathway also changed, and there were significant differences in the expression levels of *PYL*, *PP2C*, and *SnRK2* between the two varieties (Figure 4B).

3.7. Differential Expression Analysis of Genes Related to Lipid Metabolism in Peach Fruits during Low-Temperature Storage

Lipids are important compounds in living organisms, as they can participate in and regulate a variety of life activities, and they play an important physiological role in the process of plants' responses to low-temperature stress. In the transcriptome data related to the fatty acid metabolism pathway, seven genes encoding key enzymes were screened, including *PDH*, *ACC*, *MCD*, *KAR*, *SAD*, and *ENR-2*, which showed downregulated expression during the cold storage of CM fruits. This was highly positively related to the decrease in the contents of unsaturated fatty acids. During the low-temperature storage of CX peaches, the expression of *MCD*, *ENR*, and *SAD* decreased, while that of *PDH* and *KAR* first decreased and then increased, and that of *ACC* first increased and then decreased.

The contents of lipids (especially phospholipids) changed during the low-temperature storage of CX and CM fruits. Combined with the RNA-Seq data, we found that the transcription levels of *PSS*, *CDS*, *DGK*, *PAP*, *FAD*, *PLC*, *PLP ζ*, and *AAPT* were all upregulated in CX peaches, which were positively correlated with the increase in phospholipid content. The *MGDG* and *DGDG* contents increased during the low-temperature storage of CM peaches, while the related *MGD* and *DGD* contents increased at the gene level (Figure 5).

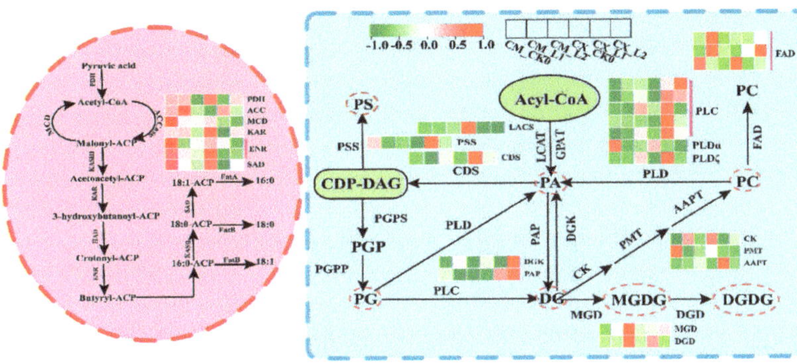

Figure 5. Gene expression analysis of lipid metabolic pathways.

3.8. Analysis of Genes Related to Different Chilling Sensitivity Traits of Peach Fruits Based on WGCNA

A co-expression network was constructed using WGCNA to determine the differentially expressed genes of different cold-sensitive peach fruits during low-temperature storage. Through WGCNA analysis, the 4290 DEGs identified were divided into nine co-expression modules. The turquoise module was highly positively correlated with ethylene, ABA-GE, total lipid content, DG, PA, PC, and PS (with correlation coefficients of 0.58, 0.91, 0.89, 0.87, 0.9, and 0.78, respectively) and negatively correlated with ABA, MGDG, and DGDG (with correlation coefficients of −0.45, 0.33, and 0.26, respectively). The correlation analysis of the red module, except for the ethylene module, showed the opposite trend to that of the turquoise module. GO enrichment analysis showed that the genes in the turquoise module were responsive to stress, contained oxides, and participated in hormone and lipid metabolism in the cell membranes, while the red module's genes were mainly involved in stress stimulation, ABA, lipids, and transcriptional regulation (Figure 6).

Combined with the PlantTFDB database, 44 differentially expressed TFs were identified in the turquoise module. These TFs were divided into 19 families. There were 10 WRKY family genes, 6 ERF family genes, 4 NAC family genes, 3 bZIP family genes, 3 HSF family genes, and 3 MYB family genes. The expression of genes in the major TF families can be found in Figure 7A. Among all TF families, the expression of ERFs was the highest, followed by WRKYs and bZIPs. *ERF5* (Prupe.5G061800), *ERF2* (Prupe.4G055500), *WRKY40* (Prupe.3G098100), and bZIP TF *HY5* (Prupe.1G478400) had the highest expression levels and low-temperature responses in the two peach varieties, and the expression levels in CX were significantly higher than those in CM. *ERF061* (Prupe.5G117800), *WRKY35* (Prupe.8G265900), and *ABF3* (Prupe.8G126600) were downregulated in CX and CM fruits during cold storage, but their transcription levels in CX were higher than those in CM fruits (Figure 7B,C). A correlation analysis showed that the expression levels of genes related to ethylene, ABA, and lipid metabolism were positively correlated with those of *ERF5* (Prupe.5G061800), *ERF2* (Prupe.4G055500), *WRKY40* (Prupe.3G098100), and *HY5*

(Prupe.1G478400), as well as negatively correlated with those of *ERF061* (Prupe.5G117800), *WRKY35* (Prupe.8G265900), and *ABF3* (Prupe.8G126600). In the red module, 34 TFs belonging to 21 families were identified. Four bHLH family genes, four bZIP genes, and four NAC family genes accounted for the largest proportion, of which the bZIP genes' expression levels were the highest, followed by the bHLH genes. The transcriptional levels of *PRE5* (Prupe.3G269500) and *bZIP44* (Prupe.8G091600) were significantly higher than those of the other TF genes, and they were positively correlated with *NCED2*, *NCED3*, and *MGD* (Figure 7D,E).

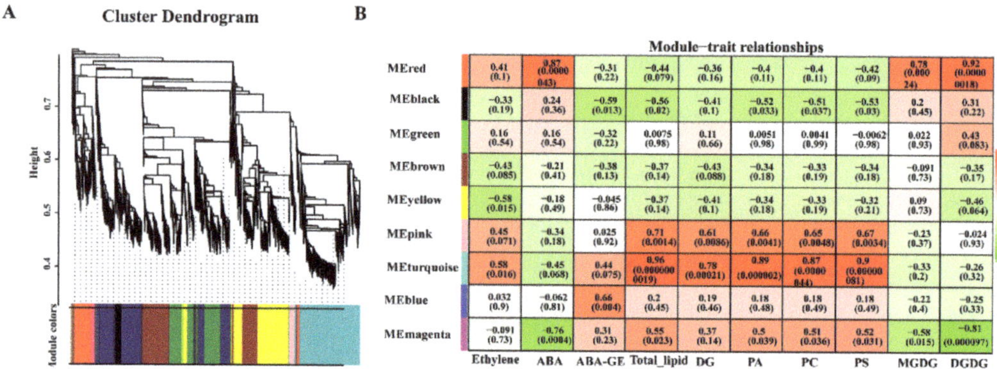

Figure 6. Correlations of metabolites with physiological indices based on WGCNA: (**A**) clustering dendrogram of the average network adjacency for the identification of metabolite co-expression modules; (**B**) module–trait relationships.

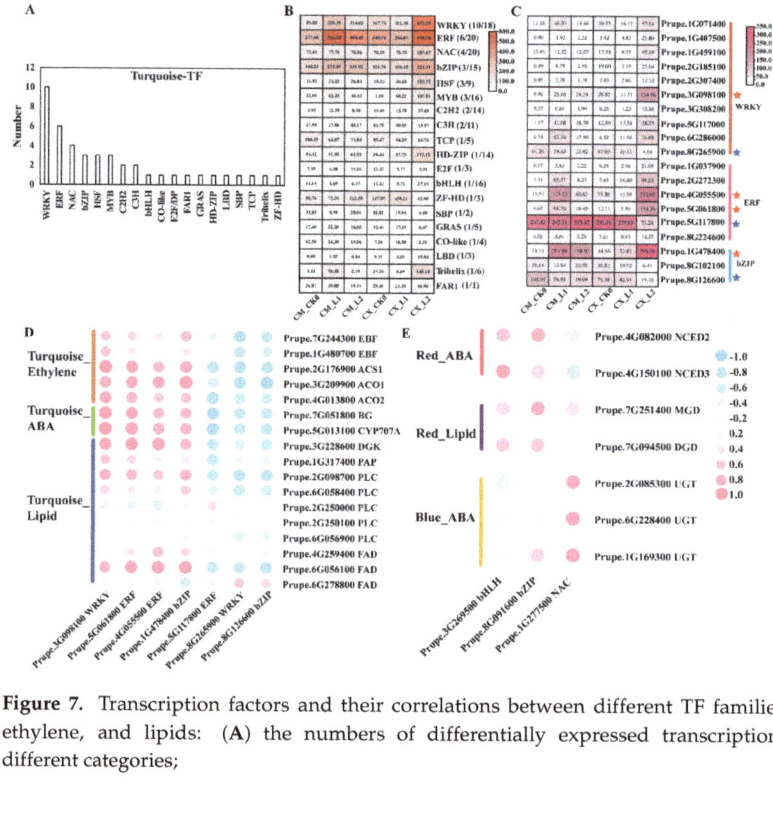

Figure 7. Transcription factors and their correlations between different TF families and ABA, ethylene, and lipids: (**A**) the numbers of differentially expressed transcription factors in different categories;

(**B**) the total FPKM heatmap of all transcription factors (TFs) of a specific TF family; (**C**) differential expression profiles of the WRKY, ERF, and bZIP families; (**D**) the correlation between differentially expressed WRKYs, ERFs, and bZIPs and differentially expressed genes related to ethylene, ABA, and lipid metabolism; (**E**) expression profiles of differentially expressed bHLH, bZIP, NAC, and ABA- and lipid-metabolism-related genes in the red and blue modules.

3.9. Analysis of Cis-Acting Elements of Key Structural Gene Promoters in Different Cold-Sensitive Peaches during Low-Temperature Storage

In order to further explore the relationships between TFs and ethylene-, ABA-, and lipid-metabolism-related genes, the cis element of the structural promoter was analyzed. The common binding sites of ERF, bZIP, and WRKY family genes were identified in the promoters of the ethylene biosynthesis genes *ACS1*, *ACO1*, and *ACO2*. The common binding sites of the bZIP, WRKY, ERF, bHLH, and NAC family genes were identified in the promoters of the ABA biosynthesis genes *CYP707A*, *NCED2*, *NCED3*, *BG1*, and *UGT*. The common binding sites of the bZIP, WRKY, ERF, and bHLH family genes were identified in the promoters of *DGK*, *PAP*, *PLC*, *FAD*, and *DGD* (Figure S8).

4. Discussion

4.1. The Levels of Endogenous Hormones in Peach Fruits Change with the Occurrence of Chilling Injury Symptoms

Low-temperature storage is the main way of prolonging the postharvest life of fresh fruits. When exposed to temperatures lower than the optimal growth temperature, cold-sensitive plants or specific plant organs may become damaged [20,22]. This study's results showed that when peach fruits were stored at about 4–5 °C, there would be severe chilling injury symptoms, such as pulp browning and flavor loss [5,23,24]. The CX and CM fruits of two different cold-sensitive peaches were picked and stored at 4 °C during commercial ripening. The phenotypic observation showed that CX fruits showed serious cold injury symptoms, but this phenotype was not obvious in CM fruits (Figure 1A). This phenomenon shows that the sensitivity of different peach varieties to low temperatures is different. Low temperatures can induce the release of endogenous ethylene, which is related to the increase in membrane permeability, which, in turn, aggravates the membrane damage, resulting in increased membrane permeability and the destruction of cell partitions [25]. We measured the ethylene release rates of CX and CM fruits during low-temperature storage, and CX began to release ethylene on the 8th day of low-temperature storage. However, ethylene was released from CM fruits only on the 24th day of low-temperature storage (Figure 1B). After low-temperature storage, the *SAMS*, *ACS1*, *ACO1*, and *ACO2* genes in CX fruits were significantly upregulated, and the expression of *ACS1* and *ACO1* in CM fruits was also upregulated, but to a significantly lower extent than that in CX fruits, which was positively correlated with the levels of ethylene production of the two varieties. It may be speculated that the great increase in endogenous ethylene during low-temperature storage may be a sign of chilling injury to some extent (Figure 4A).

Under low-temperature conditions, ABA accumulated in plants can regulate the cold-response genes [26], promote increases in the contents of some osmotic regulators in plants, and improve the stability of membranes, thus improving the cold resistance of plants [27]. The ABA contents of CM fruits were always at high levels during low-temperature storage, and there was a significant difference in ABA content between the L1 and L2 fruits of CX. (Figure 1D). Compared to CX, the expression levels of *NCED2* and *NCED3* in CM fruits were higher, indicating that the production of ABA between the two varieties was regulated by NCEDs (Figure 4B). At the same time, this result indicates that the two varieties of peach fruits have different cold resistances, as well as that the low-temperature resistance of CM fruits is significantly higher than that of CX fruits. ABA can combine with glucosyl to form inactive ABA-GE and rapidly release ABA through the action of β-glucosidase so as to regulate the contents of active ABA and make plants adapt to physiological and environmental changes [28]. During low-temperature storage, the contents of ABA-GE in

CM fruits significantly decreased, but the contents of ABA-GE in CX fruits showed the opposite trend, which was dynamically balanced with the contents of ABA in the two peach fruits during low-temperature storage so as to regulate the resistance of peach fruits to low temperatures (Figure 1D). The expression of UGT was significantly high in CM fruits at the L1 stage, while the expression of BG1 in the L2 stage was the highest in CX fruits, indicating that the two different peach varieties maintained a steady state of ABA content through different ABA metabolic processes (Figure 4B).

4.2. The Changes in the Lipid Levels of Peach Fruits with Different Cold Sensitivities Were Different during Low-Temperature Storage

In this study, a liposome analysis of CX and CM fruits during low-temperature storage was carried out. We found that the lipid composition and contents of the two different peach varieties changed after low-temperature storage. As the original site of cell injury caused by low temperatures, the cell membrane is a semi-permeable membrane composed of two layers of phospholipid molecules. Predecessors have put forward two major theories on this phenomenon: one is the "membrane lipid phase transition" theory [29], and the other is that, under stress conditions, high levels of reactive oxygen species (ROS) will aggravate membrane lipid peroxidation and membrane protein polymerization, thus destroying the membrane's structure and function [30]. The significant increase in PA contents during the low-temperature storage of CX fruits will inevitably lead to the accumulation of ROS, which will affect the integrity of the cell membranes [31]. In CX fruits, DGDG/MGDG decreased with the decrease in DGDG content (Figure 2). Increasing the ratio of DGDG/MGDG under low-temperature stress can maintain the fluidity of plasma membranes and reduce the damage caused to plants by low temperatures [32]. The MDG gene mediates the transformation from DG to MGDG, and DGD catalyzes the transformation from MGDG to DGDG. The expression level of DGD in CM fruits was significantly higher than that in CX fruits, which was consistent with the changes in DGDG contents. The contents of PA significantly increased during the L2 stage of CX fruits, and the expression of CDS and PSS in CX fruits increased during low-temperature storage, which may be related to the accumulation of phospholipids, indicating that low temperatures mediated and activated the PA biosynthesis pathway in phospholipids to promote PA accumulation, resulting in chilling injury, while the stable total lipid contents in CM peaches further showed that their cold sensitivity was lower than that of CX fruits. At the same time, we measured the fatty acid saturation of two varieties of peach fruits during low-temperature storage. In general, the shorter the fatty acid chain, the higher the degree of unsaturation, while the greater the fluidity of membrane lipids, the stronger the cold resistance of the plants [33]. Therefore, the rapid accumulation of unsaturated fatty acids in CX fruits during the L2 period represents their response to low temperatures and promotes their own protective mechanism (Figure S1), i.e., the accumulation of certain lipids to resist the damage stemming from the stressful environment [34].

4.3. Transcription Factors Play a Key Regulatory Role in Low-Temperature Stress

The peach genome was reported and has been widely used to explore the cold tolerance of peaches [35]. At present, the transcription factors known to be involved in low-temperature signal responses are AP2/ERF, NAC, WRKY, MYB, bZIP, and ZFPs [36]. The overexpression of *TabZIP60* significantly improved the tolerance of plants to drought, salt, and freezing stress [37]). *PpCBF6* participates in the mitigation of chilling injury in peach fruits by inhibiting the expression of lipoxygenase 5 (*LOX5*) and plant sulfur factor a (*PSKa*) [38]. The WRKY transcription factor increases ABA levels by directly activating NECD expression, which participates in the ABA-induced cold tolerance of banana fruits [39]. *PpBZR1* regulates interactions between BR and glucose metabolism, thus regulating the mechanism of cold tolerance in peaches [40]. Based on genome and RNA-SEQ analyses, we revealed that C_2H_2 family transcription factor *ZFP21* was involved in the response to low-temperature stress and was an important regulatory factor involved in the chilling injury mechanism of peach fruits [41]. In this study, we identified several families of transcription factors, among which several genes in the ERF, WRKY, bZIP, ABRE, and bHLH families are highly related to the ethylene, ABA, and lipid metabolism genes (Figure 7). In the promoters of key structural genes in the above three pathways, we also identified several common cis-acting elements of transcription factor families. After cold storage, the expression levels of the NCED and DGD genes in CM fruits increased with the increases in the ABA and DGDG contents. At the same time, a cis-element analysis of the NCED and DGD gene promoters revealed the existence of several bHLH and bZIP response elements (Figure S9). We speculated that transcriptional regulation may greatly affect the occurrence of chilling injury in peach fruits with different cold sensitivities during low-temperature storage. Testing this notion will also be the goal of our next study.

5. Conclusions

The results of lipogenomics and plant hormone determination showed that the CM of cold-insensitive peach fruits had strong cold resistance, and the contents of endogenous ABA and DGDG significantly increased during low-temperature storage. However, cold-sensitive peach fruits induced by low temperatures accumulated more ethylene, phospholipids, and ABA-GE than CM fruits, which also explained the severe symptoms of CI. At the same time, combined with transcriptome data, several transcription factor families, such as AP2/ERF.WRKY, were identified, and the transcriptional regulatory networks of CM and CX fruits during cold storage were constructed.

Supplementary Materials: The following supporting information can be downloaded at: https://www.mdpi.com/article/10.3390/horticulturae10010046/s1, Table S1: Statistics of RNA-seq data of 'CM' and 'CX' peach fruit sample. Table S2: Quality of reads obtained by RNA-seq analysis of 'CM' and 'CX' peach fruit sample. Table S3: Correlation matrix of RNA-seq samples. Table S4.: List of names and IDs of genes involved in ethylene and ABA biosynthesis and signaling, lipid metabolism. The abbreviations used are listed in the legend to Figure S3. Table S5: List of genes in the co-expression network of Figure S10A,B. Figure S1: Changes of firmness of CX and CM fruits during low temperature storage. Figure S2: Principal component analysis (PCA) of the lipidome data in 'CM' and 'CX' peaches stored at low temperature. Figure S3: Lipid type and saturation analysis in of 'CM' and 'CX' peaches stored at low temperature. (A) Changes in mainly lipid types; (B) Changes in lipid saturation of PC and PS. Blue, white, and red colors indicate low, medium, and high contents, respectively. The error bars represent the standard error (SE) calculated according to six independent biological replicates. Figure S4: Venn diagrams of DEGs among the different comparison groups. (A) CM responds to low temperature differential genes; (B) Spring snow responds to low temperature differential genes; (C) Spring beauty and spring beauty responds to low temperature differential genes Veen; (D) Different genes in each period of the two varieties. Figure S5: The gene significance (GS) for ethylene (y-axis), total lipid(y-axis), ABA(y-axis) and DGDG (y-axis) vs. the module membership (x-axis) in modules turquoise and red. (A,B) Gene significance (GS) for ethylene and total lipid vs. module membership in module turquoise. (C,D) Gene significance (GS) for ABA and DGDG vs. module membership in module red. Module membership (KME) is the module eigengene-based network connectivity. Figure S6: Expression heatmaps and profiles of DEGs and eigengenes in modules red, blue, turquoise and pink in response to cold storage in 'CM' and 'CX'. Red indicates upregulated genes, and green indicates downregulated genes. The bar graph of eigengene expression shows the eigengene value variance calculated from the singular value composition for each module. Figure S7: GO enrichment analysis of DEGs in turquoise (A), red (B) and blue (C). Figure S8: Transcription factors and correlation between differentially TFs family and ABA, ethylene and lipid in red and blue modules. Figure S9: Prediction of the cis-acting elements in the 2000-bp *ACS1, ACO1, ACO2, CYP707A, BG1, DGK, PAP, PLC, FAD, NCED2, NCED3, DGD* and *UGT* promoter region was performed by searching the PLACE databases. Figure S10: Transcription factor and structural gene co-expression regulatory network. (A) Turquoise module differential TF and structural gene co-expression regulatory network; (B) Red module differential TF and structural gene co-expression regulatory network.

Author Contributions: Conceptualization, funding acquisition, and writing—review and editing, W.Z. (Wenfang Zeng) and Z.W.; methodology, investigation, and writing—original draft preparation, W.Z. (Wenduo Zhan) and Y.W.; data curation and formal analysis, W.D., A.L., Y.M., H.W. and J.M.; resources, H.L., L.N., L.P., S.S. and G.C. All authors have read and agreed to the published version of the manuscript.

Funding: This work was supported by the Excellent Youth Foundation of Henan Scientific Committee of China (212300410094), the National Natural Science Foundation of China (32271930), the Central Public-Interest Scientific Institution Basal Research Fund (no. ZGS202209), the Agricultural Science and Technology Innovation Program (ASTIP) (CAAS-ASTIP-2023-ZFRI), and the Special Fund of Henan Province for Agro-Scientific Research in the Public Interest (NO. 201300110500).

Data Availability Statement: Data is contained within the article.

Conflicts of Interest: The authors declare no conflicts of interest.

References

1. Seymour, G.B.; Ostergaard, L.; Chapman, N.H.; Knapp, S.; Martin, C. Fruit development and ripening. *Annu. Rev. Plant Biol.* **2013**, *64*, 219–241. [CrossRef] [PubMed]
2. Zhu, Y.; Wang, K.; Wu, C.; Zhao, Y.; Yin, X.; Zhang, B.; Grierson, D.; Chen, K.; Xu, C. Effect of ethylene on cell wall and lipid metabolism during alleviation of postharvest chilling injury in peach. *Cells* **2019**, *8*, 1612. [CrossRef] [PubMed]
3. Brummell, D.A.; Dal Cin, V.; Crisosto, C.H.; Labavitch, J.M. Cell wall metabolism during maturation, ripening and senescence of peach fruit. *J. Exp. Bot.* **2004**, *55*, 2029–2039. [CrossRef] [PubMed]
4. Ketsa, S.; Chidtragool, S.; Lurie, S. Prestorage heat treatment and poststorage quality of mango fruit. *HortScience* **2000**, *35*, 247–249. [CrossRef]
5. Zhang, C.; Ding, Z.; Xu, X.; Wang, Q.; Qin, G.; Tian, S. Crucial roles of membrane stability and its related proteins in the tolerance of peach fruit to chilling injury. *Amino Acids* **2010**, *39*, 181–194. [CrossRef]
6. Cabrera, R.M.; Saltveit, M.E. Physiological Response to Chilling Temperatures of Intermittently Warmed Cucumber Fruit. *J. Am. Soc. Hortic. Sci.* **2019**, *115*, 256–261. [CrossRef]
7. Gong, Z.; Xiong, L.; Shi, H.; Yang, S.; Herrera-Estrella, L.R.; Xu, G.; Chao, D.Y.; Li, J.; Wang, P.Y.; Qin, F.; et al. Plant abiotic stress response and nutrient use efficiency. *Sci. China Life Sci.* **2020**, *63*, 635–674. [CrossRef]
8. Tang, J.; Zhao, Y.; Qi, S.; Dai, Q.; Lin, Q.; Duan, Y. Abscisic acid alleviates chilling injury in cold-stored peach fruit by regulating ethylene and hydrogen peroxide metabolism. *Front. Plant Sci.* **2022**, *13*, 987573. [CrossRef]
9. Deluc, L.G.; Quilici, D.R.; Decendit, A.; Grimplet, J.; Wheatley, M.D.; Schlauch, K.A.; Mérillon, J.M.; Cushman, J.C.; Cramer, G.R. Water deficit alters differentially metabolic pathways affecting important flavor and quality traits in grape berries of Cabernet Sauvignon and Chardonnay. *BMC Genom.* **2009**, *10*, 212. [CrossRef]
10. Jin, M.; Jiao, J.; Zhao, Q.; Ban, Q.; Gao, M.; Suo, J.; Zhu, Q.; Rao, J. Dose effect of exogenous abscisic acid on controlling lignification of postharvest kiwifruit (*Actinidia chinensis* cv. *hongyang). Food Control* **2021**, *124*, 107911. [CrossRef]
11. Chen, B.; Yang, H. 6-Benzylaminopurine alleviates chilling injury of postharvest cucumber fruit through modulating antioxidant system and energy status. *J. Sci. Food Agric.* **2013**, *93*, 1915–1921. [CrossRef] [PubMed]
12. Luengwilai, K.; Beckles, D.M.; Saltveit, M.E. Chilling-injury of harvested tomato (*Solanum lycopersicum* L.) cv. Micro-Tom fruit is reduced by temperature pre-treatments. *Postharvest Biol. Technol.* **2012**, *63*, 123–128. [CrossRef]
13. Cao, S.; Zheng, Y.; Wang, K.; Jin, P.; Rui, H. Methyl jasmonate reduces chilling injury and enhances antioxidant enzyme activity in postharvest loquat fruit. *Food Chem.* **2009**, *115*, 1458–1463. [CrossRef]
14. Wang, K.; Yin, X.R.; Zhang, B.; Grierson, D.; Xu, C.J.; Chen, K.S. Transcriptomic and metabolic analyses provide new insights into chilling injury in peach fruit. *Plant Cell Environ.* **2017**, *40*, 1531–1551. [CrossRef] [PubMed]
15. Guy, C.; Kaplan, F.; Kopka, J.; Selbig, J.; Hincha, D.K. Metabolomics of temperature stress. *Physiol. Plant.* **2008**, *132*, 220–235. [CrossRef] [PubMed]
16. Funnekotter, B.; Kaczmarczyk, A.; Turner, S.R.; Bunn, E.; Zhou, W.; Smith, S.; Flematti, G.; Mancera, R.L. Acclimation-induced changes in cell membrane composition and influence on cryotolerance of in vitro shoots of native plant species. *Plant Cell Tissue Organ Cult.* **2013**, *114*, 83–96. [CrossRef]
17. Yamaki, S.; Uritani, I. Mechanism of chilling injury in sweet potatoes part V biochemical mechanism of chilling injury with special reference to mitochondrial lipid components. *Agric. Biol. Chem.* **1972**, *36*, 47–55. [CrossRef]
18. Thalhammer, A.; Bryant, G.; Sulpice, R.; Hincha, D.K. Disordered cold regulated15 proteins protect chloroplast membranes during freezing through binding and folding, But do not stabilize chloroplast enzymes in vivo. *Plant Physiol.* **2014**, *166*, 190–201. [CrossRef]
19. Ma, Y.; Hu, S.; Chen, G.; Zheng, Y.; Jin, P. Cold shock treatment alleviates chilling injury in peach fruit by regulating antioxidant capacity and membrane lipid metabolism. *Food Qual. Saf.* **2022**, *6*, fyab026. [CrossRef]
20. Franzoni, G.; Spadafora, N.D.; Sirangelo, T.M.; Ferrante, A.; Rogers, H.J. Biochemical and molecular changes in peach fruit exposed to cold stress conditions. *Mol. Hortic.* **2023**, *3*, 24. [CrossRef]

21. Wang, X.; Ding, Y.; Li, Z.; Shi, Y.; Wang, J.; Hua, J.; Gong, Z.; Zhou, J.M.; Yang, S. PUB25 and PUB26 Promote Plant Freezing Tolerance by Degrading the Cold Signaling Negative Regulator *MYB15*. *Dev. Cell* **2019**, *51*, 222–235. [CrossRef] [PubMed]
22. Thomashow, M.F. Plant cold acclimation: Freezing tolerance genes and regulatory mechanisms. *Annu. Rev. Plant Biol.* **1999**, *50*, 571–599. [CrossRef] [PubMed]
23. Duan, W.; Yang, C.; Cao, X.; Wei, C.; Chen, K.; Li, X.; Zhang, B. Chilling-induced peach flavor loss is associated with expression and DNA methylation of functional genes. *J. Adv. Res.* **2022**, *53*, 17–31. [CrossRef] [PubMed]
24. Lurie, S.; Crisosto, C.H. Chilling injury in peach and nectarine. *Postharvest Biol. Technol.* **2005**, *37*, 195–208. [CrossRef]
25. Wang, C.Y. Changes of polyamines and ethylene in cucumber seedlings in response to chilling stress. *Physiol. Plant.* **1987**, *69*, 253–257. [CrossRef]
26. Barroso, C.; Romero, L.C.; Cejudo, F.J.; Vega, J.M.; Gotor, C. Salt-specific regulation of the cytosolic O-acetylserine(thiol)lyase gene from Arabidopsis thaliana is dependent on abscisic acid. *Plant Mol. Biol.* **1999**, *40*, 729–736. [CrossRef]
27. Zuo, X.; Cao, S.; Zhang, M.; Cheng, Z.; Cao, T.; Jin, P.; Zheng, Y. High relative humidity (HRH) storage alleviates chilling injury of zucchini fruit by promoting the accumulation of proline and ABA. *Postharvest Biol. Technol.* **2021**, *171*, 111344. [CrossRef]
28. Yoshida, T.; Christmann, A.; Yamaguchi-Shinozaki, K.; Grill, E.; Fernie, A.R. Revisiting the Basal Role of ABA—Roles Outside of Stress. *Trends Plant Sci.* **2019**, *24*, 625–635. [CrossRef]
29. Lyons, J.M. Chilling Injury in Plants. *Annu. Rev. Plant Physiol.* **1973**, *24*, 445–466. [CrossRef]
30. Buege, J.A.; Aust, S.D. Biomembranes—Part C: Biological Oxidations. *Methods Enzymol.* **1978**, *52*, 302–310.
31. Tan, W.J.; Yang, Y.C.; Zhou, Y.; Huang, L.P.; Xu, L.; Chen, Q.F.; Yu, L.J.; Xiao, S. Diacylglycerol acyltransferase and diacylglycerol kinase modulate triacylglycerol and phosphatidic acid production in the plant response to freezing stress. *Plant Physiol.* **2018**, *177*, 1303–1318. [CrossRef] [PubMed]
32. Moellering, E.R.; Benning, C. Galactoglycerolipid metabolism under stress: A time for remodeling. *Trends Plant Sci.* **2011**, *16*, 98–107. [CrossRef] [PubMed]
33. Munns, R.; Wallace, P.A.; Teakle, N.L.; Colmer, T.D. Measuring Soluble Ion Concentrations (Na^+, K^+, Cl^-) in Salt-Treated. In *Plants BT—Plant Stress Tolerance: Methods and Protocols*; Springer: New York, NY, USA, 2010. [CrossRef]
34. Song, C.; Wang, K.; Xiao, X.; Liu, Q.; Yang, M.; Li, X.; Feng, Y.; Li, S.; Shi, L.; Chen, W.; et al. Membrane lipid metabolism influences chilling injury during cold storage of peach fruit. *Food Res. Int.* **2022**, *157*, 111249. [CrossRef] [PubMed]
35. Cao, K.; Yang, X.; Li, Y.; Zhu, G.; Fang, W.; Chen, C.; Wang, X.; Wu, J.; Wang, L. New high-quality peach (*Prunus persica* L. Batsch) genome assembly to analyze the molecular evolutionary mechanism of volatile compounds in peach fruits. *Plant J.* **2021**, *108*, 281–295. [CrossRef] [PubMed]
36. Tweneboah, S.; Oh, S.K. Biological roles of NAC transcription factors in the regulation of biotic and abiotic stress responses in solanaceous crops. *J. Plant Biotechnol.* **2017**, *44*, 1–11. [CrossRef]
37. Zhang, L.; Zhang, L.; Xia, C.; Zhao, G.; Liu, J.; Jia, J.; Kong, X. A novel wheat bZIP transcription factor, *TabZIP60*, confers multiple abiotic stress tolerances in transgenic Arabidopsis. *Physiol. Plant.* **2015**, *153*, 538–554. [CrossRef]
38. Jiao, C. *PpCBF6* Is Involved in Phytosulfokine α-Retarded Chilling Injury by Suppressing the Expression of *PpLOX5* in Peach Fruit. *Front. Plant Sci.* **2022**, *13*, 874338. [CrossRef]
39. Luo, D.L.; Ba, L.J.; Shan, W.; Kuang, J.F.; Lu, W.J.; Chen, J.Y. Involvement of WRKY Transcription Factors in Abscisic-Acid-Induced Cold Tolerance of Banana Fruit. *J. Agric. Food Chem.* **2017**, *65*, 3627–3635. [CrossRef]
40. Zhang, S.; Cao, K.; Wei, Y.; Jiang, S.; Ye, J.; Xu, F.; Chen, Y.; Shao, X. *PpBZR1*, a BES/BZR transcription factor, enhances cold stress tolerance by suppressing sucrose degradation in peach fruit. *Plant Physiol. Biochem.* **2023**, *202*, 107972. [CrossRef]
41. Zheng, Y.; Liu, Z.; Wang, H.; Zhang, W.; Li, S.; Xu, M. Transcriptome and genome analysis to identify C_2H_2 genes participating in low-temperature conditioning-alleviated postharvest chilling injury of peach fruit. *Food Qual. Saf.* **2022**, *6*, fyac059. [CrossRef]

horticulturae

MDPI

Article

Differential Tolerance of Primary Metabolism of *Annona emarginata* (Schltdl.) H. Rainer to Water Stress Modulates Alkaloid Production

Ana Beatriz Marques Honório [1], Ivan De-la-Cruz-Chacón [2], Gustavo Cabral da Silva [1], Carolina Ovile Mimi [1,*], Felipe Girotto Campos [1], Magali Ribeiro da Silva [3], Carmen Silvia Fernandes Boaro [1] and Gisela Ferreira [1]

1 Department of Biodiversity and Biostatistics, Institute of Biosciences, São Paulo State University (Unesp), Prof. Dr. Antônio Celso Wagner Zanin Street, 250, Botucatu 18618-689, SP, Brazil; beatriz.honorio@unesp.br (A.B.M.H.); gustavo.cabral-silva@unesp.br (G.C.d.S.); felipe.girotto@unesp.br (F.G.C.); carmen.boaro@unesp.br (C.S.F.B.); gisela.ferreira@unesp.br (G.F.)

2 Laboratorio de Fisiología y Química Vegetal, Instituto de Ciencias Biológicas, Universidad de Ciencias y Artes de Chiapas, Libramiento Norte-Poniente 1150, Tuxtla Gutiérrez 29039, CHIS, Mexico; ivan.cruz@unicach.mx

3 Forest, Soil and Environmental Sciences Department, Faculty of Agronomic Science, São Paulo State University, UNESP, Avenida Universitária 3780, Botucatu 18610-034, SP, Brazil; magali.ribeiro@unesp.br

* Correspondence: carolina.mimi@unesp.br

Abstract: *Annona emarginata* produces alkaloids of ecological and pharmacological interest and is tolerant to water and biotic stress, so it is used as rootstock for other Annonaceae fruits. There are few reports in the literature on how contrasting water stress impacts the production of specialized metabolites in Annonaceae and how primary metabolism adjusts to support such production. The objective of this investigation was to evaluate how drought and flooding stress affect alkaloid concentration and the primary metabolism of young *A. emarginata* plants. Three water levels (flooding, field capacity, and drought) were studied at two moments (stress and recovery). Variables analyzed were gas exchange levels, chlorophyll *a* fluorescence, leaf sugars, total alkaloid content, alkaloid profile, and Liriodenine concentration. The photosynthetic metabolism of *A. emarginata* was affected by water stress, with plants having a greater ability to adapt to drought conditions than to flooding. During the drought, a reduction in photosynthetic efficiency with subsequent recovery, higher starch and trehalose concentrations in leaves, and total alkaloids in roots (480 $\mu g.g^{-1}$) were observed. Under flooding, there was a reduction in photochemical efficiency during stress, indicating damage to the photosynthetic apparatus, without reversal during the recovery period, as well as a higher concentration of total sugars, reducing sugars, sucrose, glucose, and fructose in leaves, and Liriodenine in roots (100 $\mu g.g^{-1}$), with a lower concentration of total alkaloids (90 $\mu g.g^{-1}$). It could be concluded that there is differential tolerance of *A. emarginata* to water stress, inducing the modulation of alkaloid production, while drought promotes a higher concentration of total alkaloids and flooding leads to an increase in the Liriodenine concentration.

Keywords: Annonaceae; Liriodenine; photosynthesis; leaf sugars; flooding; drought

Citation: Honório, A.B.M.; De-la-Cruz-Chacón, I.; da Silva, G.C.; Mimi, C.O.; Campos, F.G.; da Silva, M.R.; Boaro, C.S.F.; Ferreira, G. Differential Tolerance of Primary Metabolism of *Annona emarginata* (Schltdl.) H. Rainer to Water Stress Modulates Alkaloid Production. *Horticulturae* **2024**, *10*, 220. https://doi.org/10.3390/horticulturae10030220

Academic Editors: Adalberto Benavides-Mendoza, Yolanda González-García, Fabián Pérez Labrada and Susana González-Morales

Received: 19 January 2024
Revised: 20 February 2024
Accepted: 23 February 2024
Published: 25 February 2024

1. Introduction

The Annonaceae family has medical and economic importance and significantly contributes to the diversity of tree species in neotropical and tropical forests [1,2]. *Annona emarginata* (Schltdl.) H. Rainer is a species native to the South American continent [3–5] and is known as having two morphotypes, "terra fria" and "mirim" [6]. In Brazil, it has a prominent presence in the Atlantic Forest, a biome that has enormous water availability [3,4]. In the economic sphere, *A. emarginata* is commonly utilized as a rootstock for the large-scale production of atemoya seedlings (*Annona atemoya* Mabb) [7], as it is tolerant to pathogens that attack plant roots and stems (*Phytophthora nicotianeae* var. parasítica, *Pythium* spp., *Rhizoctonia solani*) [8–10]. The species also presents adaptations to different humidity and temperature conditions [8] and tolerance to dry soils [11]. In addition, there are personal reports from field producers that the species survives for months under excess water conditions in areas subject to flooding.

Annonaceae presents specialized metabolites of ecological and pharmacological importance like alkaloids, acetogenins, and volatile terpenes [12–15]. Alkaloids assembly the group of nitrogenous metabolites with the mayor diversification [16]. In this context, Liriodenine is the most important alkaloid, considered a chemotaxonomic marker [17–20], with antibacterial [17], antiprotozoal [21,22], cytotoxic [12,23], and antifungal activities [24,25], in addition to being a defense molecule against phytopathogens [26]. In addition, *Annona emarginata* is capable of producing alkaloids that are stored in roots, stems, and leaves, such as Liriodenine [6,19].

The biosynthesis of these metabolites depends on several environmental conditions like temperature, light and water regime. [27–29] and is related to primary metabolism, given that the precursors of the specialized metabolism come from the primary metabolism [19,28,30]. Thus, climate changes that cause irregularities in rainfall distribution and volume affect agricultural production, including in the *Annona* cultivation regions [24,25], which can impact the production of specialized metabolites such as alkaloids. Water restriction directly affects ecological and agricultural systems, given that plant reactions to this type of stress vary according to their intensity and duration [31,32]. On the other hand, flooding imposes restrictions on the cultivation of many species due to the lack of adaptations to saturated soils with low oxygen availability to the root system [33].

The literature is abundant regarding relationships between primary and specialized metabolisms under water stress. However, in *Annona*, studies are generally aimed at evaluating the effect of hydric stress on plant production, especially drought [34,35] or secondary metabolites [18,36]. Thus, there are no reports on how drought and flooding stress affect the primary metabolism and production of alkaloids in *Annona emarginata*. Therefore, this work aims to evaluate how drought and flooding affect alkaloid production (total alkaloids, Liriodenine, and profile) and how primary metabolism (gas exchange, chlorophyll *a* fluorescence, and quantification of sugars) in young *Annona emarginata* plants adapt to support or explain the production of these metabolites.

2. Materials and Methods

2.1. Plant Material

Young *Annona emarginata* (Schltdl.) H. Rainer plants, "araticum de terra-fria" morphotype, were obtained from the Seedling Production Center of São Bento do Sapucaí, CATI (Technical and Integrated Assistance Coordination), municipality of São Bento do Sapucaí—São Paulo.

Seedlings were produced with seeds obtained from ripe fruits from a collection of *Annona emarginata* matrices, used by CATI as rootstock for atemoya production. Sowing took place in April 2019 in sand beds and transplanting in June of the respective year using 2-liter plastic bags with commercial substrate (Carolina Soil®—composed of Sphagnum peat, expanded perlite, and vermiculite expanded and toasted rice husk).

To start the experiment, young plants (3 months old and with 10–15 mature leaves) were acclimatized for 15 days in the greenhouse of the Department of Forestry Science, Soil, and Environment of the Faculty of Agricultural Sciences, Unesp—Botucatu (coordinates 22°51′ latitude S and 48°26′ longitude W).

2.2. Experimental Design

The experiment was carried out in July 2019, with average temperatures in Botucatu-São Paulo varying between 12 and 23 °C and relative humidity between 10 and 30% [37].

The experimental design was in randomized blocks in a 3 × 2 factorial scheme, with three replicates of 19 plants per treatment. Treatments consisted of water levels (flooding, field capacity, and drought) and two moments (stress and recovery), totaling 342 plants. In the period before the beginning of the experiment, all plants were watered until the beginning of water percolation, then allowed to drain. As soon as the drainage stopped, indicating that the soil was at field capacity (control group), seedlings were individually weighed, and from that point, drought and flooding conditions were implemented.

In plants submitted to drought, irrigation was completely removed, and for those submitted to flooding, plants were kept constantly in trays with substrate covered in water. At the same time, to maintain field capacity, seedlings were weighed, and water was replenished daily. These treatments were maintained for thirty days. After this period, plants subjected to flooding and drought returned to field capacity for another 17 days in order to recover from water stress conditions. Plants at field capacity were maintained under this condition until the end of the experiment.

The response of the primary metabolism of *A. emarginata* to variations in water regimes was evaluated considering gas exchange, chlorophyll *a* fluorescence, and quantification of sugars, while the amount of total alkaloids, Liriodenine, and the presence of eleven other alkaloids were the specialized metabolism variables.

2.3. Gas Exchange

Gas exchange levels were weekly evaluated, with three plants from each replicate per treatment (totaling 27 plants) at each evaluation moment (stress and recovery), on the 2nd or 3rd completely expanded leaf from 09:00 a.m. to 11:00 a.m., with the aid of a CO_2 gas analyzer and water vapor by infrared radiation (Infra-Red Gas Analyzer, model GFS-3000—Heinz Walz GmbH, Effeltrich, Germany) with a saturating light of 800 m^{-2} s^{-1} determined using a light curve [38].

CO_2 assimilation rate (A_{net}, µmol CO_2 m^{-2} s^{-1}), transpiration rate (E, mmol water vapor m^{-2} s^{-1}), and stomatal conductance (gs, mmol m^{-2} s^{-1}) were determined. Instantaneous water use efficiency [WUE, µmol CO_2 (mmol H_2O^{-1})] was calculated using the ratio between assimilated CO_2 and transpiration rate (A_{net}/E). The apparent carboxylation efficiency was calculated according to the ratio between the CO_2 assimilation rate and the intercellular leaf CO_2 concentration (A_{net}/Ci, µmol m^{-2} s^{-1} Pa^{-1}).

2.4. Chlorophyll a Fluorescence

Chlorophyll *a* fluorescence was performed weekly from 09:00 a.m. to 11:00 a.m. using a fluorometer (LED-Array/PAM-Module3055-FL) on 27 plants (3 replicates with 3 plants per treatment) at each evaluation moment (stress × recovery). Leaves were acclimated for a period of 30 min in the dark, covered with aluminum foil, and then a pulse of actinic light of 4500 µmol $m^{-2}s^{-1}$ was applied to obtain Fm (maximum dark-adapted fluorescence). Subsequently, leaves were adapted to light, and the actinic light pulse was applied again to obtain Fm′ (maximum light-adapted fluorescence) [39]. In addition to maximum light- and dark-adapted fluorescence, Fo (minimum dark-adapted fluorescence) and Fo′ (minimum light-adapted fluorescence) values were also obtained. For this purpose, in the dark-adapted state (when all PSII centers are open), a light pulse is emitted. This light has an intensity low enough to induce electron transport from PSII but high enough to generate a minimum chlorophyll fluorescence value (Fo). Fo measurement and its light-adapted equivalent Fo′ are critical for fluorescence analysis. For an accurate Fo′ determination, it is necessary to use the far-red wavelength (FR) to stimulate PSI, thus extracting electrons from PSII, which ensures that quinone A (QA) remains fully oxidized during measurements [39].

Using Fm, Fo, Fm′, and Fo′, the maximum quantum yield (Fv/Fm), potential quantum efficiency (Fv′/Fm′) [40], effective quantum yield (*F*PSII) [41], photochemical quenching (qL) [42], non-photochemical quenching (NPQ) [43], and electron transport rate (ETR) were calculated, considering that 84% of light is absorbed by chlorophyll, with 50% of photons activating chlorophyll from photosystem II and 50% from photosystem I, energy from photosystem II that cannot be dissipated (*Ex*), and energy dissipated in the form of heat (D) [44].

2.5. *Qualitative and Quantitative Analysis of Leaf Sugars*

For the extraction of leaf carbohydrates, 36 samples were used, consisting of 6 replicates of 3 plants per treatment (flooding, field capacity, and drought) at two collection moments: when stress was established (18 samples) and after the recovery period (18 samples).

The extraction of total soluble sugars was carried out as proposed by Garcia et al. [45], as well as starch extraction according to Clegg [46]. The procedure to determine the concentration of total soluble sugars was performed according to Morris [47], for starch, the procedure was described by Yemm and Folkes [48], for reducing sugars, the procedure was determined by Miller [49], and for sucrose, the procedure was established by Passos [50].

Subsequently, 1 mL of total soluble sugars was separated for purification in columns containing Dowex cation and anion exchange resins eluted with 10 volumes of deionized water. The purified material had its pH neutralized with ammonium hydroxide and concentrated until complete drying in a freeze dryer. Samples were then resuspended in 5 mL of deionized water and analyzed in anion exchange liquid chromatography with pulsed amperometric detection (HPAEDPAD) (Colluna Dionex CarboPacTM PA-100, 4×250 mm) with a sodium hydroxide elution gradient (625 mM), ultrapure water (Milli Q), and sodium acetate (0.5 M). From standards, it was possible to identify the following sugars: arabinose, fructose, glucose, and trehalose.

2.6. *Extraction of Total Alkaloids*

The extraction of total alkaloids was carried out using leaf and root material from 60 samples, of which 30 corresponded to the stress moment (5 replicates per treatment, totaling 15 leaf samples and 15 root samples) and 30 to the recovery moment (5 replicates per treatment, 15 leaf samples and 15 root samples). The material was stored in an oven with forced air circulation at 30 °C for ten days, then crushed to obtain $\pm$ 5 g of dry mass for each replicate. Alkaloid extraction was carried out using the selective acid-base method, and extracts were stored in the dark [18].

2.7. *Qualitative and Quantitative Analyses of Total Alkaloids and Liriodenine*

To determine the total alkaloid content, the 60 alkaloid extracts were re-solubilized with $CHCl_3$, and their absorbance was determined by UV-Vis spectrophotometry (single-beam—model UV-M51—BEL Engineering®, Monza, Italy) at 254 nm using Liriodenine isolated from *Annnona mucosa* as reported in Sousa et al. [19], as a standard for elaborating the standard curve ($y = 0.0881x - 0.0112$, $R^2 = 0.9949$). The Liriodenine alkaloid was quantified using an ultra-high-performance liquid chromatograph with a UV-Vis detector (UHPLC—Thermo Fisher Scientific®, Waltham, MA, USA). The column was of reverse phase C18 (150×4.6 mm and 5 µm particle diameter) maintained at 30 °C; the isocratic mobile phase was water (pH 3.5 with acetic acid) and methanol (30:70) with a flow rate of 1 mL/min. Detection was carried out in UV at 254 nm. For the Liriodenine quantification, two calibration solution curves were performed ($y = 0.3595x - 0.0011$; $R^2 = 0.9989$ for samples with up to 10 µg of Liriodenine in the extract and $y = 0.3658x + 1.142$; $R^2 = 0.9992$ for samples larger than 10 µg) [18].

2.8. *Statistical Analysis*

The data were subjected to a normality test and homogeneity of variance. Subsequently, the data obtained (total alkaloids, Liriodenine, leaf carbohydrates, relative leaf water content, chlorophyll *a* fluorescence, and gas exchange) were submitted to a two-way analysis of variance (ANOVA). A two-way ANOVA was conducted to determine the effect of treatments (flooding, field capacity, and drought) with moments (stress: 30 days; recovery: 47 days) and their interaction. Means were compared using the Tukey test at the 5% significance level ($p < 0.05$).

3. Results

3.1. Effect of Water Stress on Alkaloid Metabolism

Alkaloid production in *Annona emarginata* was impacted both in stress and recovery moments (Figure 1). Drought stress stimulated almost five times more total alkaloid production (480 µg g^{-1} dry mass) than flooding and field capacity (control) (90 and 100 µg.g^{-1}, respectively).

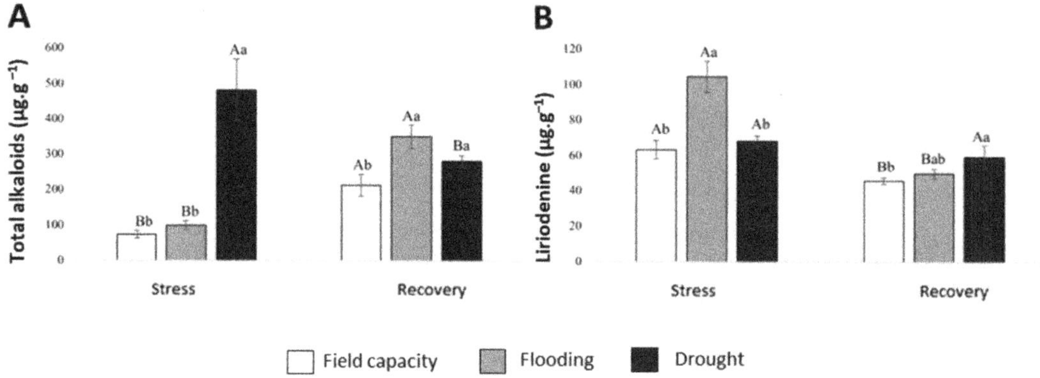

Figure 1. Concentration of (**A**) total alkaloids and (**B**) Liriodenine (means ± SD) in the roots of young *Annona emarginata* plants subjected to water stress and recovery. Means followed by capital letters indicate significant differences between assessment moments (stress × recovery), and means with lowercase letters indicate significant differences between water levels (flooding, field capacity, and drought) using the Tukey test at 5% probability (Table S1).

When plants were removed from water stress due to drought and placed at field capacity (recovery), a reduction in the concentration of total alkaloids was detected (300 µg.g^{-1}). An increase in the biosynthesis of total alkaloids was observed in plants subjected to flooding stress (350 µg.g^{-1}) without reaching the maximum production observed under drought stress conditions (480 µg.g^{-1}). Under the recovery condition, there was no difference in the amounts of total alkaloids between previously imposed drought stress and flooding; however, values were higher than those of field capacity. Thus, the results demonstrate that there is a more immediate effect at the moment of the drought imposition, with an increase in alkaloid production, while the effect of flooding occurs more slowly, being detected only during recovery.

In relation to Liriodenine, greater production was observed in flooding (100 µg.g^{-1}) than in drought stress and control (65 and 60 µg.g^{-1}). In the recovery phase, flooding and control treatments reduced the Liriodenine proportion to 50 and 45 µg.g^{-1} with no differences between them; however, in plants subjected to drought stress, the alkaloid concentration was preserved. Thus, it could be observed that flooding stress caused an increase in the Liriodenine concentration in roots, despite not having generated an increase in the total alkaloid concentration (Figure 1).

Furthermore, from the comparison through HPLC of the standards used, it was possible to identify eleven alkaloids in roots (at stress moments), nine of which are present in all water conditions (stress as recovery), namely N-Methyl-Laurotetanine, Norglaucin, Xylopinine, Xylopine, Assimilobin, Laurotetanin, Liriodenine, Oxoglaucine, and Lanulinosin. Ten alkaloids were identified in leaves, of which seven (N-Methyl-Laurotetanine, Norglaucin, Discretin, Xylopine, Laurotetanin, Liriodenine, and Lanulinosin) were present in both stress and recovery conditions. The alkaloids Norpredicentine and Xylopinine were not detected in leaves during flooding stress or recovery. Assimilobin was not detected in leaves during both stress and recovery conditions. Oxoglaucine and Reticulin were detected only in roots (Table 1). Reticulin was only detected during the recovery from water stress.

Table 1. Presence (x) and absence (-) of alkaloids in roots and leaves of *Annona emarginata* maintained at field capacity, flooding, and drought during stress and recovery.

	Roots						Leaves					
	Stress			Recovery			Stress			Recovery		
Alkaloids	Field Capacity	Flooding	Drought	Field Capacity	Flooding	Drought	Field Capacity	Flooding	Drought	Field Capacity	Flooding	Drought
Reticulin	-	-	-	x	x	-	-	-	-	-	-	-
Norpredicentine	x	x	x	x	-	-	x	-	x	x	-	-
N-Methyl-Laurotetanine	x	x	x	x	x	x	x	x	x	x	x	x
Norglaucin	x	x	x	x	x	x	x	x	x	x	x	x
Discretin	-	x	-	x	x	-	x	x	x	x	x	x
Xylopinine	x	x	x	x	x	x	x	-	x	x	-	-
Xylopine	x	x	x	x	x	x	x	x	x	x	x	x
Assimilobin	x	x	x	x	x	x	x	-	-	x	-	-
Laurotetanin	x	x	x	x	x	x	x	x	x	x	x	x
Liriodenine	x	x	x	x	x	x	x	x	x	x	x	x
Oxoglaucine	x	x	x	x	x	x	-	-	-	-	-	-
Lanulinosin	x	x	x	x	x	x	x	x	x	x	x	x

3.2. *Effect of Water Stress on Primary Metabolism*

Both water stress conditions caused changes in the photosynthesis of *A. emarginata*. The greater recovery reactions of plants in water deficit (drought) indicate greater tolerance of the species to this type of stress.

The net assimilation (A_{net}) and instantaneous water use efficiency (WUE) values were reduced during drought and flooding in relation to field capacity (control), with drought also causing significantly lower values than flooding. Stomatal conductance (gs) and transpiration (E) of plants were reduced in a similar way by treatments, differing from control. However, flooding caused a reduction in the carboxylation efficiency of the Rubisco enzyme (A_{net}/Ci), differing from drought, whose plants remained similar to control (Figure 2).

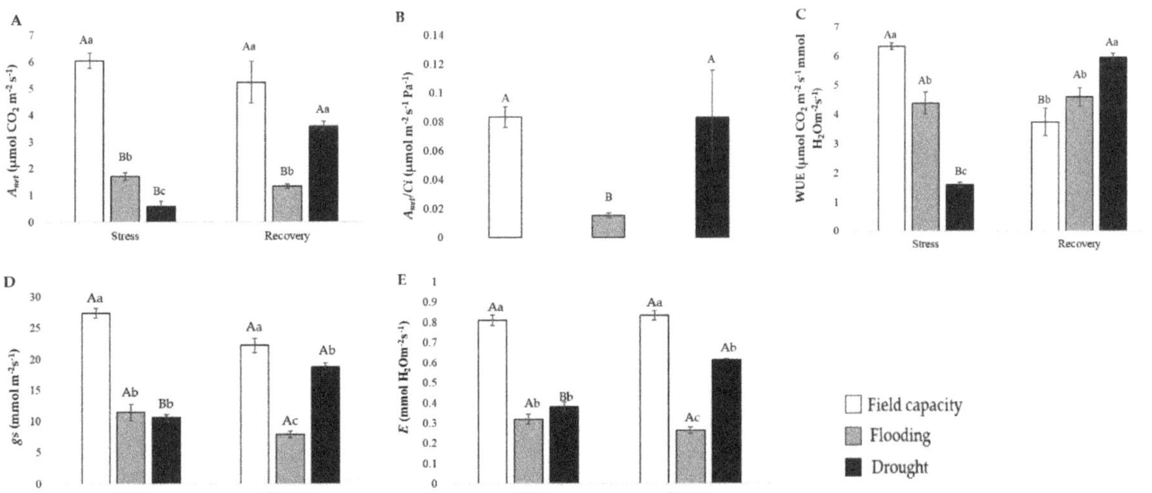

Figure 2. (**A**) Carbon assimilation rate (A_{net}), (**B**) Rubisco carboxylation efficiency (A_{net}/Ci), (**C**) instantaneous water use efficiency (WUE), (**D**) stomatal conductance (gs), and (**E**) transpiration (E) (means ± SD) in *Annona emarginata* plants maintained at field capacity, flooding, and drought during stress and subsequent recovery. Means followed by capital letters indicate significant differences between assessment moments (stress × recovery), and means with lowercase letters indicate significant differences between water levels (flooding, field capacity, and drought) using the Tukey test at 5% probability (Table S2).

At the same time, at the stress moment, the impact of treatments on chlorophyll *a* fluorescence was observed, with a reduction in maximum quantum yield (Fv/Fm) and the energy that cannot be dissipated by photosystem II (Ex), especially caused by flooding.

Furthermore, both drought and flooding caused similar reductions in potential quantum efficiency (FV′/FM′), electron transport rate (ETR), effective quantum yield (*FPSII*), and non-photochemical quenching (NPQ), and increased energy dissipated in the form of heat (D) (Figure 3).

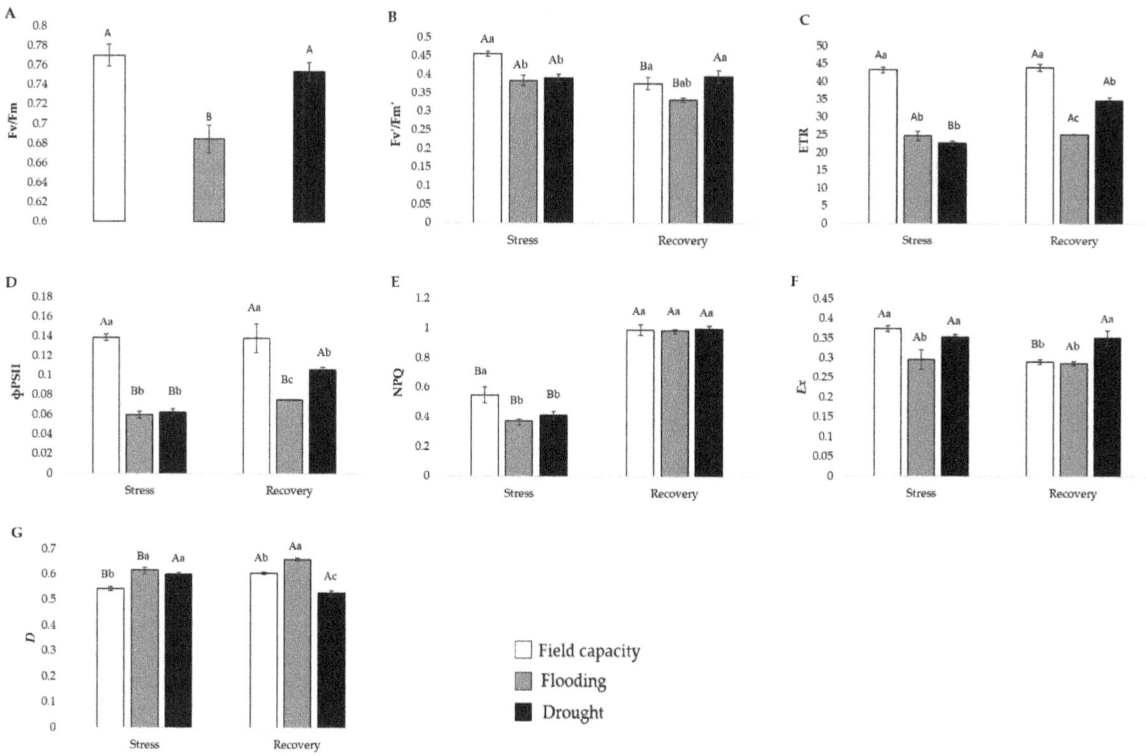

Figure 3. (**A**) Maximum quantum yield (Fv/Fm), (**B**) potential quantum efficiency (Fv′/Fm′), (**C**) electron transport rate (ETR), (**D**) effective quantum yield (ΦPSII), (**E**) non-photochemical quenching (NPQ), (**F**) energy of photosystem II that cannot be dissipated (Ex), and (**G**) energy dissipated in the form of heat (**D**) (means ± SD) of *Annona emarginata* plants maintained at field capacity, flooding, and drought during stress and subsequent recovery. Means followed by capital letters indicate significant differences between assessment moments (stress × recovery), and means with lowercase letters indicate significant differences between water levels (flooding, field capacity, and drought) using the Tukey test at 5% probability (Table S3).

The photosynthesis of plants placed in recovery reinforces the understanding of what occurred at the stress moment. Gas exchange (A_{net}, *gs*, E, WUE), chlorophyll *a* fluorescence [electron transport rate (ETR), effective quantum yield (*FPSII*), and non-photochemical quenching (NPQ)] values were higher in recovering plants (rehydration) than at the stress moment, indicating the ability of plants to recover and restore photosynthetic efficiency (Figures 2 and 3). The carboxylation efficiency (A_{net}/Ci) of plants in drought remained similar to control regardless of the assessment moment, indicating that despite stomatal closure and lower CO_2 assimilation caused by drought, no damage to the photosynthetic apparatus that would prevent CO_2 incorporation by the Rubisco enzyme was observed.

Plants submitted to flooding presented greater recovery difficulties since A_{net}, *gs*, E, and WUE values were lower than those of plants maintained in drought and field capacity, which was also observed in relation to A_{net}/Ci, regardless of whether under stress or during recovery. Similarly, *FPSII* and ETR values were maintained below values from plants submitted to drought conditions and field capacity, and Ex data were below the values of plants under drought conditions, while D values increased under flooding conditions (Figure 3), which indicates damages related to flooding stress.

Variations caused by water conditions in photosynthetic metabolism were reflected in the synthesis of leaf sugars. Under drought stress conditions, lower total sugars, reducing sugars, sucrose, glucose, and fructose concentrations, and higher trehalose concentrations were observed, both in relation to flooding and control. Arabinose and starch in drought presented higher concentrations compared to plants maintained at field capacity (control group) without, however, differing from plants under flooding stress (Figure 4).

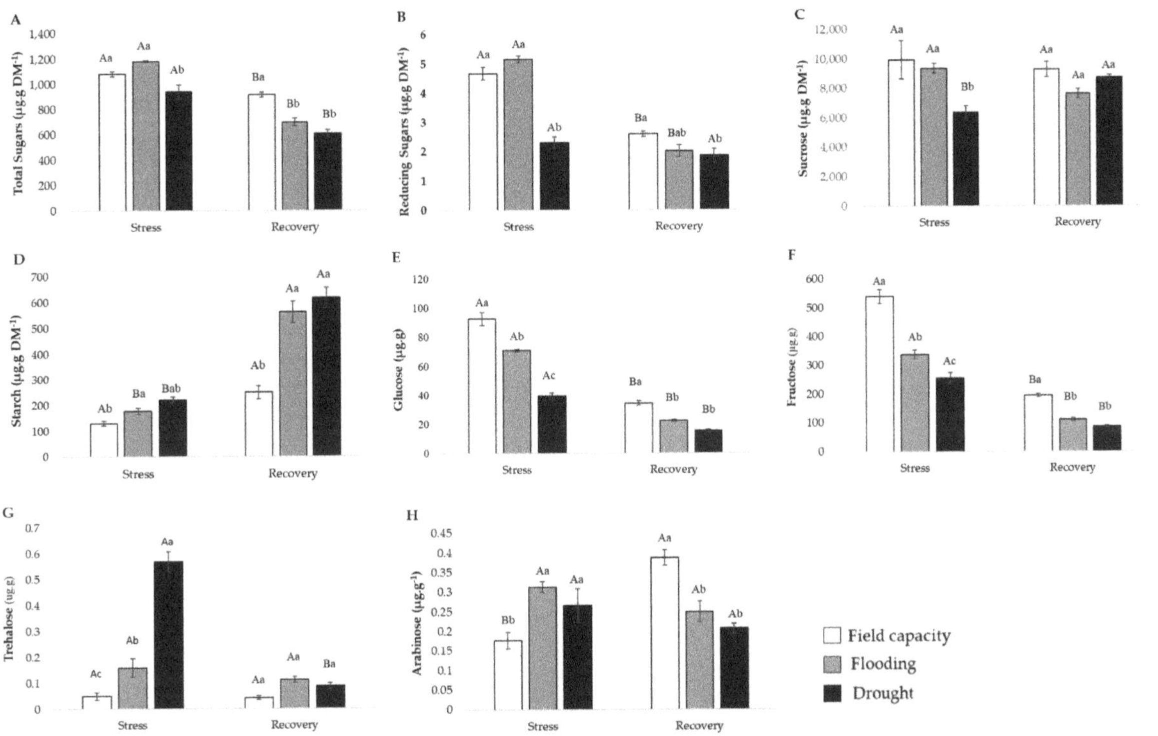

Figure 4. (**A**) Total sugars (µg.g DM^{-1}), (**B**) reducing sugars (µg.g DM^{-1}), (**C**) sucrose (µg.g DM^{-1}), (**D**) starch (µg.g DM^{-1}), (**E**) glucose (µg.g^{-1}), (**F**) fructose (µg.g^{-1}), (**G**) trehalose (µg.g^{-1}), and (**H**) arabinose (µg.g^{-1}) (means ± SD) of *Annona emarginata* plants maintained at field capacity, flooding, and drought during stress and subsequent recovery. Means followed by capital letters indicate significant differences between assessment moments (stress × recovery), and means with lowercase letters indicate significant differences between water levels (flooding, field capacity, and drought) using the Tukey test at 5% probability (Table S4).

4. Discussion

This work highlights the existence of a differential response of the primary and specialized metabolism of *Annona emarginata* to variations in water regimes.

It was observed that *A. emarginata* maintains a pool of "constitutive alkaloids" regardless of water conditions and the consequences of photosynthesis. This set is associated with at least seven alkaloids common in leaves and roots. No solid evidence of the production of alkaloids stimulated by any type of stress was detected. Plants under drought produced the greatest amount of total alkaloids, and plants under flooding produced the particular alkaloid Liriodenine in amounts that were reduced during recovery but remained at least equal to control. These variations were associated with the adjustment of the photosynthetic metabolism.

A. emarginata has been considered drought-tolerant, which is one of the reasons why it is an excellent rootstock for atemoya production [38]. Furthermore, field observations suggest that the species can grow in flooded areas, which led to the hypothesis that the species is also flood-tolerant. However, this work demonstrates that water variations modulated the photosynthetic metabolism of *A. emarginata* in a different way, with greater difficulty in maintaining primary productivity after being subjected to flooding, which indicates less tolerance for this condition.

Water stress (drought and flooding) led to a reduction in stomatal conductance (gs) and transpiration of *A. emarginata*, which was expected since stomatal closure increases resistance to the diffusion of water vapor into the atmosphere [51]. However, these mechanisms that lead to stomatal closure vary with the type of stress. In drought, stomatal closure is related to the action of abscisic acid (ABA) on guard cells as a protective mechanism to prevent water loss [52]. Under flooding, the induction of stomatal closure may occur through an increase in ABA and ethylene synthesis induced by the decrease in available O_2 levels due to soil hypoxia, in addition to leading to the accumulation of reactive oxygen species [53,54].

In addition to reducing transpiration, the decrease in gs can also interfere with the carbon assimilation rate (A_{net}) due to the increased resistance to CO_2 entry due to stomatal closure, as observed in *A. emarginata* plants subjected to water restriction [11]. However, the reduction of A_{net} may also be related to internal CO_2 accumulation (which reduces the potential for diffusion into the leaf) due to damage to the photosynthetic apparatus and the lower carboxylation efficiency of the Rubisco enzyme, which affects the generation of energy used to reduce CO_2 [55].

In this experiment, results suggest that the reduction in A_{net} in plants under flooding was due to damage to the photosynthetic apparatus since a significant reduction in the carboxylation efficiency of the Rubisco enzyme was observed (A_{net}/Ci). These results differ from those observed under drought conditions, as even with lower A_{net} and gs, plants maintained carboxylation efficiency, demonstrating greater tolerance to this condition, as demonstrated with *A. crassiflora* [56], which also maintained carboxylation efficiency under conditions of reduced CO_2 entry into leaves during drought stress.

The literature presents evidence that a water deficit leads to an increase in specialized metabolites, including alkaloids [28,57–59]. In general, water restriction events lead to stomatal closure, reduced CO_2 entry, and, as a consequence, a decrease in the NADPH + H^+ consumption for CO_2 fixation in the Calvin Cycle, which results in excess and accumulation of this reducing equivalent [28]. Thus, the increase in NADPH + H^+ alters reactions that involve its consumption, and to avoid damage caused by its excess, the reducing equivalent is involved in the synthesis of highly reduced carbon compounds, such as alkaloids [28]. This fact may explain the increase in the concentration of total alkaloids when there is less carbon assimilation under dry conditions, differing from plants under flooding. In plants under flooding, although there was an intermediate A_{net} between drought and control, a reduction in the carboxylation efficiency of the Rubisco enzyme was observed, which could also lead to NADPH + H^+ accumulation and greater production of alkaloids; however, the investment in specialized metabolism was lower, with a particular increase in Liriodenine. The alkaloid biosynthesis depends on the availability of carbon and nitrogen and the energy provided by the primary metabolism, demonstrating a high degree of connectivity between primary and specialized metabolisms [60].

The reduction in A_{net}/Ci caused by flooding, indicating damage to the photosynthetic apparatus, is proven by changes in chlorophyll *a* fluorescence patterns, changing the light energy dissipation paths, as proposed by Kalaji et al. [61]. Even without detecting damage to the photosynthetic apparatus under drought stress conditions, it was possible to verify variations in chlorophyll *a* fluorescence patterns, which may explain the differential production of alkaloids between types of stress.

Therefore, it is known that the light energy absorbed by chlorophylls can be used in three ways: (i) boosting photosynthesis through photochemical reactions; (ii) dissipating excess energy in the form of heat; and (iii) re-emitting it in the form of fluorescence [62]. These processes occur concurrently, so any increase in one will result in a decrease in the others [39].

A decline in the maximum quantum yield (Fv/Fm) of plants kept under flooding in relation to those under drought and control was observed. This reduction indicates that there was a decline in the photochemical efficiency of PSII and disturbance or damage to the photosynthetic apparatus of *A. emarginata*, since the FV/FM ratio is an estimate of the maximum quantum efficiency of the photochemical activity of PSII when the centers of reaction are open and has even been used to evaluate other types of stress, such as salt stress [63]. The same behavior was observed for *Annona crassiflora* plants under flooding [56], indicating that this variable measures the intrinsic functioning of PSII, reflecting the high sensitivity of this photosystem to environmental stresses [39]. *A. emarginata* plants in the absence of stress showed an FV/FM ratio around 0.78, similar to several species (0.83) [64].

Unlike the maximum quantum yield (FV/FM), a reduction in the potential quantum efficiency (Fv′/Fm′) of plants under flooding and drought was observed, indicating that there was a lower proportion of light absorbed under both conditions. Therefore, considering that this measurement estimates the proportion of absorbed light effectively used in the photochemical phase of PSII [64], there was less light for the photochemical phase in both water stress conditions, which resulted in a reduction in the photochemical efficiency of PSII (ΦPSII).

In this context, in stressed *A. emarginata* plants (flooding and drought), there was a reduction in the energy fraction absorbed by chlorophyll associated with PSII that was used in the photochemical activity, indicating a reduction in the amount of transported electrons and a reduction in photosynthesis. Thus, a lower electron transport rate (ETR) is observed, directly reflecting a lower effective quantum yield (ΦPSII), as proposed by Tian et al. [65]. It is noteworthy that, although in both stress conditions there was a reduction in PSII efficiency, only plants submitted to drought demonstrated recovery, achieving responses similar to control, indicating that flooding was more harmful to *A. emarginata* plants.

The reduction in light energy destined for photosynthesis provides evidence that it follows other paths [39,62,64], which is confirmed by the increase in energy dissipation in the form of heat (D) for both stress conditions. Non-photochemical extinction (NPQ) values were also expected to be high, considering that NPQ is responsible for converting a large amount of absorbed light energy into heat [66]. However, what was observed is that under stress, NPQ values are low, which seems to be characteristic of the species, as also observed by Mantoan et al. [11]. Furthermore, subtle changes in ETR values can modulate NPQ performance [64].

To avoid damage caused by excess energy, there was greater dissipation in the form of heat (D) mainly in flooding plants, similar to results obtained in *A. crassiflora* [56], a condition that continued even with recovery from stress. It is also noteworthy that the lower photosystem II energy values that cannot be dissipated (Ex) during flooding may indicate a reduction in the flow of electrons that would be destined for the production of specialized metabolites, such as alkaloids [67], which only occurred in this experiment in plants with water deficits both in stress and recovery, which showed high *Ex* values. *Ex* is energy that is in the photosystem, driving the transport of electrons and the oxidation–reduction mechanism of reaction centers, and its excess is accumulated when there is no oxidizing agent, as there is no possibility of dissipation in the form of heat. In this way, this energy can be directed to alternative drains such as the synthesis of specialized metabolites, for example, alkaloids in *A. emarginata*, which did not occur under flooding since the *Ex* value was reduced.

In summary, in both water stress conditions, there was less light available for the photochemical phase (FV′/FM′), which resulted in lower electron transport rate (ETR) and effective quantum yield values and, therefore, less photosynthesis. The energy from the unused photochemical phase accumulated (*Ex*) can be diverted to the synthesis of total alkaloids in drought plants and only to Liriodenine in flooding plants, where Ex is more reduced. During recovery, drought plants began to divide their energy flow between photosynthesis (effective quantum yield values increased) and the production of alkaloids, which reduced the production of these metabolites. In contrast, flooding plants maintained the lowest photosynthesis values and high D values, and although they maintained low *Ex* in relation to drought, there is evidence that energy has been targeted to the production of specific alkaloids, such as Liriodenine.

Increased alkaloid production in plants subjected to drought has been reported in several species, such as *Catharanthus roseus* [68], *Senecio jacobaea* [69], *Nicotiana tabacum* [70], and *Phellodendron amurense*, in which an increase in the concentration of benzylisoquinoline alkaloids was observed [71], the same group of alkaloids commonly found in species of the genus *Annona* [26].

In this experiment, regardless of drought or flooding stress, the presence of the alkaloids N-Methyl-Laurotetanine, Norglaucin, Laurotetanin, Lanulinosin, Liriodenine, and Xylopine was detected in the roots and leaves of all treatments, both in stress and recovery. These alkaloids may have played an antioxidant role, and, in the specific case of isoquinoline alkaloids, there is the potential to inhibit lipid peroxidation through the elimination of singlet oxygen [72]. Ovile Mimi et al. [6] worked with two *A. emarginata* morphotypes, "terra-fria" and "mirim", and also detected the presence of these alkaloids in the root material.

Particularly in the genus *Annona*, it has been described that drought increases Liriodenine production [36], an oxoaporphinic alkaloid recorded in more than 250 species [18], which has potential antibacterial [17], antiprotozoal [21,22], cytotoxic [22,23], and antifungal activities against phytopathogens [24]. In *A. emarginata*, the increase in Liriodenine production only occurred in plants submitted to drought stress and then returned to field capacity, while flooding stimulated Liriodenine production already at the time of stress establishment. Flooding caused an increase in Liriodenine concentration (around 110 μg^{-1}) in roots, around twice as much as in plants maintained in drought, which is the first report of the presence and quantification of Liriodenine in species of the genus *Annona* under flooding.

The increase in alkaloid production under drought conditions may indicate plants' strategies to perform osmotic adjustment and ensure tissue hydration. This osmotic adjustment occurs due to the accumulation of osmolytes, such as amino acids and sugars, in roots [59], in addition to total alkaloids production, which increased in roots. With the increase in alkaloids in roots, reductions in root water potential could occur, which could favor water absorption even under water stress or maintenance of cellular water status. Thus, in this experiment, under reduced water conditions, an increase in alkaloid production was observed, perhaps with the aim of mitigating the effect of drought.

In relation to the presence of osmolytes, plants subjected to drought showed a higher trehalose concentration (found only in some plants). This reducing disaccharide acts on the stabilization of enzymes, proteins, and lipid membranes against denaturation in situations of water restriction [73,74]. According to Paul et al. [74], trehalose is also related to starch accumulation, as the trehalose-6-phosphate enzyme (T6P) can reflect conditions of high supply of assimilates, and, in this case, it activates the AGPase enzyme through a redox activation mechanism, boosting starch formation.

Thus, drought increased the starch content in *A. emarginata* leaves, as reported by Matos Filho and Carvalho [75], however, not differing from flooding plants, corroborating Honório et al. [56]. This carbohydrate is part of the carbon "pool" within the plant [75], and it is worth mentioning that, in photosynthetic cells, starch can be synthesized and temporarily stored in chloroplasts, which is a "transitory" starch; therefore, it can be synthesized and degraded in other sugars within a 24-hour window, influencing carbon allocation throughout the plant and thus mitigating the negative effect of stress [76–78].

Under flooding, *A. emarginata* plants accumulate total sugars, reducing sugars and sucrose similar to results found by Henrique et al. [79], and, according to these authors, in this type of stress, there is a reduction in the translocation rate of carbohydrates from leaves to roots, in addition to a reduction in plant growth and metabolic activities; thus, the demand for carbohydrates decreases, leading to the accumulation of these photosynthates in leaves. On the other hand, the sucrose concentration was lower in plants submitted to drought, and this situation can be explained, according to Kuai et al. [80], due to the fact that this sugar can be used in two ways: (i) directly via glycolysis; or (ii) translocated within the plant through the phloem to draining tissues, so that when recovery was imposed on plants, the sucrose concentration did not differ between treatments.

With regard to arabinose, the high concentration of this monosaccharide in plants subjected to stress, whether due to excess or lack of water, corroborates the results found by Moore et al. [81]. These authors argue that the accumulation of this sugar induces the flexibilization of the cell walls of photosynthetic cells during water stress, facilitating subsequent plant recovery, and this may be one of the reasons why the concentration of this sugar was lower when plants were submitted to recovery.

In summary, it could be concluded that stress (drought and flooding) caused changes in variables such as gas exchange and chlorophyll *a* fluorescence more markedly in flooded *Annona emarginata* plants "terra fria" morphotype. What draws attention is that plants maintained in drought did not reduce carboxylation efficiency even with a reduction in stomatal conductance and A_{net}, indicating that there was no damage to the photosynthetic apparatus, which was confirmed by the maintenance of maximum quantum yield (FV/FM) at levels similar to plants kept at field capacity. It is noteworthy that except for FV/FM and *Ex*, all other fluorescence variables were reduced or increased by stress treatments, but only plants subjected to drought showed recovery of variables such as effective quantum efficiency, ETR, and gas exchange, which generally represent greater photosynthesis.

During stress establishment, it was also evident that the energy not used in photochemistry was dissipated in the form of heat (increase in D) by plants in both treatments. Furthermore, a reduction in non-photochemical extinction in the antenna complexes was observed (reduction in NPQ), releasing thermal energy in both treatments. However, the maintenance of high *Ex* values only in drought plants suggests that this energy was used to increase the synthesis of total alkaloids, and during flooding, it was directed to the synthesis of Liriodenine. However, in the recovery, *Ex* values remained high in drought plants, D was reduced, and the quantum yield was increased to intermediate values (higher than that of flooding plants and lower than control), indicating that part of the energy was used to maintain photosynthesis and part was used for the synthesis of specialized metabolites, such as alkaloids.

Therefore, the results suggest that the use of the *Annona emarginata* "terra fria" morphotype should be better evaluated as rootstock for regions with high water regimes while confirming its suitability for soils with greater water restrictions. Furthermore, the variation in alkaloid production can be managed depending on the water regime, which can be interesting in the search for molecular biodiversity, which deserves to be further investigated.

5. Conclusions

The primary metabolism of *Annona emarginata* is differentially affected by drought and flooding conditions, which results in the modulation of the alkaloid metabolism. Flooding in general causes greater damage to gas exchange as well as to chlorophyll *a* fluorescence, which results in greater plant recovery difficulties. However, flooding promotes an increase in the concentration of Liriodenine but not of total alkaloids. On the other hand, water stress due to drought causes less damage to the photosynthetic apparatus and greater plant recovery capacity. Under this condition, there is an increase in the biosynthesis of total alkaloids. Some alkaloids are constant in *Annona emarginata* plants regardless of stress (N-Methyl-Laurotetanine, Norglaucin, Xylopine, Laurotetanin, Liriodenine, and Lanulinosin). However, other alkaloids (Reticulin, Norpredicentine, Discretin, Xylopinine, Assimilobin, and Oxoglaucine) are observed differently depending on water levels during stress and recovery periods. Although it is a species native to the Atlantic Forest, a biome with high water availability, *Annona emarginata* adapts better to drought than flooding periods.

Supplementary Materials: The following supporting information can be downloaded at: https://www.mdpi.com/article/10.3390/horticulturae10030220/s1. Table S1: *p* and f values (two-way ANOVA) are indicated. Analyzed the interaction between water levels (field capacity, flooding, drought) × moments (stress and recovery) on the concentration of total alkaloids and Liriodenine; Table S2: *p* and f values (two-way ANOVA) are indicated. Analyzed the interaction between water levels (field capacity, flooding, drought) × moments (stress and recovery) on the carbon assimilation rate (A_{net}), Rubisco carboxylation efficiency (A_{net}/Ci), instantaneous water use efficiency (WUE), stomatal conductance (*gs*), and transpiration (E); Table S3: *p* and f values (two-way ANOVA) are indicated. Analyzed the interaction between water levels (field capacity, flooding, drought) × moments (stress and recovery) on the maximum quantum yield (Fv/Fm), potential quantum efficiency (Fv′/Fm′), electron transport rate (ETR), effective quantum yield (ΦPSII), non-photochemical quenching (NPQ), photosystem II energy that cannot be dissipated (Ex), and energy dissipated in the form of heat (D); Table S4: *p* and f values (two-way ANOVA) are indicated. Analyzed the interaction between water levels (field capacity, flooding, drought) × moments (stress and recovery) on the total sugars, reducing sugars, sucrose, starch, glucose, fructose, trehalose, and arabinose.

Author Contributions: Conceptualization, A.B.M.H. and G.F.; formal analysis, A.B.M.H., F.G.C. and G.F.; investigation, A.B.M.H., G.C.d.S., M.R.d.S. and F.G.C.; resources, M.R.d.S., C.S.F.B. and G.F.; writing—original draft preparation, A.B.M.H. and G.F.; writing—review and editing, A.B.M.H., I.D.-l.-C.-C., C.O.M., M.R.d.S., C.S.F.B. and G.F.; supervision, G.F.; funding acquisition, G.F. All authors have read and agreed to the published version of the manuscript.

Funding: This research was funded by Coordenação de Aperfeiçoamento de Pessoal de Nível Superior—Brasil (CAPES)—Finance Code 001, CAPESPrint—Unesp and National Council for Scientific and Technological Development (CNPq) (140073/2022−1).

Data Availability Statement: Data are contained with in the article and Supplementary Materials.

Acknowledgments: To Seedling Production Center of São Bento do Sapucaí, CATI (Technical and Integrated Assistance Coordination), municipality of São Bento do Sapucaí—São Paulo, for the donation of young *Annona emarginata* plants ("araticum de terra-fria" morphotype).

Conflicts of Interest: The authors declare no conflicts of interest.

References

1. Chatrou, L.W.; Erkens, R.H.J.; Richardson, J.E.; Saunders, R.M.K.; Fay, M.F. The Natural History of Annonaceae. *Bot. J. Linn. Soc.* **2012**, *169*, 1–4. [CrossRef]
2. Couvreur, T.L.P.; Helmstetter, A.J.; Koenen, E.J.M.; Bethune, K.; Brandão, R.D.; Little, S.A.; Sauquet, H.; Erkens, R.H.J. Phylogenomics of the Major Tropical Plant Family Annonaceae Using Targeted Enrichment of Nuclear Genes. *Front. Plant Sci.* **2019**, *9*, 1941. [CrossRef] [PubMed]
3. Mendes-Silva, I.; Lopes, J.C.; Silva, L.V.; Bazante, M.L. Annona in Flora Do Brasil 2020. Available online: https://floradobrasil2020.jbrj.gov.br/reflora/floradobrasil/FB110243 (accessed on 20 July 2023).
4. Maas, P.J.M.; de Kamer, H.M.; Junikka, L.; de Mello-Silva, R.; Rainer, H. Annonnaceae from Central-Eastern Brazil. *Rodriguésia* **2001**, *52*, 65–98. [CrossRef]
5. Chatrou, L.W.; Turner, I.M.; Klitgaard, B.B.; Maas, P.J.M.; Utteridge, T.M.A. A Linear Sequence to Facilitate Curation of Herbarium Specimens of Annonaceae. *Kew Bull.* **2018**, *73*, 39. [CrossRef] [PubMed]
6. Ovile Mimi, C.; De-la-Cruz-Chacón, I.; Caixeta Sousa, M.; Aparecida Ribeiro Vieira, M.; Ortiz Mayo Marques, M.; Ferreira, G.; Silvia Fernandes Boaro, C. Chemophenetics as a Tool for Distinguishing Morphotypes of *Annona emarginata* (Schltdl.) H. Rainer. *Chem. Biodivers.* **2021**, *18*, e2100544. [CrossRef] [PubMed]
7. De Lemos, E.E.P. A Produção de Anonáceas No Brasil. *Rev. Bras. Frutic.* **2014**, *36*, 77–85. [CrossRef]
8. Bettiol Neto, J.E.; Pio, R.; Bueno, S.C.S.; Bastos, D.C.; Scarpare Filho, J.A. Enraizamento de Estacas Dos Porta-Enxertos Araticum-de-Terra-Fria (*Rollinia* Sp.) e Araticum-Mirim (*Rollinia emarginata* Schltdl.) Para Anonáceas. *Ciência Agrotecnologia* **2006**, *30*, 1077–1082. [CrossRef]
9. Junqueira, N.T.V.; Junqueira, K.P. Principais Doenças de Anonáceas No Brasil: Descrição e Controle. *Rev. Bras. Frutic.* **2014**, *36*, 55–64. [CrossRef]

10. São José, A.R.; Pires, M.D.M.; de Freitas, A.L.G.E.; Ribeiro, D.P.; Perez, L.A.A. Atualidades e Perspectivas Das Anonáceas No Mundo. *Rev. Bras. Frutic.* **2014**, *36*, 86–93. [CrossRef]
11. Mantoan, L.P.B.; Rolim de Almeida, L.F.; Macedo, A.C.; Ferreira, G.; Boaro, C.S.F. Photosynthetic Adjustment after Rehydration in *Annona emarginata*. *Acta Physiol. Plant* **2016**, *38*, 157. [CrossRef]
12. de Lima, J.P.S.; Pinheiro, M.L.B.; Santos, A.M.G.; Pereira, J.L.d.S.; Santos, D.M.F.; Barison, A.; Silva-Jardim, I.; Costa, E.V. In Vitro Atileishmanial and Cytotoxic Activities of *Annona mucosa* (Annonaceae). *Rev. Virtual Química* **2012**, *4*, 692–702. [CrossRef]
13. Quílez, A.M.; Fernández-Arche, M.A.; García-Giménez, M.D.; De la Puerta, R. Potential Therapeutic Applications of the Genus *Annona*: Local and Traditional Uses and Pharmacology. *J. Ethnopharmacol.* **2018**, *225*, 244–270. [CrossRef]
14. Makabe, H.; Konno, H.; Miyoshi, H. Current Topics of Organic and Biological Chemistry of Annonaceous Acetogenins and Their Synthetic Mimics. *Curr. Drug Discov. Technol.* **2008**, *5*, 213–229. [CrossRef] [PubMed]
15. Chowdhury, S.S.; Tareq, A.M.; Tareq, S.M.; Farhad, S.; Sayeed, M.A. Screening of Antidiabetic and Antioxidant Potential along with Phytochemicals of *Annona* Genus: A Review. *Futur. J. Pharm. Sci.* **2021**, *7*, 144. [CrossRef]
16. Debnath, B.; Singh, W.S.; Das, M.; Goswami, S.; Singh, M.K.; Maiti, D.; Manna, K. Role of Plant Alkaloids on Human Health: A Review of Biological Activities. *Mater. Today Chem.* **2018**, *9*, 56–72. [CrossRef]
17. Costa, E.V.; da Cruz, P.E.O.; de Lourenço, C.C.; de Souza Moraes, V.R.; de Lima Nogueira, P.C.; Salvador, M.J. Antioxidant and Antimicrobial Activities of Aporphinoids and Other Alkaloids from the Bark of *Annona salzmannii* A. DC. (Annonaceae). *Nat. Prod. Res.* **2013**, *27*, 1002–1006. [CrossRef]
18. De La Cruz Chacón, I.; González-Esquinca, A.R. Liriodenine Alkaloid in *Annona diversifolia* during Early Development. *Nat. Prod. Res.* **2012**, *26*, 42–49. [CrossRef]
19. Sousa, M.C.; De-la-Cruz-Chacón, I.; Campos, F.G.; Vieira, M.A.R.; Corrêa, P.L.C.; Marques, M.O.M.; Boaro, C.S.F.; Ferreira, G. Plant Growth Regulators Induce Differential Responses on Primary and Specialized Metabolism of *Annona emarginata* (Annonaceae). *Ind. Crops Prod.* **2022**, *189*, 115789. [CrossRef]
20. Leboeuf, M.; Cavé, A.; Bhaumik, P.K.; Mukherjee, B.; Mukherjee, R. The Phytochemistry of the Annonaceae. *Phytochemistry* **1982**, *21*, 2783–2813. [CrossRef]
21. Costa, E.V.; Pinheiro, M.L.B.; Xavier, C.M.; Silva, J.R.A.; Amaral, A.C.F.; Souza, A.D.L.; Barison, A.; Campos, F.R.; Ferreira, A.G.; Machado, G.M.C.; et al. A Pyrimidine-β-Carboline and Other Alkaloids from *Annona foetida* with Antileishmanial Activity. *J. Nat. Prod.* **2006**, *69*, 292–294. [CrossRef]
22. Cota, L.G.; Vieira, F.A.; Melo Júnior, A.F.; Brandão, M.M.; Santana, K.N.O.; Guedes, M.L.; Oliveira, D.A. Genetic Diversity of *Annona crassiflora* (Annonaceae) in Northern Minas Gerais State. *Genet. Mol. Res.* **2011**, *10*, 2172–2180. [CrossRef] [PubMed]
23. Khan, T.-M.; Gul, N.S.; Lu, X.; Wei, J.-H.; Liu, Y.-C.; Sun, H.; Liang, H.; Orvig, C.; Chen, Z.-F. In vitro and in vivo Anti-Tumor Activity of Two Gold(III) Complexes with Isoquinoline Derivatives as Ligands. *Eur. J. Med. Chem.* **2019**, *163*, 333–343. [CrossRef] [PubMed]
24. De-la-Cruz-Chacón, I.; López-Fernández, N.Y.; Riley-Saldaña, C.A.; Castro Moreno, M.; González-Esquinca, A.R. Antifungal Activity in Vitro of *Sapranthus microcarpus* (Annonaceae) against Phytopathogens. *Acta Bot. Mex.* **2018**, *126*, e1420. [CrossRef]
25. Chen, C.-Y. Review on Pharmacological Activities of Liriodenine. *Afr. J. Pharm. Pharmacol.* **2013**, *7*, 1067–1070. [CrossRef]
26. Lúcio, A.S.S.C.; da Silva Almeida, J.R.G.; Da-Cunha, E.V.L.; Tavares, J.F.; Barbosa Filho, J.M. Alkaloids of the Annonaceae: Occurrence and a Compilation of Their Biological Activities. *Alkaloids Chem. Biol.* **2015**, *74*, 233–409. [CrossRef]
27. Yang, L.; Wen, K.-S.; Ruan, X.; Zhao, Y.-X.; Wei, F.; Wang, Q. Response of Plant Secondary Metabolites to Environmental Factors. *Molecules* **2018**, *23*, 762. [CrossRef]
28. Selmar, D.; Kleinwächter, M. Stress Enhances the Synthesis of Secondary Plant Products: The Impact of Stress-Related over-Reduction on the Accumulation of Natural Products. *Plant Cell Physiol.* **2013**, *54*, 817–826. [CrossRef]
29. Zhao, Y.H.; Jia, X.; Wang, W.K.; Liu, T.; Huang, S.P.; Yang, M.Y. Growth under Elevated Air Temperature Alters Secondary Metabolites in *Robinia pseudoacacia* L. Seedlings in Cd- and Pb-Contaminated Soils. *Sci. Total Environ.* **2016**, *565*, 586–594. [CrossRef]
30. Jan, R.; Asaf, S.; Numan, M.; Lubna; Kim, K.-M. Plant Secondary Metabolite Biosynthesis and Transcriptional Regulation in Response to Biotic and Abiotic Stress Conditions. *Agronomy* **2021**, *11*, 968. [CrossRef]
31. Shao, H.-B.; Chu, L.-Y.; Jaleel, C.A.; Zhao, C.-X. Water-Deficit Stress-Induced Anatomical Changes in Higher Plants. *Comptes Rendus Biol.* **2008**, *331*, 215–225. [CrossRef]
32. Sengupta, D.; Guha, A.; Reddy, A.R. Interdependence of Plant Water Status with Photosynthetic Performance and Root Defense Responses in *Vigna radiata* (L.) Wilczek under Progressive Drought Stress and Recovery. *J. Photochem. Photobiol. B* **2013**, *127*, 170–181. [CrossRef]
33. Martinazzo, E.G.; Perboni, A.T.; de Oliveira, P.V.; Bianchi, V.J.; Bacarin, M.A. Atividade Fotossintética Em Plantas de Ameixeira Submetidas Ao Déficit Hídrico e Ao Alagamento. *Ciência Rural.* **2012**, *43*, 35–41. [CrossRef]
34. de Oliveira, J.D.S.; de Lemos, E.E.P.; Filho, R.V.d.C.; dos Santos, E.F.; Gallo, R.B.S.C.M. Alterações Fisiológicas No Crescimento Inicial de Pinheira (*Annona squamosa* L.) Submetida Ao Stresse Hídrico Physiological Changes in the Initial Growth of Sugar Apple (*Annona squamosa* L.) Submitted to Water Stress. *Rev. Ciências Agrárias* **2020**, *43*, 53–63. [CrossRef]
35. Moreira, R.C.L.; Brito, M.E.B.; Fernandes, P.D.; da Silva Sá, F.V.; Silva, L.D.A.; Oliveira, C.J.A.; Veloso, L.L.d.S.A.; de Queiroga, T.B. Growth and Physiology of *Annona squamosa* L. Under Different Irrigation Depths and Phosphate Fertilization. *Biosci. J.* **2019**, *35*, 389–397. [CrossRef]

36. Castro-Moreno, M.; Tinoco-Ojangurén, C.L.; Cruz-Ortega, M.D.R.; González-Esquinca, A.R. Influence of Seasonal Variation on the Phenology and Liriodenine Content of *Annona lutescens* (Annonaceae). *J. Plant Res.* **2013**, *126*, 529–537. [CrossRef]
37. Da Cunha, A.R.; Martins, D. Classificação Climática Para os Municípios de Botucatu e São Manuel, SP. *Irriga* **2009**, *14*, 1–11. [CrossRef]
38. Baron, D.; Amaro, A.C.E.; Campos, F.G.; Ferreira, G. Leaf Gas Exchanges Responses of Atemoya Scion Grafted onto *Annona* Rootstocks. *Theor. Exp. Plant Physiol.* **2018**, *30*, 203–213. [CrossRef]
39. Murchie, E.H.; Lawson, T. Chlorophyll Fluorescence Analysis: A Guide to Good Practice and Understanding Some New Applications. *J. Exp. Bot.* **2013**, *64*, 3983–3998. [CrossRef] [PubMed]
40. Kitajima, M.; Butler, W.L. Quenching of Chlorophyll Fluorescence and Primary Photochemistry in Chloroplasts by Dibromothymoquinone. *Biochim. Biophys. Acta (BBA) Bioenerg.* **1975**, *376*, 105–115. [CrossRef]
41. Genty, B.; Briantais, J.-M.; Baker, N.R. The Relationship between the Quantum Yield of Photosynthetic Electron Transport and Quenching of Chlorophyll Fluorescence. *Biochim. Biophys. Acta (BBA) General. Subj.* **1989**, *990*, 87–92. [CrossRef]
42. Schreiber, U.; Schliwa, U.; Bilger, W. Continuous Recording of Photochemical and Non-Photochemical Chlorophyll Fluorescence Quenching with a New Type of Modulation Fluorometer. *Photosynth. Res.* **1986**, *10*, 51–62. [CrossRef]
43. Bilger, W.; Björkman, O. Role of the Xanthophyll Cycle in Photoprotection Elucidated by Measurements of Light-Induced Absorbance Changes, Fluorescence and Photosynthesis in Leaves of Hedera Canariensis. *Photosynth. Res.* **1990**, *25*, 173–185. [CrossRef]
44. Demmig-Adams, B.; Adams, W.W., III; Barker, D.H.; Logan, B.A.; Bowling, D.R.; Verhoeven, A.S. Using Chlorophyll Fluorescence to Assess the Fraction of Absorbed Light Allocated to Thermal Dissipation of Excess Excitation. *Physiol. Plant* **2008**, *98*, 253–264. [CrossRef]
45. Garcia, I.S.; Souza, A.; Barbedo, C.J.; Dietrich, S.M.C.; Figueiredo-Ribeiro, R.C.L. Changes in Soluble Carbohydrates during Storage of *Caesalpinia echinata* LAM. (Brazilwood) Seeds, an Endangered Leguminous Tree from the Brazilian Atlantic Forest. *Braz. J. Biol.* **2006**, *66*, 739–745. [CrossRef]
46. Clegg, K.M. The Application of the Anthrone Reagent to the Estimation of Starch in Cereals. *J. Sci. Food Agric.* **1956**, *7*, 40–44. [CrossRef]
47. Morris, D.L. Quantitative Determination of Carbohydrates with Dreywood's Anthrone Reagent. *Science* **1948**, *107*, 254–255. [CrossRef] [PubMed]
48. Blenkinsopp, A. A Good Year. *Int. J. Pharm. Pract.* **2011**, *7*, 197. [CrossRef]
49. Miller, G.L. Use of Dinitrosalicylic Acid Reagent for Determination of Reducing Sugar. *Anal. Chem.* **1959**, *31*, 426–428. [CrossRef]
50. Passos, L.P. *Métodos Analíticos e Labora Toriais em Fisiologia Vegetal*; EMBRAPA-CNPGL: Coronel Pacheco, MG, Brazil, 1996; ISBN 8585748087.
51. Wan, J.; Griffiths, R.; Ying, J.; McCourt, P.; Huang, Y. Development of Drought-Tolerant Canola (*Brassica napus* L.) through Genetic Modulation of ABA-Mediated Stomatal Responses. *Crop Sci.* **2009**, *49*, 1539–1554. [CrossRef]
52. Liu, H.; Song, S.; Zhang, H.; Li, Y.; Niu, L.; Zhang, J.; Wang, W. Signaling Transduction of ABA, ROS, and Ca2+ in Plant Stomatal Closure in Response to Drought. *Int. J. Mol. Sci.* **2022**, *23*, 14824. [CrossRef] [PubMed]
53. De Oliveira, A.K.M.; Gualtieri, S.C.J. Trocas Gasosas E Grau De Tolerância Ao Estresse Hídrico Induzido Em Plantas Jovens De *Tabebuia aurea*; (PARATUDO) Submetidas A Alagamento. *Ciência Florest.* **2017**, *27*, 181–191. [CrossRef]
54. Chen, S.; ten Tusscher, K.H.W.J.; Sasidharan, R.; Dekker, S.C.; de Boer, H.J. Parallels between Drought and Flooding: An Integrated Framework for Plant Eco-physiological Responses to Water Stress. *Plant-Environ. Interact.* **2023**, *4*, 175–187. [CrossRef] [PubMed]
55. Ashraf, M.; Harris, P.J.C. Photosynthesis under Stressful Environments: An Overview. *Photosynthetica* **2013**, *51*, 163–190. [CrossRef]
56. Honório, A.B.M.; De-la-Cruz-Chacón, I.; Martínez-Vázquez, M.; da Silva, M.R.; Campos, F.G.; Martin, B.C.; da Silva, G.C.; Fernandes Boaro, C.S.; Ferreira, G. Impact of Drought and Flooding on Alkaloid Production in *Annona crassiflora* Mart. *Horticulturae* **2021**, *7*, 414. [CrossRef]
57. Kleinwächter, M.; Selmar, D. New Insights Explain That Drought Stress Enhances the Quality of Spice and Medicinal Plants: Potential Applications. *Agron. Sustain. Dev.* **2015**, *35*, 121–131. [CrossRef]
58. Liu, Y.; Meng, Q.; Duan, X.; Zhang, Z.; Li, D. Effects of PEG-Induced Drought Stress on Regulation of Indole Alkaloid Biosynthesis in *Catharanthus roseus*. *J. Plant Interact.* **2017**, *12*, 87–91. [CrossRef]
59. Ghorbanpour, M. Role of Plant Growth Promoting Rhizobacteria on Antioxidant Enzyme Activities and Tropane Alkaloids Production of Hyoscyamus Niger under Water Deficit Stress. *Turk. J. Biol.* **2013**, *37*, 350–360. [CrossRef]
60. Dudareva, N.; Klempien, A.; Muhlemann, J.K.; Kaplan, I. Biosynthesis, Function and Metabolic Engineering of Plant Volatile Organic Compounds. *N. Phytol.* **2013**, *198*, 16–32. [CrossRef]
61. Kalaji, H.M.; Jajoo, A.; Oukarroum, A.; Brestic, M.; Zivcak, M.; Samborska, I.A.; Cetner, M.D.; Łukasik, I.; Goltsev, V.; Ladle, R.J. Chlorophyll *a* Fluorescence as a Tool to Monitor Physiological Status of Plants under Abiotic Stress Conditions. *Acta Physiol. Plant* **2016**, *38*, 102. [CrossRef]
62. Wilhelm, C.; Selmar, D. Energy Dissipation Is an Essential Mechanism to Sustain the Viability of Plants: The Physiological Limits of Improved Photosynthesis. *J. Plant Physiol.* **2011**, *168*, 79–87. [CrossRef]
63. De Azevedo Neto, A.D.; Pereira, P.P.A.; Costa, D.P.; Santos, A.C.C. dos Fluorescência Da Clorofila Como Uma Ferramenta Possível Para Seleção de Tolerância à Salinidade Em Girassol. *Rev. Ciência Agronômica* **2011**, *42*, 893–897. [CrossRef]
64. Maxwell, K.; Johnson, G.N. Chlorophyll Fluorescence—A Practical Guide. *J. Exp. Bot.* **2000**, *51*, 659–668. [CrossRef] [PubMed]

65. Tian, Y.; Ungerer, P.; Zhang, H.; Ruban, A.V. Direct Impact of the Sustained Decline in the Photosystem II Efficiency upon Plant Productivity at Different Developmental Stages. *J. Plant Physiol.* **2017**, *212*, 45–53. [CrossRef] [PubMed]
66. Nilkens, M.; Kress, E.; Lambrev, P.; Miloslavina, Y.; Müller, M.; Holzwarth, A.R.; Jahns, P. Identification of a Slowly Inducible Zeaxanthin-Dependent Component of Non-Photochemical Quenching of Chlorophyll Fluorescence Generated under Steady-State Conditions in Arabidopsis. *Biochim. Biophys. Acta Bioenerg.* **2010**, *1797*, 466–475. [CrossRef] [PubMed]
67. Corrêa, P.L.C.; De-la-Cruz-Chacón, I.; Sousa, M.C.; Vieira, M.A.R.; Campos, F.G.; Marques, M.O.M.; Boaro, C.S.F.; Ferreira, G. Effect of Nitrogen Sources on Photosynthesis and Biosynthesis of Alkaloids and Leaf Volatile Compounds in *Annona sylvatica* A. St.-Hil. *J. Soil. Sci. Plant Nutr.* **2022**, *22*, 956–970. [CrossRef]
68. Jaleel, C.A.; Manivannan, P.; Kishorekumar, A.; Sankar, B.; Gopi, R.; Somasundaram, R.; Panneerselvam, R. Alterations in Osmoregulation, Antioxidant Enzymes and Indole Alkaloid Levels in *Catharanthus roseus* Exposed to Water Deficit. *Colloids Surf. B Biointerfaces* **2007**, *59*, 150–157. [CrossRef]
69. Kirk, H.; Vrieling, K.; Van Der Meijden, E.; Klinkhamer, P.G.L. Species by Environment Interactions Affect Pyrrolizidine Alkaloid Expression in *Senecio jacobaea*, *Senecio aquaticus*, and Their Hybrids. *J. Chem. Ecol.* **2010**, *36*, 378–387. [CrossRef]
70. Çakir, R.; Çebi, U. The Effect of Irrigation Scheduling and Water Stress on the Maturity and Chemical Composition of *Virginia tobacco* Leaf. *Field Crops Res.* **2010**, *119*, 269–276. [CrossRef]
71. Xia, L.; Yang, W.; Xiufeng, Y. Effects of Water Stress on Berberine, Jatrorrhizine and Palmatine Contents in Amur Corktree Seedlings. *Acta Ecol. Sin.* **2007**, *27*, 58–63. [CrossRef]
72. Arango, O.; Pérez, E.; Granados, H.; Rojano, B.; Sáez, J. Inhibición de la Peroxidación lipídica y capacidad de atrapadora de radicales libres de alcaloides aislados de dos annonaceae, Xylopia amazonica cf. y Duguetia vallicola. *Actualidades Biológicas* **2017**, *26*, 105–110. [CrossRef]
73. Cortina, C.; Culiáñez-Macià, F.A. Tomato Abiotic Stress Enhanced Tolerance by Trehalose Biosynthesis. *Plant Sci.* **2005**, *169*, 75–82. [CrossRef]
74. Paul, M.J.; Primavesi, L.F.; Jhurreea, D.; Zhang, Y. Trehalose Metabolism and Signaling. *Annu. Rev. Plant Biol.* **2008**, *59*, 417–441. [CrossRef] [PubMed]
75. Matos Filho, H.A.; Carvalho, R.D.C.M. Análise de Carboidratos Solúveis em Plantas de Arroz. *Científic@ Multidiscip. J.* **2020**, *7*, 1–8. [CrossRef]
76. Thalmann, M.; Santelia, D. Starch as a Determinant of Plant Fitness under Abiotic Stress. *New Phytol.* **2017**, *214*, 943–951. [CrossRef]
77. Liu, K.; Zou, W.; Gao, X.; Wang, X.; Yu, Q.; Ge, L. Young Seedlings Adapt to Stress by Retaining Starch and Retarding Growth through ABA-Dependent and -Independent Pathways in Arabidopsis. *Biochem. Biophys. Res. Commun.* **2019**, *515*, 699–705. [CrossRef]
78. Krasensky, J.; Jonak, C. Drought, Salt, and Temperature Stress-Induced Metabolic Rearrangements and Regulatory Networks. *J. Exp. Bot.* **2012**, *63*, 1593–1608. [CrossRef]
79. Henrique, P.D.C.; Alves, J.D.; Goulart, P.d.F.P.; Deuner, S.; Silveira, N.M.; Zanandrea, I.; de Castro, E.M. Características Fisiológicas e Anatômicas de Plantas de Sibipiruna Submetidas à Hipoxia. *Ciência Rural.* **2009**, *40*, 70–76. [CrossRef]
80. Kuai, J.; Chen, Y.; Wang, Y.; Meng, Y.; Chen, B.; Zhao, W.; Zhou, Z. Effect of Waterlogging on Carbohydrate Metabolism and the Quality of Fiber in Cotton (*Gossypium hirsutum* L.). *Front. Plant Sci.* **2016**, *7*, 877. [CrossRef]
81. Moore, J.P.; Nguema-Ona, E.E.; Vicré-Gibouin, M.; Sørensen, I.; Willats, W.G.T.; Driouich, A.; Farrant, J.M. Arabinose-Rich Polymers as an Evolutionary Strategy to Plasticize Resurrection Plant Cell Walls against Desiccation. *Planta* **2013**, *237*, 739–754. [CrossRef]

horticulturae

MDPI

Article

Changes in Growth Parameters, C:N:P Stoichiometry and Non-Structural Carbohydrate Contents of *Zanthoxylum armatum* Seedling in Response to Five Soil Types

Tao Gu [1], Hongyu Ren [1], Mengying Wang [1], Wenzhang Qian [1], Yunyi Hu [1], Yao Yang [1], Ting Yu [2], Kuangji Zhao [1,3] and Shun Gao [1,3,*]

[1] Department of Forestry, Faculty of Forestry, Sichuan Agricultural University, Chengdu 611130, China; gutao1@stu.sicau.edu.cn (T.G.); renhongyu@stu.sicau.edu.cn (H.R.); wangmengying1@stu.sicau.edu.cn (M.W.); 202001247@stu.sicau.edu.cn (W.Q.); huyunyi@stu.sicau.edu.cn (Y.H.); yangyao@stu.sicau.edu.cn (Y.Y.); zhaokj@sicau.edu.cn (K.Z.)

[2] School of Life Science and Engineering, Southwest Jiaotong University, Chengdu 611756, China; yuting212916@my.swjtu.edu.cn

[3] National Forestry and Grassland Administration Key Laboratory of Forest Resources Conservation and Ecological Safety on the Upper Reaches of the Yangtze River, Sichuan Agricultural University, Chengdu 611130, China

* Correspondence: shungao@sicau.edu.cn

Abstract: *Zanthoxylum armatum* (*Z. armatum*) is an economic crop widely planted for both spice and medicinal purposes in Southwest China. Soil is a key environmental condition that affects seedling growth and development, and screening suitable soil types is of great significance for the large-scale cultivation of crops. This study designed growth experiments of *Z. armatum* seedlings in red soil (RS), yellow soil (YS), acidic purple soil (ACPS), alkaline purple soil (ALPS), and alluvial soil (AS) to screen for more suitable soil types. The growth traits of *Z. armatum* seedlings and the carbon (C), nitrogen (N), phosphorus (P), C:N:P stoichiometry, and non-structural carbohydrate (NSC) content of different organs were comparatively analyzed. The results showed that the morphological indexes of *Z. armatum* seedlings cultured in AS were better than those in the other four soils.AS and RS may be beneficial for the culture of *Z. armatum* seedlings due to higher nutrient levels in three organs. Two-factor ANOVA and PCA analysis showed that C, N, and P and their proportions would affect the uptake and distribution of NSC in various organs of *Z. armatum* seedlings. These results showed that soil types and plant organs significantly affected the accumulation and distribution of N, P, and NSC in *Z. armatum* seedlings. These results are conducive to screening soil types suitable for the growth and development of *Z. armatum* and provide data support for further large-scale cultivation of *Z. armatum* in suitable areas.

Keywords: *Zanthoxylum armatum*; soil type; organs; growth parameters; stoichiometry; non-structural carbohydrates

Citation: Gu, T.; Ren, H.; Wang, M.; Qian, W.; Hu, Y.; Yang, Y.; Yu, T.; Zhao, K.; Gao, S. Changes in Growth Parameters, C:N:P Stoichiometry and Non-Structural Carbohydrate Contents of *Zanthoxylum armatum* Seedling in Response to Five Soil Types. *Horticulturae* **2024**, *10*, 261. https://doi.org/10.3390/horticulturae10030261

Academic Editors: Adalberto Benavides-Mendoza, Yolanda González-García, Fabián Pérez Labrada and Susana González-Morales

Received: 4 February 2024
Revised: 6 March 2024
Accepted: 7 March 2024
Published: 8 March 2024

1. Introduction

Soil is the basic substrate for plant growth and development, and its physicochemical properties are important external influencing factors for plant growth and development [1,2]. Different types of soil represent different physical and chemical characteristics due to differences in their formation processes, nutritional conditions, etc., which will lead to differences in water absorption efficiency, nutrient absorption, and transportation efficiency during plant growth and development [3,4]. In plants, the responses of physiology and biochemicals, such as photosynthesis, mineral metabolism, etc., are closely related to the growth environment, especially soil types [1,5,6]. Variations in plant growth status and chemical composition content are evidence that plants could be affected by environmental conditions such as soil.

Significant differences in plant growth parameters and oil yield per plant are observed between *Mentha arvensis* grown in different soil types [7]. Different soil textures influence the phytochemical contents and antioxidant properties of *Solanum nigrum* L. cultivated in variable soil types [8]. These reports suggested that different soil types do have an impact on plant growth and development.

Carbon (C), nitrogen (N), and phosphorus (P) are considered essential elements in the processes of plant growth [9]. C can form the basic structure of plants and accounts for about 50% of their biomass [10]. N and P play vital roles in protein and nucleic acid formation and are also prominent mineral nutrients affecting plant biomass production [10,11]. Thus, the levels of C, N, and P serve as indicators for assessing nutrient absorption efficiency, utilization effectiveness, and adaptability to environmental stress in plants. C, N, and P are strongly coupled in plant biochemical functions, and their balance and interaction relationship usually alter physiological activity and growth rates [12]. C:N:P chemometrics has turned into the most explored factor in the interactional mechanism between plant growth and the eco-environment [13,14]. Some studies have found that plant N:P is related to growth rate, which means that N:P can be used as an index to predict plant growth rate [11,15]. Previous studies have shown that both N and P are key limiting elements controlling plant development and primary production, and N:P helps to understand the limiting types of ecosystems, which is crucial for screening and modifying soil conditions [16]. The external supply rate of C, N, and P limits the uptake of nutrients, and the functional divergence of plant tissues leads to internal differences in plant organs [17]. The nutrient distribution and their stoichiometry in plants represent the choices that plants make when acquiring and distributing above-ground and subsurface resources [18]. Plants can be curtailed due to insufficient soil nutrition, and they will constantly adjust the physiological and ecological processes of vegetative organs to ensure their normal growth and development [19]. Soil types, representing changes in the soil environment, will directly affect the adaptation strategies of plants and change the stoichiometric ratio of plants [20]. As a consequence, it is very necessary to broaden our knowledge about C, N, and P and their ecological stoichiometric characteristics in plants.

Non-structural carbohydrates (NSC), generally including glucose, fructose, sucrose, and starch, are prominent energy sources for plant growth and metabolism and play a crucial role in plant resistance to external adverse environmental stresses [21,22]. NSC is the product of plant photosynthesis. When the amount of carbon obtained by the plant is greater than the amount of carbon required for growth, its photosynthates are deposited in the form of NSC, and these NSCs can play specific roles at the appropriate time for the survival, growth, and/or other physiological functions of plants [23,24]. In plant organs, the quantity of NSCs commonly reflects C supply as well as the balance between C acquisition and consumption [25]. Higher NSC concentrations can improve the drought resistance and cold stress resistance of plants, which is beneficial for the survival and growth of plants in adverse environments [26,27]. Plants may adjust their biomass, nutrient content distribution, and photosynthetic characteristics to meet their own growth and development needs in response to different soil environments, which is an adaptation strategy to variable environments [21,28]. Therefore, studying the changes inplant nutrient content and stoichiometry in different soils provides a basis for exploring the relationship between plant functions and environmental adaptation mechanisms and is of great significance for analyzing and evaluating the soil type most suitable for plant growth.

Zanthoxylum armatum DC. (*Z. armatum*) is a small deciduous tree of *Zanthoxylum*, belonging to the Rutaceae family, and has enormous value for development, utilization, and long-term cultivation in Southwest China, Pakistan, India, etc. [29]. In Southwest China, the fruit of *Z. armatum* is usually harvested in its immature state, and the pericarp of *Z. armatum* is green, so it is called Qinghuajiao or Tengjiao, etc. The pericarp of its fruits can be used directly as traditional seasoning condiments or ground into powder for cooking in Sichuan cuisine, offering a unique, numb taste [30]. Moreover, different partsof *Z. armatum* include amides, essential oils, alkaloids, lignans, flavonoids, polyphenols, etc., which show antiviral, anti-tumor, antibacterial, anti-inflammatory, analgesic, antioxidant, and other pharmacological activities [31,32]. Thus, *Z. armatum* is also used in traditional Chinese medicine in China and can be used for the treatment of fever, toothache, stomachache, bruises, rheumatoid arthritis, etc. [32–34]. Due to its higher yield, stronger adaptability, and higher economic values, people have started to commercially cultivate this plant for rural livelihood improvement in Southwest China. Recently, the planting area of *Z. armatum* in China has been expanded and increased, and there is a wide range of soil types in these cultivation areas [29,35]. Although *Z. armatum* can grow on a variety of soil types, its production and quality vary significantly between different geographical locations. The seedling stage is the most critical life-history stage in plants, while the demands of different seedlings have diversified demands for a suitable soil environment [6,36]. The growth status of seedlings is important for large-scale cultivation of *Z. armatum* under different soil conditions, which directly and indirectly affects the growth cycle and the potential yield of *Z. armatum*. Thus, it is particularly important to evaluate and analyze the adaptability of *Z. armatum* seedling cultivation to different soil types in order to get the maximum yield of this plant. Here, the current study was designed to analyze growth parameters, nutrient content, and NSC content of *Z. armatum* seedlings in response to five soil types. These findings may offer an effective way to screen for suitable soil type and nutrient management during seedling cultivation.

2. Materials and Methods

2.1. Experimental Site

The experimental site lies at Sichuan Agricultural University in Chengdu, China, with geographical coordinates of 103°52′ E, 30°42′ N. It belongs to a subtropical monsoon climate with four distinct seasons, a mild climate, and abundant rainfall. The annual rainfall varied from 801.4 to 1445.5 mm. The rainy season lasts from June to September, with an average annual relative humidity of 84%. The annual average temperature is 17.5 °C, the monthly average highest temperature is 27.0 °C (July), the monthly average lowest temperature is 6.0 °C (January), and the annual average Sunshine duration is 1104.5 h.

2.2. Soil Basic Information

The RS, YS, ALPS, ACPS, and AS were obtained from Sanxing Village, Fengle Township, Shimian County, Ya'an, China (N 29°32′, E 102°54′, H 878 m), Baisheng Village, Baolin Town, Qionglai, Chengdu, China (N 30°21′, E 103°30′, H 552 m), Jifeng Town, Zhongjiang County, Deyang, China (N 31°03′, E 104°68′, H 900 m), Laobanshan Reading Park (N 29°58′, E 102°58′) in District Yucheng, Ya'an, China, and the Village Jing-shan, Town Yongan, District Shuangliu, Chengdu, China (30°365954′ N, 104°004114′ W, H 425 m), respectively. These soils were classified, crushed, sieved, and air-dried in the sun for further use. Soil pH was measured viaan electrode pH meter in a suspension of soil and water (1:2). The TN content was measured using the Kjeldahl method [37]. The organic matter content was measured by potassium dichromate oxidation—the external heating method [37]. TP content was measured by the Mo-Sb colorimetric method [37]. The main physico-chemical properties of five soils are displayed in Table 1.

Table 1. Main physico-chemical properties of five types of soil.

Soil Type	pH	Organic Matter ($g\cdot kg^{-1}$)	Total Nitrogen ($g\cdot kg^{-1}$)	Total Phosphorus ($g\cdot kg^{-1}$)
Red soil (RS)	4.9	16.64	1.14	0.067
Yellow soil (YS)	4.8	33.50	1.89	0.080
Acidic purple soil (ACPS)	4.3	42.53	2.33	0.235
Alkaline purple soil (ALPS)	8.7	20.99	1.65	0.301
Alluvial soil (AS)	8.1	38.38	2.35	0.300

2.3. Seedling Culture

Mature *Z. armatum* fruits (red) were collected in October 2021 from Lezhi County, Ziyang, China, and were dried in the shade at room temperature. The seeds were separated from the pericarps and deposited in damp sand until used. In February 2022, 500 seeds were sown in a box with nutrient soil in a green house. After 25–30 days, the seeds began to sprout, and the seedlings grew up to fully extended cotyledons (Stage BBCH10, about 2 cm high). One hundred seedlings were selected and planted in pots containing 10 kg of different soils (diameter and height, 28 × 22 cm, and the soil layer height, 20 cm). These pots were placed in the greenhouse of Sichuan Agricultural University from March 2022 to August 2022. During the seedling cultivation, the relative water moisture of the soils was checked and maintained at 60–80%, and weed and pest management were performed uniformly.

2.4. Sample Collection and Growth Parameter Measurement

After cultivation, the basic growth parameters of 10 seedlings in each group, such as plant height and ground diameter, were measured and recorded. Five seedlings were randomly collected from each group and sectioned into roots, stems, and leaves. Three parts were cleaned with deionized water, and the fresh weight (FW) was measured. They were first dried at 105 °C for 30 min, and then dried the second time for 12 h at 70 °C until they reached constant weight. The dry weight (DW) of each sample was recorded, and the relative water content (RWC) was calculated.

2.5. Determination of Nutrient Element Content

The dried samples of leaves, stems, and roots were grinded and sieved using a 60-mesh sieve, and then the content of C, N, P, and NSC was measured. The total C content was measured using the potassium dichromate sulfuric acid oxidation method. The total N and total P were measured by the semi-micro Kjeldahl method, and the total P was measured using molybdenum antimony anti-colorimetry [37].

2.6. Measurement of NSC Content

NSC content was measured using the anthrone colorimetric method. In brief, 100 mg of sample was added to 10 mL of 80% ethanol and heated in a boiling water bath for 10 min. The extracts were centrifuged at 4000 rpm for 10 min at room temperature, and the supernatant was collected. These steps were repeated three times, and the supernatants were combined and diluted with deionized water to 50 mL. The precipitate was suspended in 10 mL of 30% perchloric acid, and the mixtures sat overnight. The suspension was extracted in a water bath at 80 °C for 10 min and centrifuged at 4000 rpm for 10 min at 4 °C. The supernatants were harvested and diluted with deionized water to 50 mL for starch extraction. For glucose determination, 0.1 mL of extract was blended with 5 mL of anthrone solution. After 15 min in a water bath at 90 °C, the reaction solutions were measured at a wavelength of 620 nm when they were cooled. For sucrose determination, 0.1 mL of 7.6 M KOH and 0.1 mL of sugar extraction were mixed and placed for 10 min in boiling water, and then 5 mL of anthrone solution was added. The reaction mixtures were placed for 15 min in 90 °C water. After cooling, the absorbance was measured at a wavelength of 620 nm. For fructose determination, 0.1 mL of extract was blended with 5 mL of anthrone solution and placed in a water bath at 25 °C for 90 min. The absorbance was recorded at a wavelength of 620 nm. For starch determination, the test steps were the same as for glucose. The contents of glucose, fructose, sucrose, and starch were estimated based on the standard curves of glucose, sucrose, fructose, and starch, respectively. The data were shown as μg/mg dry matter.

2.7. Statistical Analysis

Data analyses were processed using Excel and SPSS 27.0, and the significant differences in *Z. armatum* seedling growth parameters, non-structural sugar, and nutrient element content in different soil types were examined by one-way ANOVA and Duncan methods. A correlation heatmap was established using the Pearson correlation coefficient to study the correlation among C content, N content, P content, C:N:P stoichiometry, and NSC content. A two-factor analysis of variance was adopted to examine the effects of different soil types and organs and their interactions on C, N, P, its stoichiometric ratio, and NSC in leaf, stem, and root. The above significance levels are set as $\alpha = 0.05$. The drawings were drawn using Origin 2021. To get a good idea of the relationships between tested parameters and treatments, principal component analysis (PCA) was performed using OriginPro 2021.

3. Results

3.1. Influences of Soil Types on the Growth Parameters

As exhibited in Table 2, the plant height of seedlings cultured in different types of soil is AS > ALPS > RS > ACPS > YS. There were no significant differences in seedling height among ACPS, RS, and YS ($p > 0.05$), but there was a significant difference compared to seedlings in other soil types ($p < 0.05$). The ground diameter showed a similar trend compared to the plant height. When seedlings were cultured in ALPS and RS, there was no significant difference in ground diameter ($p > 0.05$), while there was a significant difference in other soil types ($p < 0.05$). These results showed that plant height and ground diameter are related to soil types. The fresh weight of leaves, stems, and roots in the AS remarkably exceeds that in the other soil types ($p < 0.05$), but there were no obvious differences when seedlings were cultured in RS, ACPS, and ALPS ($p > 0.05$). Moreover, the fresh weight of three organs cultured in YS had the lowest levels. The highest fresh biomass of roots was observed when seedlings were cultured in the AS, and there were no significant differences between cultures in other soil types ($p > 0.05$). The dry biomass of leaves, stems, and roots cultured in AS was remarkably higher than that of other soil types, but there were no obvious differences among ALPS, RS, and ACPS (Table 2).

Table 2. Effects of five soil types on growth indexes of *Z. armatum* seedlings.

Soil Type		RS	YS	ACPS	ALPS	AS
Plant height (cm)		56.1 ± 10.78 ^{c}	44.72 ± 7.3 ^{d}	54.14 ± 5.01 ^{cd}	66.46 ± 4.56 ^{b}	77.22 ± 9 ^{a}
Ground diameter (mm)		8.56 ± 0.64 ^{b}	6.82 ± 0.53 ^{d}	7.76 ± 0.19 ^{c}	9.09 ± 0.69 ^{b}	10.62 ± 0.7 ^{a}
Fresh weight (g)	Leaf	28.47 ± 8.64 ^{b}	16.25 ± 2.35 ^{c}	26.91 ± 1.22 ^{b}	31.25 ± 5.08 ^{b}	48.25 ± 3.93 ^{a}
	Stem	16.26 ± 5.01 ^{b}	8.92 ± 0.95 ^{c}	12.94 ± 0.66 ^{bc}	18.09 ± 1.25 ^{b}	27.64 ± 4.15 ^{a}
	Root	36.3 ± 9.04 ^{ab}	22.42 ± 0.74 ^{b}	26.25 ± 6.76 ^{b}	24.63 ± 10.19 ^{b}	54.28 ± 19.21 ^{a}
Dry weight (g)	Leaf	9.21 ± 2.83 ^{b}	5.53 ± 0.92 ^{c}	8.79 ± 0.52 ^{bc}	10.66 ± 1.65 ^{b}	32.17 ± 2.62 ^{a}
	Stem	6.94 ± 2.1 ^{b}	3.98 ± 0.56 ^{c}	5.59 ± 0.22 ^{bc}	8.05 ± 0.43 ^{b}	12.44 ± 2.18 ^{a}
	Root	8.17 ± 2.03 ^{b}	5.08 ± 0.78 ^{b}	6.93 ± 0.86 ^{b}	7.05 ± 2.52 ^{b}	14.91 ± 3.5 ^{a}
RWC (%)	Leaf	67.69 ± 0.27	66.06 ± 0.8	67.35 ± 0.43	65.86 ± 0.46	66.66 ± 2.2
	Stem	57.28 ± 0.55	55.42 ± 2.4	56.73 ± 2.42	55.44 ± 1.09	55.11 ± 1.13
	Root	77.39 ± 2.15	77.33 ± 3.6	72.86 ± 4.48	70.96 ± 1.73	71.47 ± 4.8

Different lowercase letters represent the obvious differences of various indicators in different soil types ($p < 0.05$).

3.2. Influences of Soil Type on C, N, and P Contents

From Figure 1 and Table 3, it can be concluded that soil types and the interaction of soil type and organs had no significant impact on the C content in the different organs of *Z. armatum* seedlings, while soil type, organs, and their interaction had a remarkable impact on N and P contents ($p < 0.01$). As exhibited in Figure 1A, the C content in the organs of the seedlings planted under five soil types was the stem, root, and leaf. From the highest to the lowest, the C contents of the leaves, stems, and roots ranged from 350.33 to 406.28 g/kg, from 449.71 to 458.61 g/kg, and from 387.62 to 430.58 g/kg (Figure 1A), respectively. As shown in Figure 1B, the N contents in the different organs have great differences, and the N content of leaves was variably and considerably greater than that of the roots and stems. The N content of the leaves, stems, and roots ranged from 18.65 to 24.61 g/kg, from 6.02 to 12.76 g/kg, and from 11.89 to 17.72 g/kg, respectively. Moreover, the highest N content in seedlings was observed in RS, and the lowest value was found in AS. The N contents of roots, stems, and leaves in RS were 17.72 g/kg, 12.76 g/kg, and 24.61 g/kg, respectively, and the contents of roots, stems, and leaves in AS were 11.89 g/kg, 6.02 g/kg, and 18.65 g/kg, separately. These results showed significant differences in the N content of different organs cultured in the two soil types. As exhibited in Figure 1C, the P content of the leaves, roots, and stems ranged from 0.35 to 0.44 g/kg, 0.34 to 0.42 g/kg, and 0.28 to 0.34 g/kg, respectively. Among the five soil types, ACPS seedlings had the highest P content in their roots, stems, and leaves; the contents were 0.42 g/kg, 0.34 g/kg, and 0.44 g/kg, respectively. Moreover, when these seedlings were cultured in AS, the P content in their leaves and roots was the lowest, and the contents were 0.35 g/kg and 0.34 g/kg, respectively.

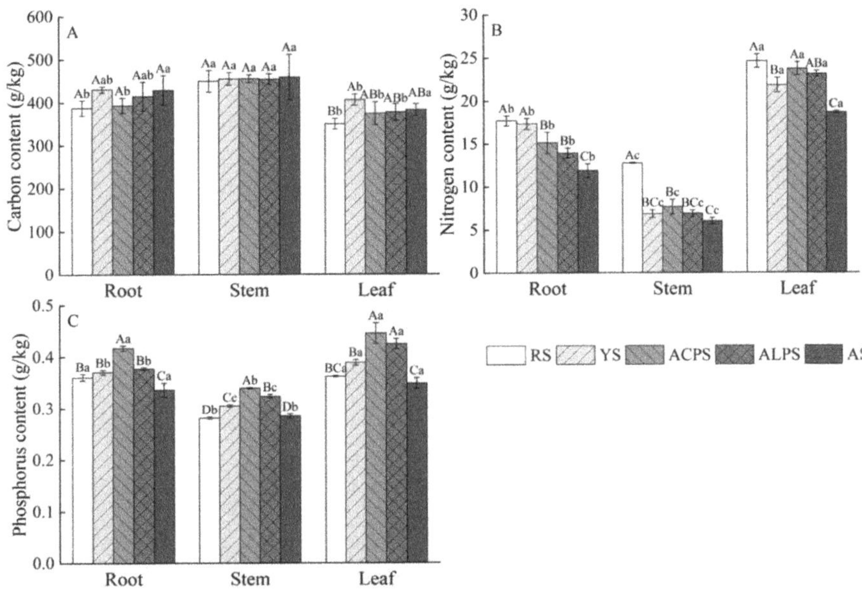

Figure 1. Infuences of five soil types on the C (**A**), N (**B**), and P (**C**) contents in leaves, stems, and roots of *Z. armatum* seedlings. The data are shown as means ± SE (n = 3). Different capital letters represented the differences among five soil types in the same tissues, and different lower-case letters represented the differences among different tissues in the same soil type ($p < 0.05$).

Table 3. Two-factor variance analysis of C, N, and P content, stoichiometric ratio, and non-structural sugar in leaves, stems, and roots of *Z. armatum* seedlings in five soil types.

Target	Source of Variation		
	Soil Type	**Organ**	**Soil Type × Organ**
C content	1.947	26.113 **	0.452
N content	66.641 **	1172.280 **	9.021 **
P content	92.034 **	321.649 **	4.104 **
C/N	27.251 **	425.359 **	8.883 **
C/P	13.862 **	133.170 **	0.796
N/P	79.003 **	562.228 **	6.687 **
Glucose content	4.829 **	21.706 **	5.362 **
Fructose content	10.346 **	55.789 **	5.927 **
Sucrose content	49.574 **	68.128 **	14.443 **
Starch content	9.454 **	70.579 **	2.376 *

Note: The value is the F-value of the analysis of variance. * indicates significant ($p < 0.05$); ** indicates highly significant ($p < 0.01$).

3.3. *Infuences of Soil Types on the Accumulation of C, N, and P*

As exhibited in Figure 2, the total accumulation of C, N, and P showed a similar trend cultured in five soil types, and the total accumulation of C, N, and P in *Z. armatum* seedlings reached the maximum of 24.4 g/plant, 852.04 g/plant, and 19.75 mg/plant cultured in AS, respectively. Moreover, the values reached the significance level with other soils ($p < 0.05$). However, the minimum accumulation of C, N, and P was observed when these seedlings were cultured in YS, and the values represented 6.25 g/plants, 236.15 mg/plant, and 5.24 mg/plant, respectively. Moreover, the total accumulations of C in roots and stems were little different, and were generally lower than those in leaves (Figure 2A). As shown in Figure 2B, in most soil types, the accumulation of N in various organs showed a significant difference ($p < 0.05$), with leaves > roots > stems. In various soil types, the trend of P accumulation in various organs is the same as that of N accumulation (Figure 2C). These results show that the accumulation of C, N, and P is related to soil types and tissues, which may be due to the differences in nutrient requirements in *Z. armatum* seedlings.

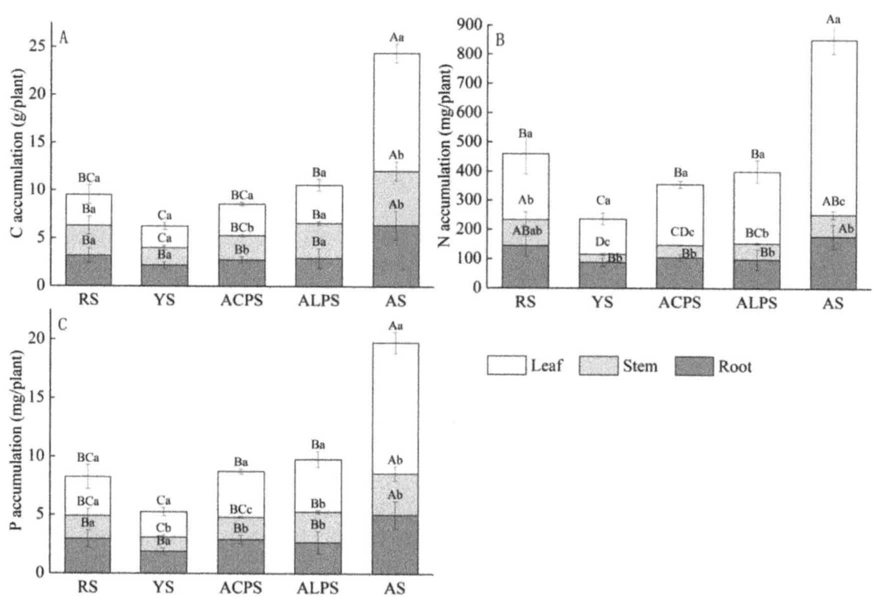

Figure 2. Influences of five soil types on the accumulation of C (**A**), N (**B**), and P (**C**) in *Z. armatum* seedlings. The data are shown as means ± SE ($n = 3$). Different capital letters represented the differences among five soil types in the same tissues, and different lower-case letters represented the differences among different tissues in the same soil type ($p < 0.05$).

3.4. *Influences of Five Soil Types on Stoichiometry*

As shown in Figure 3 and Table 3, soil types and organs have extremely significant influences on the C/N, C/P, and N/P ratios in *Z. armatum* seedlings ($p < 0.01$), and their interaction has an extremely noticeable impact on the ratios of C/N and N/P ($p < 0.01$). However, the interaction of soil types and organs has no obvious effects on the C/P ratios. The mean C/N and C/P of seedlings cultured in five soil types showed significant differences in each organ, with stems significantly higher than roots and roots significantly higher than leaves ($p < 0.05$). However, mean N/P ratios showed the opposite trend. Moreover, the C/N ratios of leaves, stems, and roots cultured in the RS are the smallest, but the N/P ratios are the largest. The leaves, stems, and roots cultured with AS have the highest C/N and C/P. The C/P of the leaves, stems, and roots cultured in the ACPS are the smallest. The N/P ratios of leaves and stems do not show significant differences in YS,

ACPS, ALPS, and AS, while the N/P ratios of roots do not show significant differences in ACPS, ALPS, and AS.

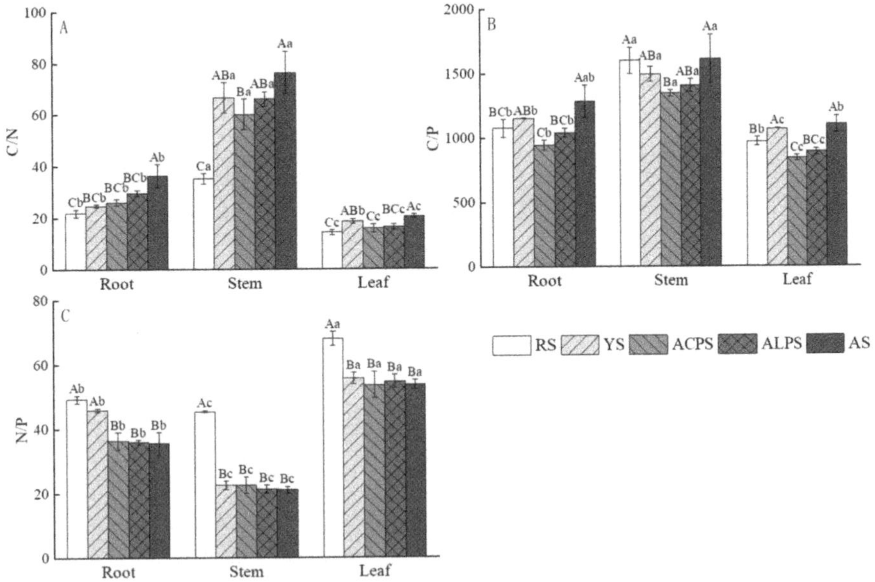

Figure 3. Effects of five soil types on C:N (**A**), C:P (**B**), and N:P (**C**) of leaves, stems, and roots in *Z. armatum* seedlings. The data are shown as means ± SE ($n = 3$). Different capital letters represented the differences among five soil types in the same tissues, and different lower-case letters represented the differences among different tissues in the same soil type ($p < 0.05$).

3.5. Effects of Five Soil Types on Non-Structural Carbohydrates (NSC)

As shown in Figure 4 and Table 3, soil type and organ have extremely significant effects on non-structural sugar ($p < 0.01$). Their interaction has extremely significant effects on soluble sugar ($p < 0.01$) and significant effects on starch ($p < 0.05$). Among the five soil types, the seedlings in ACPS have the lowest glucose content in their roots and leaves, while the seedlings in YS have the highest glucose content in their leaves. In addition, only seedlings planted on YS and AS show significantly higher glucose content in leaves compared to roots and stems (Figure 4A). Among the five soil types, the fructose content of seedlings planted on ACPS is the lowest in stems and leaves, and the fructose content in the roots is the lowest in ALPS. In most soil types, the fructose content of seedlings is shown as leaf > stem > root in each organ (Figure 4B). The sucrose contents of leaves and stems cultured in the AS were significantly higher than those in other soil types. However, the values in two purple soils were significantly lower than those in other soil types. Similarly, the sucrose contents in the roots cultured with AS were remarkable higher than those of in other soil types. However, the sucrose content of seedlings planted in two purple soils was markedly greater than that of those cultured in YS and RS (Figure 4C). The starch contents of leaves, stems, and roots cultured in the YS were relatively higher than those in the other soil types. The starch content of leaves cultured in the RS was the lowest, the starch content of stems cultured in the AS was the lowest, and the starch content of roots cultured in the RS, ACPS, and ALPS was relatively lower than those of the other soil types (Figure 4D).

As shown in Figure 5, in five soil types, the total accumulation of glucose, fructose, sucrose, and starch showed a different trend, and the total accumulation of glucose, fructose, sucrose, and starch in *Z. armatum* seedlings reached the maximum of 3531.66, 1668.85, 2372.08, and 3424.59 mg/plant cultured in AS, respectively. These values reached the significance level with other soils ($p < 0.05$). However, the minimum accumulation of glucose, fructose, sucrose, and starch was observed when these seedlings were cultured in YS, and the values represented 868.07, 371.19, 406.41, and 980.57 mg/plant, respectively. These results show that the accumulation of glucose, fructose, sucrose, and starch isrelated to soil types and tissues in *Z. armatum* seedlings.

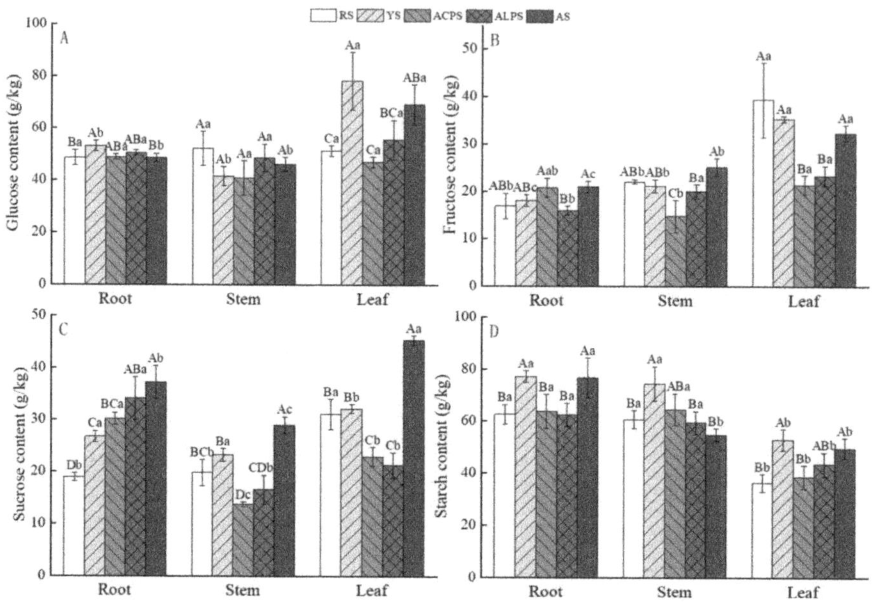

Figure 4. Effects of five soil types on non-structural carbohydrates (NSC) contents in *Z. armatum* seedlings. The data are shown as means ± SE (n = 3). (**A**), glucose content. (**B**), fructose content. (**C**), sucrose content. (**D**), starch content. Different capital letters represented the differences amongfive soil types in the same tissues, and different lower-case letters represented the differences among different tissues in the same soil type ($p < 0.05$).

3.6. Correlation Analysis

Figure 6 reveals the correlation between various measurement results of *Z. armatum* seedlings. C content had significant positive correlations with C:N (r = 0.86, $p < 0.05$), C:P (r = 0.89, $p < 0.05$), and starch (r = 0.67, $p < 0.05$), and noticeable negative associations with N content (r = −0.91, $p < 0.05$), P content(r = −0.73, $p < 0.05$), N:P (r = −0.86, $p < 0.05$), fructose (r = −0.49, $p > 0.05$), glucose (r = −0.37, $p > 0.05$), and sucrose (r = −0.35, $p > 0.05$), respectively. N content was positively related to P content(r = 0.74, $p < 0.05$), N:P (r = 0.96, $p < 0.05$), glucose (r = 0.53, $p < 0.05$), fructose (r = 0.51, $p > 0.05$), and sucrose (r = 0.27, $p > 0.05$), and negatively related to C:N (r = −0.94, $p < 0.05$), C:P (r = −0.84, $p < 0.05$), and starch (r = −0.64, $p < 0.05$), respectively. P content showed a certain degree of positive correlation with N:P (r = 0.53, $p < 0.05$), glucose (r = 0.24, $p > 0.05$), fructose (r = 0.045, $p > 0.05$), and sucrose (r = 0.095, $p > 0.05$), and a negative correlation with C:N (r = −0.72, $p < 0.05$), C:P (r = −0.95, $p < 0.05$), and starch content (r = −0.41, $p > 0.05$), respectively. C:N showed positive correlation with C:P (r = 0.83, $p < 0.05$) and starch (r = 0.42, $p > 0.05$), and negative correlation with N:P (r = −0.92, $p < 0.05$), glucose (r = −0.53, $p < 0.05$), fructose (r = −0.35, $p > 0.05$), and sucrose (r = −0.35, $p > 0.05$). C:P showed a positive correlation with starch (r = 0.48, $p > 0.05$) and a negative correlation with N:P (r = −0.69, $p < 0.05$), glucose (r = −0.32, $p > 0.05$), fructose (r = −0.20, $p > 0.05$), and sucrose (r = −0.25, $p > 0.05$). N:P showed a positive correlation with glucose (r = 0.57, $p < 0.05$), fructose (r = 0.61, $p < 0.05$), and sucrose (r = 0.32, $p > 0.05$), but was negatively related to starch (r = −0.61, $p < 0.05$). Glucose was positively related to fructose (r = 0.64, $p < 0.05$) and sucrose (r = 0.54, $p < 0.05$), but was negatively related to starch content (r = −0.30, $p > 0.05$). Fructose was positively related to sucrose (r = 0.51, $p < 0.05$), but negatively related to starch(r = −0.61, $p < 0.05$). As shown in Table 3, soil type showed statistical significance for the tested parameters, except for C contents. There were significant organs for all tested parameters. Significant interactions of soil type × organ were observed for the N, P, C:N, N: P, glucose, fructose, sucrose, and starch content.

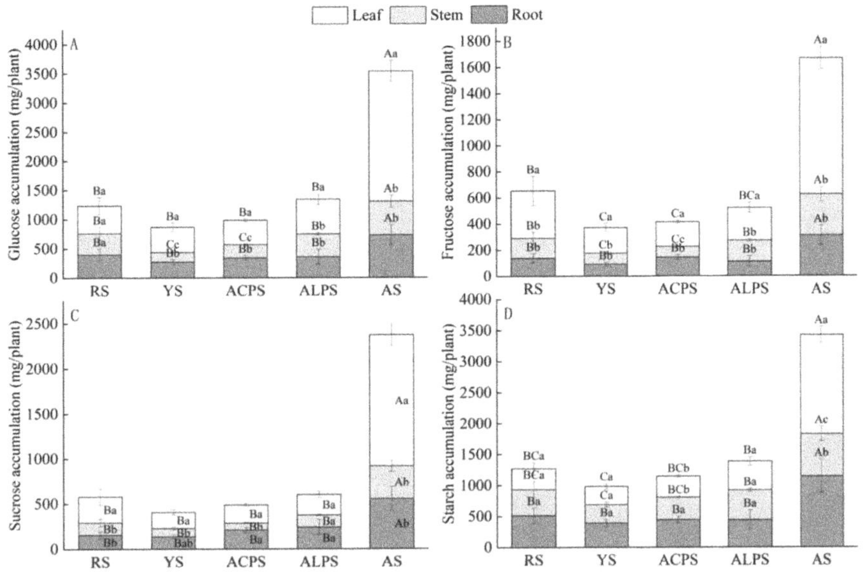

Figure 5. Effects of five soil types on the accumulation of glucose (**A**), fructose (**B**), sucrose (**C**), and starch (**D**) in *Z. armatum* seedlings. The data are shown as means ± SE (n = 3). Different capital letters represented the differences among five soil types in the same tissues, and different lower-case letters represented the differences among different tissues in the same soil type ($p < 0.05$).

3.7. Principal Component Analysis (PCA)

As shown in Figure 7, significant differences were observed when *Z. armatum* seedlings were cultured in five soil types. The contribution rate of the principal component 1 (PC1) is 62.0%, with the highest weight. The contribution rate of the principal component 2 (PC2) is 16.9%, and the two principal components account for a cumulative 78.9% of the data variation. Between various chemical components and stoichiometric ratios, NC, PC, and N/P will have a strong positive response to PC1, while CC, C/N, and C/P will have a strong negative response to PC1. Glc, Fru, and Suc have a strong positive response to PC2, while PC will have a strong negative response to PC2. Analysis shows that there is a positive correlation between CC, C/N, C/P, and Sta, while there is also a positive correlation between Glc, Fru, Suc, NC, and N/P. Meanwhile, there is a negative correlation between CC, C/N, C/P, and Sta, and NC, PC, and N/P. These results indicated that the variations of C, N, P, and NSC contents, C/N, C/P, and N/P, can be used to explain and better understand the growth and nutrient uptake in *Z. armatum* seedlings in response to five soil types.

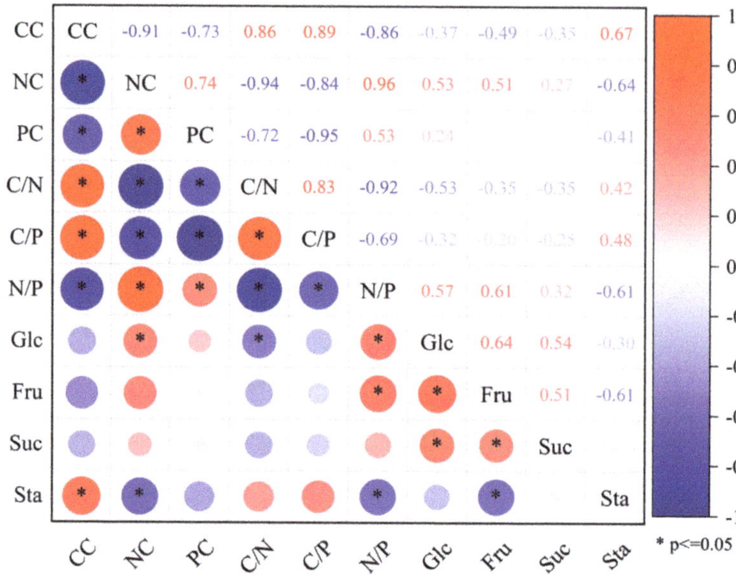

Figure 6. Correlation analysis among different tested parameters in *Z. armatum* seedlings. * Indicated significant ($p \leq 0.05$). The red area indicated positive correlations. The blue area represented negative correlations. CC, carbon content. NC, nitrogen content. PC, phosphorus content. C/N, C/P, and N/P. Glu, glucose content. Fru, fructose content. Suc, sucrose content. Sta, starch content.

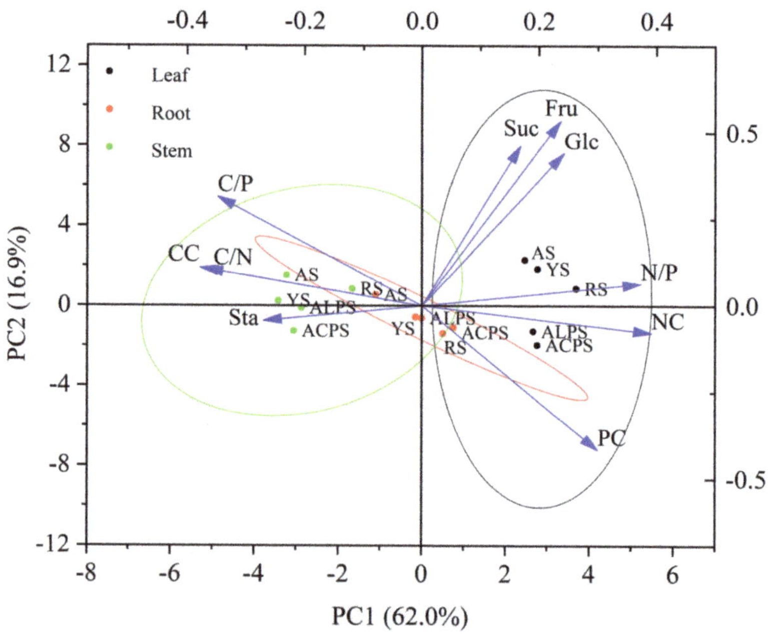

Figure 7. PCA of chemical constituents of *Z. armatum* seedlings in different organs under five soil types. CC, carbon content. NC, nitrogen content. PC, phosphorus content. C/N, C/P, and N/P. Glu, glucose content. Fru, fructose content. Suc, sucrose content. Sta, starch content. The percentage change described for each component is given next to the axis. Location of a trait in the diagram closest to the intersection of 0 on the x-axis (PC1) and y-axis (PC2).

4. Discussion

Soil, as a basic condition for agricultural production, has a significant impact on plant growth, yield, and biochemical contents [6,8]. Growth parameters, like seedling height, ground diameter, and biomass, can reflect the growth status of plants in response to various environmental conditions, like soil [6,7]. Previous reports showed that plant height, base diameter, and biomass of *Jatropha curcas* plants are related to soil types, and the growth status of different species varies significantly in different types of soils [37]. Previous studies also suggested that *Phellodendron chinense, Ricinus communis,* and *Firmiana simplex* seedlings show significant differences in seedling height, basal diameter, and total biomass when they are cultured in RS, YS, ALPS, and ACPS [6,38,39]. These variations are due to different soil types with different physical, chemical, and microbial community structures. The content of various nutrients provided by soil, soil texture, water holding capacity, and changes in microbial and soil enzyme activities can all have a certain impact on the nutrient absorption and utilization efficiency of plant growth and development. Therefore, different plants adapt to different soil types.

Z. armatum plant is a perennial and economic crop that is widely distributed in Southwest China and has strong adaptability to variable growth conditions [40]. Studies have shown that its chemical composition varies with changes in climate and soil conditions and is significantly affected by the species and geographical variation of *Zanthoxylum* [32,41,42]. Although *Z. armatum* may adapt to a variety of soil types, the soil can offer the foundation of nutrition for better seedling growth, maximum fruit yield, and optimal quality, which will be considered the optimal soil type. As shown in Table 2, the AS displays the best comprehensive performances in terms of the agronomic and nutrient characteristics, like seedling height, base diameter, and biomass, as well as the nutrient contents of *Z. armatum*, followed by those of ALPS, ACPS, RS, and YS in the order from high to low. The reason may be that *Z. armatum* prefers to grow and proliferate in clay or loamy soil enriched with high organic matter [43]. AS possesses the above conditions, and the pH of AS is alkaline and contains a higher level of organic matter, total nitrogen, and total phosphorus. Moreover, its texture is loose, and its air permeability is good (Table 1). Thus, this type of AS is suitable for seedling cultivation of *Z. armatum*. The results show that YS has good organic matter content, but its growth is still poor. This may be due to its poor water permeability, poor air permeability, and heavy clay texture. The soil properties, like organic matter, water-holding capacity, texture, water permeability, air permeability, etc., had significant influences on plant growth and development [1,8,44]. Soil pH is also a limiting factor, implying that an appropriate pH value is more conducive to plant growth [29]. Good soil structure, loose soil, good ventilation performance, and high organic matter content are conducive to the growth and development of seedlings (Table 2). Table 2 shows that *Z. armatum* growing in different soil types has certain differences in growth status and nutrient composition. Therefore, screening the appropriate soils is a useful way to expand the suitable planting area of *Z. armatum,* cultivate high-quality seedlings, and improve fruit yield.

Mineral elements, such as C, N, and P, play an indispensable role in the life of plants. C is the structural foundation in different organs of plants, while N and P are the necessary components for the metabolism of nucleotides, proteins, sugars, and lipids [13,45]. In plants, C:N and C:P ratios may reflect the utilization capacity of C assimilation in the process of nutrient uptake. The N:P ratio is widely regarded as an effective index of nutrient supply in plant systems, suggesting that plant growth is limited by N and/or P [17]. Thus, we know that the changes in C, N, and P contents and their stoichiometric ratios in *Z. armatum* seedlings cultured in five soil types are of great significance for screening suitable soil for further widespread cultivation in Southwest China.

The present study shows that significant differences among the N and P contents of *Z. armatum* were observed in response to five soil types, and organ types have significant effects on the content of C, N, and P in *Z. armatum* seedlings (Figure 1, Table 3). Moreover, our report also suggests that the soil type-organs interactions have remarkable impacts on the N and P contents of *Z. armatum* seedlings (Figure 1, Table 3). The N content in different organs of *Z. armatum* cultured in RS is significantly higher than thatin other soils, and the P contents cultured in ACPS and ALPS are greater than those in other soils (Figure 1). The contents of N and P of seedlings in AS soil are lower in the five soil types. One of the reasons for this may be that the diversity of physico-chemical properties in the five soil types and the higher values of P are observed in ACPS and ALPS (Table 1), which might result in differences in the content absorbed by *Z. armatum* seedlings. It may also be because the different growth rates of *Z. armatum* seedlings lead to inconsistent stages, and the elements absorbed and utilized are different [11]. Since the biomass of *Z. armatum* planted on different soil types showed obvious trends, the C, N, and P of the whole seedlings also showed the same trends (Figure 2). Many studies have found that the nutrient stoichiometry varies in different organs in response to various environmental conditions, including soil types. Some reports indicated that the N and P contents and N:P ratios in the leaves are significantly greater than those of stems and roots, while the C content and C:N and C:P ratios of stems are significantly higher than those of leaves and roots [14,46,47]. Our results are consistent with these above-mentioned studies (Figure 2). In most cases, leaves had the lowest C:N ratio and stems had the highest C:N ratio, which might be connected with the sugar transfer from photosynthetic tissues to structural tissues. As an important organ of photosynthesis, leaves need higher nutrient concentrations to improve their photosynthetic and metabolic capacity. This also indicates that leaves preferentially distribute nutrients relative to stems and roots, and the nutrient content of stems decreases with individual development. As a supporting structure, the stem has a higher content of C-rich structural compounds such as lignin and cellulose, so the values of C:N, C:P, and N:P exceed those of leaves [21,23,47,48]. Based on the above results, this clearly shows that the C:N:P stoichiometry of different organs varies depending on their function. The present results showed that the relationships of C,N, and P contents and ratios among the three organs responded to different soil types. Among the five soils, the values of C:N and C:P in seedlings cultivated in AS were greater than those of other soil types, which meant that the support structure of *Z. armatum* was better in AS, which was conducive to the early growth of seedlings. The maximum N:P of seedlings growing in RS indicated that *Z. armatum* had a higher nutrient concentration in RS. Therefore, AS and RS are more suitable for the cultivation of *Z. armatum* seedlings.

NSC, which mainly includes glucose, fructose, sucrose, and starch, are products of plant photosynthesis, and its levels in plant organs can reflect the correction of C uptake. Moreover, NSC levels are also necessary for transport metabolism and osmotic regulation, which can be used to resist adverse external environments in plants [22,27]. Soil, as one of the environmental factors, will affect the normal physiological metabolism of trees and then change the storage of carbohydrates in trees and their distribution in various organs [27]. Therefore, NSC can be used as one of the indexes to evaluate whether a certain soil is suitable for the cultivation of *Z. armatum* [49]. NSC is stored in different organs in the form of mobile carbon and participates in physiological metabolic processes such as material transport, energy metabolism, osmotic regulation, and stress adaptation [49]. The present study indicated that soluble sugars, such as glucose, fructose, and sucrose, have higher content in leaves, while starch has a lower content in leaves than in stems and roots (Figure 4), which was similar to the results of previous studies and reflected the performance of plants adapting to the environment by adjusting NSC usage strategies [50]. In addition, the C contents, N contents, C:N, and N:P were apparently associated with the NSC contents in *Z. armatum* seedlings (Figures 6 and 7). Notably, the C, N, and P contents in the leaves arethe key factors affecting the changes in NSC storage (Figures 6 and 7). This may be due to the fact that photosynthesis and NSC accumulation in leaves are closely related to N concentration and P concentration. With increasing N contents in leaves, the photosynthesis of plants significantly increases. In addition, increasing nitrogen can improve the fixation and assimilation abilities of CO_2, as well as the production of NSC [20,21]. In this study, N:P was remarkably positively related to glucose and fructose and negatively correlated with the concentrations of starch. This indicated that changes in

N:P can affect the reciprocal conversion of soluble sugars and starch, playing a vital role in the fluctuation of NSC contents. Our results are consistent with those of other plants [51]. Among the five soil types, the NSC content of *Z. armatum* seedlings cultured in ACPS and ALPS is lower, indicating that these two soil types had fewer stress effects.

5. Conclusions

In summary, this study selected the soil type that is more suitable for the growth and cultivation of *Z. armatum* seedlings in five different soils and measuring their growth and nutrient absorption. Research has shown that the changes in experimental parameters of *Z. armatum* seedlings are related to soil types with different physical and chemical properties. Although there are certain differences in growth parameters, nutrient accumulation and distribution, and NSC content among seedlings, *Z. armatum* seedlings can grow well in all five soil types. Among the five representative soil types, the seedlings cultured in AS had higher C/P and C/P, while those cultured in RS had higher N/P. AS and RS were beneficial for cultivating *Z. armatum* seedlings through higher nutrient levels. Two-factor ANOVA and PCA analysis showed that C and N contents and their proportions would affect the accumulation and distribution of NSC in *Z. armatum* plants, reflecting the adaptation of *Z. armatum* plants to different soil types. The results showed that ACPS and ALPS had less stress effect, while YS and AS had more NSC accumulation. These findings will help us understand how this plant develops different nutrient acquisition strategies under different soil conditions and provide directions for finding effective ways for farmers to choose suitable soil types for planting *Z. armatum* seedlings before these plants are widely planted.

Author Contributions: Conceptualization, visualization, and supervision, S.G.; methodology, conduct research, collect data, analyze data, T.G., H.R., M.W., W.Q. and Y.H.; investigation, T.G., H.R. and M.W.; writing—drafting articles, T.G. and S.G.; writing—review and editing, T.G., H.R., M.W., W.Q., Y.H., Y.Y., T.Y., K.Z. and S.G. All authors have read and agreed to the published version of the manuscript.

Funding: This work was financially supported by the National Undergraduate Training Program for Innovation and Entrepreneurship (No. 202310626038) and the Natural Science Foundation of Sichuan Province (No. 2022NSFSC1014).

Data Availability Statement: Data are contained within the article.

Acknowledgments: We are grateful to all of the group members and workers for their assistance in the field experiment.

Conflicts of Interest: The authors declare no conflicts of interest.

References

1. Dan, T.H.; Brix, H. Effects of soil type and water saturation on growth, nutrient and mineral content of the perennial forage shrub *Sesbania sesban*. *Agroforest. Syst.* **2017**, *91*, 173–184. [CrossRef]
2. Lee, E.H.; Lee, B.E.; Kim, J.G. Effects of water levels and soil nutrients on the growth of *Iris laevigata* seedlings. *J. Ecol. Environ.* **2018**, *42*, 5–13. [CrossRef]
3. Pei, J.; Li, H.; Li, S.; An, T.; Farmer, J.; Fu, S.; Wang, J. Dynamics of Maize carbon contribution to soil organic carbon in association with soil type and fertility level. *PLoS ONE* **2015**, *10*, e120825–e120840. [CrossRef] [PubMed]
4. Yang, X.; Zhang, Y.; Liang, J.; Zhang, X. Effect of soil configuration on alfalfa growth under drought stress. *Sustainability* **2023**, *15*, 5400. [CrossRef]
5. Zhao, Y.; Li, T.; Liu, J.; Sun, J.; Zhang, P. Ecological stoichiometry, salt ions and homeostasis characteristics of different types of halophytes and soils. *Front. Plant Sci.* **2022**, *13*, 990246. [CrossRef]
6. Yang, Y.; Hu, Y.; Qian, W.; Wang, Y.J.; Ren, H.Y.; Gao, S.; Cao, G.X. Early growth characterization and antioxidant responses of *Phellodendron chinense* seedling in response to four soil types at three growth stages. *Forests* **2023**, *14*, 1746. [CrossRef]
7. Kahkashan, P.; Najat, B.; Iram, S.; Iffat, S. Influence of soil type on the growth parameters, essential oil yield and biochemical contents of *Mentha arvensis* L. *J. Essent. Oil-Bear. Plants* **2016**, *19*, 76–81. [CrossRef]
8. Ogundola, A.F.; Bvenura, C.; Ehigie, A.F.; Afolayan, A.J. Effects of soil types on phytochemical constituents and antioxidant properties of *Solanum nigrum*. *S. Afr. J. Bot.* **2022**, *151*, 325–333. [CrossRef]

9. Sardans, J.; Rivas-Ubach, A.; Penuelas, J. The elemental stoichiometry of aquatic and terrestrial ecosystems and its relationships with organismic lifestyle and ecosystem structure and function: A review and perspectives. *Biogeochemistry* **2012**, *111*, 1–39. [CrossRef]
10. Ye, Y.; Liang, X.; Chen, Y.; Li, L.; Ji, Y.; Zhu, C. Carbon, nitrogen and phosphorus accumulation and partitioning, and C:N:P stoichiometry in late-season Rice under different water and nitrogen managements. *PLoS ONE* **2014**, *9*, e101776. [CrossRef]
11. Sun, H.; Li, Q.; Lei, Z.; Zhang, J.; Song, X.; Song, X. Ecological stoichiometry of nitrogen and phosphorus in Moso bamboo (*Phyllostachys Edulis*) during the explosive growth period of new emergent shoots. *J. Plant Res.* **2019**, *132*, 107–115. [CrossRef]
12. Niklas, K.J.; Owens, T.; Reich, P.B.; Cobb, E.D. Nitrogen/phosphorus leaf stoichiometry and the scaling of plant growth. *Ecol. Lett.* **2005**, *8*, 636–642. [CrossRef]
13. Gusewell, S. N:P ratios in terrestrial plants: Variation and functional significance. *New Phytol.* **2004**, *164*, 243–266. [CrossRef]
14. Jing, H.; Zhou, H.; Wang, G.; Xue, S.; Liu, G.; Duan, M. Nitrogen addition changes the stoichiometry and growth rate of different organs in *Pinus Tabuliformis* seedlings. *Front. Plant Sci.* **2017**, *8*, 1922–1932. [CrossRef]
15. Reef, R.; Ball, M.C.; Feller, I.C.; Lovelock, C.E. Relationships among RNA: DNA ratio, growth and elemental stoichiometry in mangrove trees. *Funct. Ecol.* **2010**, *24*, 1064–1072. [CrossRef]
16. Zhan, S.; Wang, Y.; Zhu, Z.; Li, W.; Bai, Y. Nitrogen enrichment alters plant N:P stoichiometry and intensifies phosphorus limitation in a steppe ecosystem. *Environ. Exp. Bot.* **2017**, *134*, 21–32. [CrossRef]
17. Minden, V.; Kleyer, M. Internal and external regulation of plant organ stoichiometry. *Plant Biol.* **2014**, *16*, 897–907. [CrossRef] [PubMed]
18. Ding, D.; Arif, M.; Liu, M.; Li, J.; Hu, X.; Geng, Q.; Yin, F.; Li, C. Plant-soil interactions and C:N:P stoichiometric homeostasis of plant organs in Riparian plantation. *Front. Plant Sci.* **2022**, *13*, 979023–979040. [CrossRef] [PubMed]
19. Schreeg, L.A.; Santiago, L.S.; Wright, S.J.; Turner, B.L. Stem, root, and older leaf N:P ratios are more responsive indicators of soil nutrient availability than new foliage. *Ecology* **2014**, *95*, 2062–2068. [CrossRef] [PubMed]
20. Liu, Q.; Huang, Z.; Wang, Z.; Chen, Y.; Wen, Z.; Liu, B.; Tigabu, M. Responses of leaf morphology, NSCs contents and C:N:P stoichiometry of *Cunninghamia lanceolata* and *Schima superba* to shading. *BMC Plant Biol.* **2020**, *20*, 354–364. [CrossRef]
21. Xie, H.; Yu, M.; Cheng, X. Leaf non-structural carbohydrate allocation and C:N:P stoichiometry in response to light acclimation in seedlings of two subtropical shade-tolerant tree species. *Plant Physiol. Bioch.* **2018**, *124*, 146–154. [CrossRef]
22. Long, R.W.; Adams, H.D. The osmotic balancing act: When sugars matter for more than metabolism in woody plants. *Global Change Biol.* **2023**, *29*, 1684–1687. [CrossRef]
23. Luo, G.; Li, J.; Guo, S.; Li, Y.; Jin, Z. Photosynthesis, nitrogen allocation, non-structural carbohydrate allocation, and C:N:P stoichiometry of *Ulmus elongata* seedlings exposed to different light intensities. *Life* **2022**, *12*, 1310. [CrossRef] [PubMed]
24. Zhang, M.; Zhu, J.; Li, M.; Zhang, G.; Yan, Q. Different light acclimation strategies of two coexisting tree species seedlings in a temperate secondary forest along five natural light levels. *Forest Ecol. Manag.* **2013**, *306*, 234–242. [CrossRef]
25. Ai, Z.; Xue, S.; Wang, G.; Liu, G. Responses of non-structural carbohydrates and C:N:P stoichiometry of *Bothriochloa ischaemum* to nitrogen addition on the Loess Plateau, China. *J. Plant Growth Regul.* **2017**, *36*, 714–722. [CrossRef]
26. Li, M.H.; Jiang, Y.; Wang, A.; Li, X.; Zhu, W.; Yan, C.F.; Du, Z.; Shi, Z.; Lei, J.; Schönbeck, L.; et al. Active summer carbon storage for winter persistence in trees at the cold Alpine treeline. *Tree Physiol.* **2018**, *38*, 1345–1355. [CrossRef] [PubMed]
27. Han, H.; He, H.; Wu, Z.; Cong, Y.; Zong, S.; He, J.; Fu, Y.; Liu, K.; Sun, H.; Li, Y.; et al. Non-structural carbohydrate storage strategy explains the spatial distribution of treeline species. *Plants* **2020**, *9*, 384. [CrossRef]
28. Smith, M.G.; Arndt, S.K.; Miller, R.E.; Kasel, S.; Bennett, L.T. Trees use more non-structural carbohydrate reserves during Epicormic than Basal resprouting. *Tree Physiol.* **2018**, *38*, 1779–1791. [CrossRef]
29. Xu, D.; Zhuo, Z.; Wang, R.; Ye, M.; Pu, B. Modeling the distribution of *Zanthoxylum armatum* in China with Maxent modeling. *Glob. Ecol. Conserv.* **2019**, *19*, e691. [CrossRef]
30. Devkota, K.P.; Wilson, J.; Henrich, C.J.; McMahon, J.B.; Reilly, K.M.; Beutler, J.A. Isobutylhydroxyamides from the pericarp of Nepalese *Zanthoxylum Armatum* inhibit Nf1-defective tumor cell line growth. *J. Nat. Prod.* **2013**, *76*, 59–63. [CrossRef]
31. Agnihotri, S.; Wakode, S.; Ali, M. Chemical constituents isolated from *Zanthoxylum armatum* stem bark. *Chem. Nat. Comp.* **2017**, *53*, 880–882. [CrossRef]
32. Kumar, V.; Kumar, S.; Singh, B. Quantitative and structural analysis of amides and lignans in *Zanthoxylum armatum* by UPLC-DAD-ESI-QTOF-MS/MS. *J. Pharmaceut. Biomed.* **2014**, *94*, 23–29. [CrossRef]
33. Nooreen, Z.; Kumar, A.; Bawankule, D.U.; Tandon, S.; Ali, M.; Xuan, T.D.; Ahmad, A. New chemical constituents from the fruits of *Zanthoxylum armatum* and its in vitro anti-inflammatory profile. *Nat. Prod. Res.* **2019**, *33*, 665–672. [CrossRef]
34. Bhatt, V.; Sharma, S.; Kumar, N.; Singh, B. A new lignan from the leaves of *Zanthoxylum armatum*. *Nat. Prod. Commun.* **2017**, *12*, 99–100. [CrossRef] [PubMed]
35. Hu, Y.; Qian, W.; Yi, L.; Mao, Y.-D.; Ye, Y.-L.; Ren, H.-Y.; Gu, T.; Zhang, D.-J.; Cao, G.-X.; Gao, S. Chemical composition and antioxidant activity of *Zanthoxylum armatum* leaves in response to plant age, shoot type and leaf position. *Forests* **2023**, *14*, 1022. [CrossRef]
36. Zhi, X.; Song, Y.; Yu, D.; Qian, W.; He, M.; Lin, X.; Zhang, D.; Gao, S. Early growth characterization and C:N:P stoichiometry in *Firmiana simplex* seedlings in response to shade and soil types. *Forests* **2023**, *14*, 1481. [CrossRef]
37. Shu, X.; Zhang, K.; Zhang, Q.; Wang, W.B. Ecophysiological responses of *Jatropha curcas* L. seedlings to simulated acid rain under different soil types. *Ecotox. Environ. Safe* **2019**, *185*, 109705–109717. [CrossRef]

38. Li, Z.; Qiu, X.; Sun, Y.; Liu, S.; Hu, H.; Xie, J.; Chen, G.; Xiao, Y.; Tang, Y.; Tu, L. C:N:P stoichiometry responses to 10 years of nitrogen addition differ across soil components and plant organs in a subtropical *Pleioblastus amarus* forest. *Sci. Total Environ.* **2021**, *796*, 148925–148937. [CrossRef] [PubMed]
39. Zhi, X.; Yang, Y.; Zou, J.; Ma, N.; Liu, T.; Wang, H.; Hu, Y.; Gao, S. Responses of the growth and nutrient stoichiometry in *Ricinus communis* seedlings on four soil types. *J. Elementol.* **2022**, *27*, 223–238. [CrossRef]
40. Tang, N.; Cao, Z.; Wu, P.; Liu, Y.; Lou, J.; Hu, Y.; Sun, X.; Si, S.; Chen, Z. Comparative transcriptome analysis reveals hormone, transcriptional and epigenetic regulation involved in prickle formation in *Zanthoxylum armatum*. *Gene* **2023**, *871*, 147434–147448. [CrossRef]
41. Ma, Y.; Tian, J.; Wang, X.; Huang, C.; Tian, M.; Wei, A. Fatty acid profiling and chemometric analyses for *Zanthoxylum* pericarps from different geographic origin and genotype. *Foods* **2020**, *9*, 1676. [CrossRef]
42. Chen, X.; Wang, W.; Wang, C.; Liu, Z.; Sun, Q.; Wang, D. Quality evaluation and chemometric discrimination of *Zanthoxylum bungeanum* maxim leaves based on flavonoids profiles, bioactivity and HPLC-fingerprint in a common garden experiment. *Ind. Crop. Prod.* **2019**, *134*, 225–233. [CrossRef]
43. Agnihotri, S.; Dobhal, P.; Ashfaqullah, S.; Chauhan, H.K.; Tamtal, S. Review of the botany, traditional uses, pharmacology, threats and conservation of *Zanthoxylum armatum* (Rutaceae). *S. Afr. J. Bot.* **2022**, *150*, 920–927. [CrossRef]
44. Yousefi, M.; Hajabbasi, M.; Shariatmadari, H. Cropping system effects on carbohydrate content and water-stable aggregates in a calcareous soil of central Iran. *Soil Till. Res.* **2008**, *101*, 57–61. [CrossRef]
45. Chen, X.; Chen, H. Plant mixture balances terrestrial ecosystem C:N:P stoichiometry. *Nat. Commun.* **2021**, *12*, 4562–4571. [CrossRef] [PubMed]
46. Kleyer, M.; Minden, V. Why Functional ecology should consider all plant organs: An allocation-based perspective. *Basic Appl. Ecol.* **2015**, *16*, 1–9. [CrossRef]
47. Li, H.; Li, J.; He, Y.; Li, S.; Liang, Z.; Peng, C.; Polle, A.; Luo, Z. Changes in carbon, nutrients and stoichiometric relations under different soil depths, plant tissues and ages in Black Locust plantations. *Acta Physiol. Plant.* **2013**, *35*, 2951–2964. [CrossRef]
48. Maathuis, F.J. Physiological functions of mineral macronutrients. *Curr. Opin. Plant Biol.* **2009**, *12*, 250–258. [CrossRef]
49. Zhai, P.F.; Guan, J.X.; He, P.; Liu, H.Y.; Man, L.; Jiang, Y.; Ma, C.C. Changes of non-structural carbohydrates and nitrogen contents of needles and twigs in *Pinus sylvestris* var. *mongolica* plantations along an aridity gradient. *Ying Yong Sheng Tai Xue Bao* **2022**, *33*, 1518–1524. (In Chinese) [CrossRef]
50. Wang, K.; Shen, C.; Cao, P.; Song, L.N.; Yu, G.Q. Changes of non-structural carbohydrates of *Pinus sylvestris* var. *mongolica* seedlings in the process of drought-induced mortality. *Ying Yong Sheng Tai Xue Bao* **2018**, *29*, 3513–3520. (In Chinese) [CrossRef]
51. Wang, Y.; Han, X.; Ai, W.; Zhan, H.; Ma, S.; Lu, X. Non-structural carbohydrates and growth adaptation strategies of *Quercus mongolica* Fisch. ex Ledeb. seedlings under drought stress. *Forests* **2023**, *14*, 404. [CrossRef]

MDPI AG
Grosspeteranlage 5
4052 Basel
Switzerland
Tel.: +41 61 683 77 34

Horticulturae Editorial Office
E-mail: horticulturae@mdpi.com
www.mdpi.com/journal/horticulturae

www.ingramcontent.com/pod-product-compliance
Lightning Source LLC
Chambersburg PA
CBHW061616210926
51850CB00009B/2346